VISTAS IN
ELECTRIC POWER

In Three Volumes

VOLUME 2

VISTAS IN ELECTRIC POWER

BY

PHILIP SPORN

Volume 2

PERGAMON PRESS

OXFORD · LONDON · EDINBURGH · NEW YORK
TORONTO · SYDNEY · PARIS · BRAUNSCHWEIG

Pergamon Press Ltd., Headington Hill Hall, Oxford
4 & 5 Fitzroy Square, London W.1

Pergamon Press (Scotland) Ltd., 2 & 3 Teviot Place, Edinburgh 1

Pergamon Press Inc., 44–01 21st Street, Long Island City, New York 11101

Pergamon of Canada Ltd., 207 Queen's Quay West, Toronto 1

Pergamon Press (Aust.) Pty. Ltd., 19a Boundary Street, Rushcutters Bay,
N.S.W. 2011, Australia

Pergamon Press S.A.R.L., 24 rue des Écoles, Paris 5e

Vieweg & Sohn GmbH, Burgplatz 1, Braunschweig

First edition 1968

Library of Congress Catalog Card. No. 66–18232

08 011742 2

CONTENTS

VOLUME 1

VOLUME 2

SECTION II. DYNAMIC TECHNOLOGY

VOLUME 3

SECTION III. POWER FOR HEAT

SECTION IV. POWER FOR DEFENSE   951

SECTION V. SOVIET POWER   977

SECTION VI. THE SEARCH FOR TRUTH AND ITS REWARDS

SECTION II

DYNAMIC TECHNOLOGY

CHAPTER 1

CONVENTIONAL STEAM GENERATION

CONTENTS

1. PHILO STATION DESIGN †

AN AMERICAN power station that has operated for one month on 14,300 Btu per kwhr, sets an economy record for a public-serving company plant using coal for fuel. This new station of the Ohio Power Company—Philo—embodying several departures from past design practices successfully met the design specifications of operation at 15,000 Btu per kwhr. It is the result of careful planning both as regards station design and its use in connection with the large interconnected system to which it supplies power.

Company experience—six years' operating record of its Windsor plant with its comparatively high steam pressure (250 psi), high load factor, combination mine operation and large coal storage and high-tension outdoor substations— led the management to seek a location for its new plant that would make possible advanced development of these arrangements.

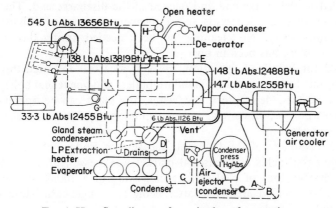

FIG. 1. Heat flow diagram for unit plan of generation.

Located on the Muskingum River the site is right in the center of the southern Ohio coal fields; a number of the mines are on the river, thus making possible direct river haulage from mine to plant.

A dam across the river permits gravity flow through the condensers at all times except during high-flood. An old canal leaving the river above the dam created an artificial island of just about the proper size (115 acres), for use as a condenser intake.

Thus water may be taken to the screen house through an intake tunnel (one for each unit) running the entire width of the boiler room to the condenser, and continuing under the turbine room to the discharge tunnel leading to the river. In addition this arrangement allows full utilization of a portion of the old canal beyond the remotest future limits of the plant as a 300,000-ton wet-storage basin. It also permitted the high-tension yard to be so located that the 132-kv circuits could take off without any crossing or recrossing of the yard.

† *Electrical World* (with M. L. Sindeband), August 22, 1925.

12a*

Philo consists of two 35,000-kw Curtis 19-stage turbines designed for 530-psi, 725 F operation with provision for extraction and reheating to 725 F between the seventh and eighth stages, and extraction at the twelfth and sixteenth stages for feed-water heating. This size and type of unit seemed to offer the most attractive, and hitherto unattained economy when operating under design conditions. The reheat and extraction arrangement seemed particularly promising from the economy standpoint, besides offering a solution to the problem of blading erosion at the low-pressure end.

Not only have the advance studies and predictions of fuel economy been confirmed but operation has very definitely proved that practically as great economy can be obtained from this unit at one-quarter, one-half and three-quarter load as at full load, due principally to the combination of reheat and extraction.

Eight boilers are employed with 14,080 sq ft heating surface each, six being standard and two being reheat. The condensers are two-pass vertical, the tubes being rolled at the intake end and packed at the discharge end. The unusual feature about these condensers is an arrangement for reversal of flow to make cleaning unnecessary. No cleaning has been needed during the more than 8 months the plant has been in operation.

Designed for base-load operation, the unit idea was followed throughout. Thus each unit has its own three standard boilers and one reheat boiler, its own auxiliaries, heaters, de-aerators, etc. Thus, the plant can be operated either with a single unit or, ultimately, with six units on the line with practically the same thermal economy.

To take advantage of the favorable location of Philo with respect to the central Ohio coal fields and seasonal market weaknesses, provision has been made for 400,000 tons of coal storage consisting of a 100,000-ton dry and a 300,000-ton submerged area, the latter being for longer-term storage of the cheaper grades of fuel.

All motors driving the crushers and conveyors are automatically controlled, standard contactors and control panels being employed. A master push button in the control house starts the various motors singly and in proper order to reduce the starting load and to prevent coal accumulating on the belts. Emergency-stop push buttons are provided in the crusher house and along the conveyors, so that in an emergency all conveyers carrying coal toward the push-button station that has been actuated will shut down but the conveyors carrying coal away will remain in operation.

By a proper manipulation of switches on the master control panel, four general schemes of automatic operation can be obtained:

1. Through operation—conveying coal from the track hoppers to the coal bunkers in the plant.

2. Hoppers to storage yard—conveying coal from track to yard, the coal being deflected from the belt conveyor by a tripper in the crane housing.

3. Storage yard to bunkers—conveying coal from the yard to the bunkers in the plant.

4. Crushers idle—all belts running to clear slack coal.

Double 132-kv Bus

The unit idea was extended to the design of the electrical end. Each unit is connected through a low-voltage oil circuit breaker to a transformer bank and then to the 132-kv bus, each generator and transformer constituting a unit; neither component operates without the other. From the standpoint of maintenance, it was believed that a minimum loss of time and capacity would result if the entire unit were shut down and every part of it gone over at the same time.

Each machine has its own transformer that feeds the auxiliaries of the particular generator only, but a reserve transformer is provided for each pair of units in case one transformer is out of service. Switching arrangement contemplates no paralleling of generators on the 11,000-v side, and, except for synchronizing, no low-voltage breaker would have been employed but it was not felt safe to synchronize a generator on a 132-kv breaker. Elimination of the low-voltage bus obviously eliminated the problem of heavy short-circuit currents with their concomitant high stresses which would necessitate special phase arrangements, etc. The high rupturing capacity needed in the oil breakers was taken care of much more easily at 132 kv.

The auxiliary transformer is tapped to a by-pass connection of the low-voltage breaker, thus making it possible to start the plant cold. After the unit is in operation, closing the lower of the two by-pass switches and opening the upper one will bring about a connection equivalent to the auxiliary transformer being tied in solidly on the generator terminals. Such an arrangement will permit clearing everything beyond the generator low-tension oil switch without in any way affecting the auxiliaries of the generator as long as the steam end and the generator proper remain unaffected.

Two separate 132-kv buses have been provided and the entire plant can be operated on either one. It was felt that for a plant of this magnitude no reliance should be placed on less than two buses; there should always be a complete bus in reserve. This is accomplished with only one oil breaker per circuit by utilizing motor-operated disconnecting selector switches arranged in sets of six on a common frame which is floor mounted to aid inspection and adjustment. To permit oil-switch inspection and repair, a transfer or inspection bus has been provided; this allows a single extra oil switch to be substituted for any oil circuit breaker, whether the circuit with which it is connected is on the main bus or on the reserve.

All low-tension breakers, auxiliary transformers and main transformers have been placed close to the main building, making necessary only a very short cable run from the generator to the main transformers and from the auxiliary transformers back to the 2300-v bus inside.

Placing the transfer or inspection bus on each side of the two main buses makes it possible to utilize a single bay for two feeders leaving in opposite directions, or for a feeder and transformer, or for a feeder and transfer switch. This gives ample room for clearances and operation without spreading the structure out too far.

Auxiliary Drive and Control

All auxiliaries with the exception of a duplicate steam-driven boiler-feed pump are electrically driven; the large units are 2300-v motors which have been designed to start by throwing directly on the line. Exceptions are the motors driving the forced-draftfans and generator blowers and the few synchronous motors for which fractional-voltage starting has been provided. Since the initial operation of the plant, eliminating this complicated fan and blower set-up so far seems to indicate that such a procedure will not harm the fans.

Each boiler fan has two motors mounted on the same shaft, one low-speed and the other high-speed. Driving the forced-draft fan are two 2300-v squirrel-cage motors; the low-speed motor starts directly off the bus and the high-speed motor starts by means of an autotransformer. Driving the induced-draft fan are two 2300-v wound-rotor induction motors, the starting and speed adjustments of which are obtained by the use of standard rotor resistances and motor-operated drum controllers. It was found that the full range of speed required for the induced-draft fan motors could not be covered economically by a single motor; this was the reason for using two motors.

The low-speed motors are used for light boiler loads and the high-speed motors for heavy loads. The circuit breakers for both types are interlocked so that only one motor can be run at a time. The control is so arranged that the following combinations are the only ones possible:

1. Either the low-speed motor or the high-speed motor of the induced-draft fan may operate alone. This combination is obviously not applicable to the forced-draft fan.

2. The high-speed motor of the forced-draft fan can operate only in conjunction with the high-speed motor of the induced draft fan.

3. The low-speed motor of the forced-draft fan can operate in conjunction with either the low-speed motor or the high-speed motor of the induced-draft fan.

In case of overload on either of the forced-draft fan motors, a multi-contact relay shuts down all fans at which point the low-speed motor of the induced-draft fan, if not already running, is started up. In case of overspeed on the main turbines, the throttle and intercepting valves close, causing the reheat-boiler doors to open, which in turn actuate the multi-contact relay.

In case of undervoltage on either of the motors used with the induced-draft fans (which would necessarily be running in conjunction with a forced-draft fan), a definite-time delay relay between the undervoltage and the multi-contract relays prevents the fans from shutting down due to momentary fluctuations in supply voltage.

Overload on the high-speed motor of the induced-draft fan would shut it down, after which the definite-time and multi-contact relays being energized in turn would shut down all the fans and start the low-speed motor of the induced-draft fan.

In case of overload on the low-speed motor of the induced-draft fan (which is relied upon in abnormal conditions), the motor shuts down, and if the low-speed motor of the forced-draft fan is running at the same time, it too is shut down.

Field and Auxiliary Transformer Interlocks

To insure generator-field excitation, the generator oil circuit breaker cannot be closed unless the field circuit breaker, or the circuit breaker controlling the feed from the excitation bus, or both, are closed, as well as the synchronizing switch. For the same reason, neither of these breakers can be tripped unless the generator oil circuit breaker is open or both the former circuit breakers are closed. In the latter case, tripping one of the circuit breakers still will leave potential on the generator field from the other circuit breaker.

The auxiliary-transformer breaker cannot be closed unless its generator breaker is closed, and the latter cannot be closed except through a synchronizing receptacle. This makes it practically impossible to feed in on a dead machine. In case of internal trouble in the generator, a multi-contract relay is energized by the generator PQ-6 differential relays. The same multi-contact relay is also energized by PQ-6 overload relays in the generator neutral leads. The following oil circuit breakers are tripped when the multi-contact relay is energized: generator breaker, auxiliary transformer breaker, power-transformer breaker, excitation-bus breakers after the generator breaker has first tripped.

The generator breaker is also tripped, in case of internal trouble in the 45,000-kva bank of power transformers, by means of a second multi-contact relay which is energized by the differential relays of the power-transformer bank.

To make it impossible to tie the two sections of the 11-kv bus together, or open the motor-operated bus-tie disconnects under load, the latter cannot be opened or closed unless both the reserve auxiliary-transformer oil circuit breaker and the other bus-tie disconnects are open. Conversely, the other tie disconnects cannot be opened or closed unless both the tie disconnects and the auxiliary-transformer breaker are open.

Protective Arrangements

Motors.—All auxiliary motors are protected by instantaneous overload relays with current settings of approximately six to eight times normal. Various other elaborate arrangements were investigated but the present scheme was found to be the simplest and the least likely to cause trouble due to improper operation.

Generator.—The generators utilize outside blowers to supply cooling air in a closed system of air circulation with surface coolers. A bimetallic thermostatic relay is placed in the intake chamber to give an alarm in case the air temperature reaches a dangerous value. In addition there is the usual balance scheme of protection, utilizing for the purpose instantaneous plunger-type relays.

Transformer.—Difficulties are encountered with a balanced scheme of protection for the transformers if the balancing current-transformers are placed in the terminals of the high-voltage side switches and low-voltage side switches of the transformer bank. Besides that due to tapping the auxiliary transformer beyond the low-tension switch, there is also the difficulty brought about by utilizing the transfer switch in place of the normal transformer switch. All these problems were cleared away by placing the bushing transformers in the terminals of the transformer proper.

Potential transformers on the 11,000-v side are all fused using a series resistance; the 132-kv potential transformers, which were necessary for synchronizing between various plants, have all been connected solidly on the line side of the oil switch without any disconnect of fuse.

Lightning.—Each of the 132-kv lines is protected by an oxide-film arrester. The station, which has been in service considerably over half a year and which has gone through one of the most severe lightning seasons, has so far experienced no trouble due to lightning, although the insulation of the station bus and of a good deal of the equipment on the buses is considerably lower than that of the line. This does not necessarily prove the efficacy of the lightning arresters, but it is an interesting result nevertheless.

Line.—The Crooksville and St. Albans lines are both protected by straight overload relays. The two Canton lines, which are on the same towers and are

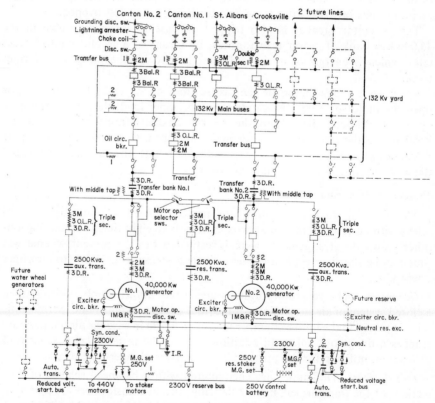

Fig. 2. One-line diagram of extension of unit idea to electrical at Philo.

of exactly the same length, have separate overload protection on each of the lines, with an arrangement to give instantaneous balance protection when both lines are in circuit. As originally laid out, a lock-out feature was provided that would delay one line for a period of about five seconds when the other line tripped on balanced protection. Six months' operation with this arrangement during particularly heavy lightning storms showed the desirability of removing this lock-out feature, as cases occurred where both lines apparently flashed over at the same time and the delay in clearing the second line due to this relay arrangement was invariably long enough to burn the conductor badly.

Grounding of Neutrals

The 132-kv transformers are all grounded solidly at Philo. Since the removal of the lock-out feature, relay functioning has been entirely satisfactory; sufficient ground current is obtained under all conditions and the current is apparently not heavy enough to cause any conductor burning.

The 11,000-v neutrals of the generators, however, are grounded through a 2-ohm, 2500-amp, 1-min outdoor resistor. A short-circuiting switch is provided for this resistor, and is automatically actuated by a thermal relay when the resistor temperature reaches 350 C. It was felt not only that this arrangement would protect the resistor but, should the grounding current last for any length of time yet be insufficient to cause relay operation, that the short-circuiting of the resistor would increase the ground current and give the necessary increase in current for relay functioning.

Communication

For plant intercommunication a private automatic telephone is employed. For communication between Philo and Canton and Windsor, and between Philo, St. Albans and Crooksville, carrier-current equipment of the interphase type is employed. The coupling to the Crooksville and St. Albans circuits is straight wire. In the case of the Philo–Canton lines, however, the arrangement of the two circuits on the towers is asymmetrical, making wire coupling impossible, and condensers had to be used. Various types were considered, the type finally chosen being a mica condenser built in 22,000-v units with six in series to give the necessary insulation and two such stacks in parallel to give the proper capacity.

The plans of Philo were worked out jointly by Sargent & Lundy and by the engineers of the American Gas and Electric Company.

2. SUPERPOSITION AND HYDROGEN COOLING FOR LOGAN†

ALMOST continuously the American Gas and Electric system has been carefully studied to determine where new capacity could best be added when needed. Always the answer has been to extend the present plant as against building a large new plant up to the point where the total load would be approximately

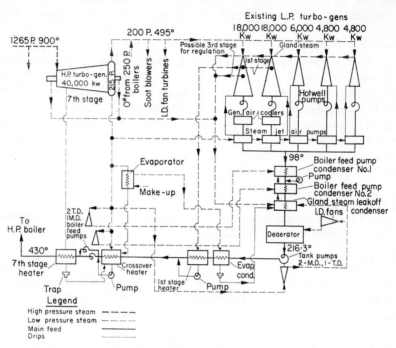

Simple steam cycle of Logan superposition to nearly double thermal Efficiency.

twice its present value. This was true not only from the standpoint of cost per kw of additional capacity, but particularly so from the standpoint of transmission system and service-area protection. Hence, when it became evident toward the end of last year that additional capacity would be needed to meet expected loads during the low-water period of the 1937 hydro season, capacity-extension studies were confined to purchase considerations and to studies of enlargement or modification of existing plants. The final decision was to superpose at the Logan station of Appalachian Electric Power Company.

Economical superposition would give a block of power at Logan which possibly would serve for several decades and definitely would complete the station from a capacity standpoint.

† *Electrical World*, April 11, 1936.

Choosing Pressure and Temperature

In choosing the pressure, temperature, unit size, boiler and heat cycle certain factors gathered from previous experience, or developed and brought to light during the past six years, were kept in mind. Among these were:

Generally satisfactory operation of 1250-psi boilers and turbines as shown by Deepwater and other plants;

Improved thermodynamic performance made possible by increased temperature;

Elimination of reheat made possible by going to a 900–925 F initial range to bring about piping simplification;

Conviction that a single high-pressure unit was absolutely necessary to save capital cost;

Space limitations which compelled adoption of a single boiler unit for the generation of a million pounds of steam;

Lack of low-pressure capacity which, of necessity, called for a maximum ratio between high-pressure and low-pressure capacity to develop the block of power desired and incrementally available on an attractive basis; and finally,

Conviction that the art of power generation had progressed to the point where a single boiler and generator installation could be made to give an availability of 95%.

In short, the goal was an economical superposition scheme yielding the maximum amount of incremental capacity on an attractive basis. To do this, resort was had to new developments in turbines, alternators, switching and boiler pressure and temperature conditions.

Superposition Cycle

Logan, at present, consists of two 20,000-kva turbo-generators operating at 260 psi and 580 F. It also has two 5000-kva turbo-generators designed for 200 psi and 480 F, which have not been operated for about six years. The first-stage nozzles of the 18,000-kw units will be replaced to increase the steam flow required by the future lower steam pressure and temperature and the two small units will be placed in first-class condition. A 40,000-kw high-pressure unit will exhaust at 200 psi absolute pressure into the present four low-pressure units and into a fifth 6000-kw machine which is to be moved to Logan from one of the other plants.

A new high-pressure 1,000,000 lb per hour boiler will take care of the entire plant.

When revamped the station will have a total capacity of 96,100 kw (44,500 kw in the high-pressure turbine and 51,600 kw in the low-pressure turbines). The usual ratio between the high-pressure and the low-pressure section is roughly 2:3. However, in the case of Logan, the ratio will be considerably higher— 1.1 : 16. This high ratio will be obtained by using steam-driven boiler-feed pumps and fans and by liberal bleeding. The cycle of steam flow is extremely simple. A single crossover pressure of 200-psi supplies all five low-pressure turbines, the fan turbines, boiler-feed pumps, steam-jet air pumps, evaporator and the crossover heater. The high-pressure turbine is bled at the seventh stage into the only high-pressure heater used. This cycle, it is expected, will give a heat reduction from an average of 23,000 Btu per kwhr output to 12,000 Btu or a decrease of about 48%.

Turbo-generator Details

Rated at 50,000 kva (or 40,000 kw at 0.8 power factor) the new turbine will be three phase, 60 cycles, 3600 rpm with a trottle pressure of 1250 psi at a temperature of 925 F. The turbine will have an inner and outer casing and the exhaust steam at about 200 psi will surround the inner casings so that the outer casings will be subjected to only about 200 psi. The unit will have eleven stages, with bleeding at the seventh. The six first-stage nozzles will be distributed around the periphery of the nozzle head and fed by six control valves coming off a common manifold below the turbine room floor, and supplied by two main feeds coming from the superheater to the two throttles at each end of the manifold. This gives the necessary evenness of steam temperature to the entire group of control valves. The oil pump will be driven from the generator shaft and will be located in the basement to reduce fire hazard.

Governing will center on maintenance of a fixed back-pressure under the control of a pressure regulator so that the high-pressure turbine will follow low-pressure unit loadings instead of having the latter controlled by the high-pressure governor. This will minimize any possibility of hunting between the low-pressure units.

In obtaining a large 3600 rpm machine, the difficulty has always been the limitation in the maximum size of alternator capable of operating satisfactorily at this speed. Air was used as the cooling medium, but with the introduction of hydrogen this difficulty automatically disappeared. The present limit is possibly 75,000 kva. Hence there was no difficulty in obtaining the 40,000-kw 0.8 power-factor machine needed to carry out a proper superposition job. With the background of experience on our eight hydrogen-cooled synchronous condensers, going back to 1928, we did not hesitate to adopt hydrogen cooling for this alternator. In addition, more direct cooling of the iron and windings in the stator will be resorted to in this machine by the use of so-called "water pads", so that hydrogen and water really will be the cooling media.

The stator will have the conventional air-duct spaces in the armature coil but instead of being empty they will be filled by hollow pads through which water will circulate to take away the stator heat. Armature coil and windings and rotor windings will be cooled by the circulating hydrogen gas which will be propelled by fans at each end of the rotor. To remove the heat in the hydrogen a sufficient number of water pads at the central portion of the armature core will have radial fins welded to them over which gas will pass. Hydrogen cooling on this machine will reduce the heating and windage losses by about 450 kw under that of a conventional air-cooled machine of the same rating. Generator shell construction will be similar to that employed in the hydrogen-cooled synchronous condensers, with emphasis laid upon minimizing hydrogen leakage and making the shell explosion-proof.

Boiler Details

Three requirements were given the boiler manufacturers as a basis for boiler design, with their relative importance stressed in the order given below:

Reliability and Availability.—A request was made for a boiler design which would have an availability of 50 weeks a year, with some leeway as to when it would be necessary or desirable to utilize the two weeks' annual outage.

Space Economy.—This definitely restricted design to a single unit.

Capital Cost Economy.—This was particularly important since Logan is located in a low coal-freight zone, which meant that Btu economy, justifiable in areas where the freight item runs from $2.50 to $3 per ton, could not be justified here with coal available at a 7-cent (switching charge) freight rate.

The boiler chosen will have a rating of 1,000,000 lb of steam per hour at 1325 psi and 925 F. Superheater temperature will be regulated by a by-pass with an expected tolerance of ± 10 degrees. The boiler will be a six-drum, double V-type unit using tangent-tube construction for the water walls, tangential firing and dry bottom. Superheaters will be of the convection type, vertically supported; economizers of the return-bend tube type.

The decision to go to a dry-bottom furnace was based upon the fact that, with the development of other superposed plants, more and more would it become necessary for such installations to operate at comparatively low rating or face the necessity of being shut down completely, a procedure totally unsuited to sound operation of high-temperature, high-pressure, large-size boiler units. Hence, a load range of 5:1 was requested. No wet-bottom design available at the present time seemed capable of fulfilling this range for the ash-fusion temperatures expected on this particular job.

Ash-fusion temperatures received particular attention in the study of gas temperatures in the upper reaches of the boilers. It was felt that in order to make available to the plant all the economical sources of fuel, a greater than normal range of fusion temperatures would have to be handled. The range chosen was from 2300 F to 3000 F. The boiler dimensions, including the relative distribution of surface between water walls and the rest of the boiler, gave the maximum assurance of keeping liquid or viscous ash out of the upper boiler reaches over the entire range of operation up to and including full guaranteed rating.

Boiler efficiency frankly was not the maximum that might have been attained if additional heat-absorbing surface had been utilized. In this particular design, however, it was felt that the most economical balance was the one which gave less weight to the Btu efficiency, but gave maximum weight to overall economy and to reliability and availability.

3. WINDSOR MODERNIZATION†

SINCE the formulation and statement of the basic ideas behind the Logan superposition, a great many other topping installations have been undertaken and are now under way. Among these are Waterside, Springdale, West End, Fiske Street, Schuylkill, Dayton and a number of others. Basically they all follow the Logan pattern, i.e. the steam pressure is of the order of 1250 psi, the steam temperature is around 900 F and a unit of from 30,000 to 50,000 kw for topping purposes. Logan's lead in the direction of one boiler, however, has not always been followed. Rather, a more conservative practice—usually with two boilers—has generally been employed. So far as is known, all the 60-cycle topping machines have a speed of 3600 rpm, and those with a rating of 30,000 kw or over, employ hydrogen as a cooling medium. But basically the same ideas developed for Logan have been employed in all the subsequent installations.

In the turbine a size of unit was undertaken that was at that time over twice the size of the largest unit previously attempted at 3600 rpm; the throttle temperature was the highest attempted on a regular commercial unit of US manufacture. Also it was the first turbine to employ the double-casing construction on the steam end for handling the high-temperature stress problem. The generator embodied not only new principles and new materials in the construction of the rotor through the use of aluminum windings, but was the first commercial alternator to be hydrogen-cooled. The stator employed a new method of cooling the iron and the embedded portion of the armature coils, and for the first time hydrogen was used for the cooling of the rest of the winding. A totally new method of sealing against hydrogen leakage was employed.

The Logan boiler was the largest attempted in that temperature and pressure range and was definitely laid out with the idea that service availability would be high enough to make such a size economically feasible. As against the more prevailing practice of handling the ash in the wet condition, the Logan boiler, because of the broad load range that had to be handled, was definitely designed for a dry bottom. Further, the rate of heat liberation and the radiant absorbing surfaces were purposely laid out with the idea of making the upper reaches of the boiler substantially free from slag deposits under all operating ranges up to full load.

The cycle at Logan was notable not only for its direct and simple flow arrangement but for the unusually high ratio between high-pressure and topped capacity, the ratio developed being 1.1:16 against the more common one of roughly 2 to 3. This was accomplished by dropping the cross-over pressure and by the introduction of turbine-drive for auxiliaries that had heretofore become more or less standardized on an electrical-drive basis.

A number of problems, however, were not completely solved by Logan. The floating throttle valves on the turbine, for instance, introduced complicated problems in turbine and turbine piping behaviour. Again, since the work on the Logan turbine definitely showed that the various design principles and ideas

† *Combustion*, March, 1937.

employed in this unit could be projected satisfactorily into much larger sizes there was a natural conclusion that where a proper opportunity presented itself, the development of a larger topping unit would make possible very attractive economies. On the boiler end, the solution of several problems which were satisfactory for Logan did not have general application. Thus, the dry bottom, it was felt, might present many difficulties in the handling of coals having much lower fusion temperatures than those encountered at Logan. Again, it was felt that the solution of the problem of superheat control at Logan was only approximate and that further development of higher temperatures and pressures would call for a closer and more definite working out of superheat regulation.

The Problem at Windsor

The Windsor plant, which has had an effective life of close to twenty years, was originally erected in 1916 and extended to its present size in 1922. Since that time it has been operating with six 30,000-kw turbines, four at 250-psi throttle pressures, and two at 325 psi, all with a steam temperature of approximately 625 F.

Under the normal, but at present particularly intense pressure to reduce costs, something more creative than mere "tightening", so to speak, of the old plant would have to be done, their superposition began to be studied.

New Record Superposition Size

The turbine will be a 12-stage unit rated 60,000 kw at 90 % power factor and operating at a speed of 3600 rpm. It will have a throttle pressure of 1250 psi, a throttle temperature of 925 F and exhaust against a back-pressure of 235 psi gage.

Thus, the unit will be larger than any heretofore attempted in a superposition. The need for such a size is the divided ownership of the plant. If ownership of one of the high-pressure units was not to be mixed it was a question of topping one-half of the plant with either one or two turbines. The additional economies that one turbine offered, plus the satisfactory development of the design work on the Logal unit, prompted the decision to go up to the 60,000-kw size.

The turbine uses well-known double-casing construction. The throttle valve, however, is definitely anchored to the turbine foundation; this, it is felt, is an advance in turbine design. In contrast with the Logan unit having distributing valves below the turbine-floor, the Windsor turbine has valve-in-head construction, thus considerably simplifying the piping problem from construction and operating standpoints. The generator is substantially similar to the Logan unit.

Heat Cycle

Except for a number of minor details, the cycle chosen is similar to that adopted for Logan. Two high-pressure boilers supply the necessary steam for the 60,000-kw high-pressure unit which, in turn delivers its exhaust to three low-pressure turbines having a combined rating of 90,000 kw. Use of turbines for boiler-feed pump and induced-draft fan drives, and rather limited bleeding followed the Logan setup but the pump condensers were advanced somewhat in the cycle. The expected performance on full load under normal conditions will be better than at Logan, due to better Rankine efficiency of the Windsor low-pressure units and their more efficient low-pressure cycle.

Boilers

Two 750,000-lb per hour boilers will supply steam at 1,350 psi and 925 F for a total capacity of 150,000 kw. The decision to use two boilers was influenced by the requirements of $1\frac{1}{2}$ million lb of steam per hour and the limited experience to date with boilers of even 1 million of rating at 1350 psi pressure, such as the Logan design. A single boiler would definitely involve experimentation with too large a number of unknown factors. On the other hand, no more than two boilers would be justified in view of the satisfactory operation of the Logan and other boilers of which it was a forerunner.

In the Windsor design, the final decision was to adopt a dry bottom for one boiler and a wet bottom for the second boiler. It has always been recognized that the dry bottom gave the utmost flexibility in range of operation. The problem has been that to keep temperatures at the bottom at a point where no difficulty would be experienced with fused slag, a low superheat temperature might result. The independently fired radiant superheater, however, offered the ideal solution in this case. If the wet bottom has any inherent advantage it always comes into play at higher ratings when fusion of ash is never a problem. Thus, what is not possible normally in a single boiler—obtaining the advantages of dry bottom at low load and of wet bottom at high load—are entirely possible for a plant with one boiler operating dry and the other wet.

A further point of interest is the full use being made of existing foundation caissons. Only three of the existing 12 boilers will have to be removed to make room for additional steam capacity, equivalent to 150% of the present capacity. Thus 75% of the present low-pressure boiler capacity will be left available to back up the high-pressure boilers for low-pressure turbine operation. Some of this gain in space has been made at the expense of additional height, but it is nevertheless a very striking example of efficient utilization of floor space.

Summary

In addition to carrying forward many of the ideas first developed at Logan, many new steps have been taken, including:

The Windsor high-pressure turbo-generator, larger than any of its predecessors by about 20%.

Two different methods for close regulation of superheat will be utilized, each being suited to or inherently deriving from the particular boiler design.

An unusually wide and flexible range of successful station operation has been provided by combining the wet and dry-bottom designs in a manner that sacrifices none of the inherent advantages of the particular type of bottom chosen.

Although not touched upon in the body of the article it is worthy of note that at Windsor there will be attempted the first full supervisory operation of a steam turbine. This is a phase of operation in which steam practice has lagged behind straight electrical and hydraulic practice, but the need for it is as definite in steam work as in the other fields.

4. TWIN BRANCH TO USE 2400 psi†

THE 67,500-kw extension at the Twin Branch plant of the Indiana and Michigan Electric Company will be the first American installation at 2400 psi and 940 F. Size and type of turbo-generator unit, size of boiler, pressures and temperatures, both initial and crossover, and the cycle chosen were all selected because they seemed technically feasible and economically desirable.

As the load developed it became practical to consider adding a reasonable block of new capacity but before making a definite decision the economics of transmission were reconsidered, although this problem has been very thoroughly explored only some ten years previously. Such considerable progress had been made on the transmission art of late that it was thought best to reconsider

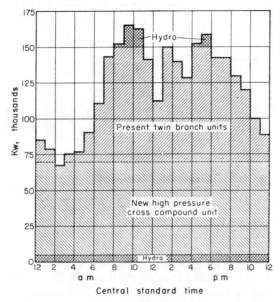

FIG. 1. Typical winter daily load curve expected.

the present status of the entire problem. Results of these studies showed that while the previously entertained economic limits of transmission had moved out somewhat, basically they were of the same order. Thus, at 70% load factor, the utmost distance that power can be transmitted electrically on the same economic basis as coal can be transmitted by freight is still around 100 miles, and this distance decreases as the load factor decreases. Consequently, the possibility of a major strengthening of the transmission system to permit electrical access to more economical sources of fuel than are available in the immediate territory was rejected. It was decided to continue to utilize the present

† *Electrical World*, October 9, 1937.

transmission system for general integration, co-ordination and emergency protection purposes for such transmission as can be carried on within its present capacity limits, and to supply the balance of the requirements from local—territorial—generation.

Superposition Considered

Twin Branch appeared to be too young a plant to be superposed, particularly if economical use in its present condition could be obtained. Besides, the economics of superposition press for a limited number of boiler units; within the

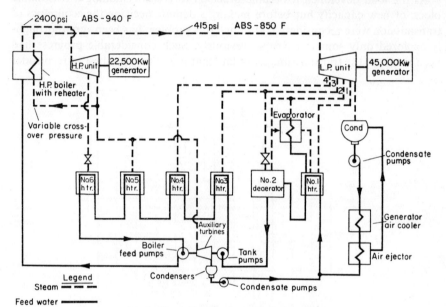

FIG. 2, Flow diagram in preliminary stage.

size of plant considered possible at Twin Branch, it called for a single high-pressure reducing unit. Keeping in mind the fact that Twin Branch is located at the most expensive fuel point on an interconnected system, having a present Peak of 810,000 kw, it definitely appeared unsound to commit such a plant to a layout that would permit economies during future recession in loads only by backing down on the load at the most expensive fuel points. The net evaluation of all of these considerations was a decision to proceed with a condensing unit.

Reason Behind High Pressures

There were three possible and practical ways of adding condensing capacity at Twin Branch. The first and the easiest would have been to continue the present pressure of 650 psi. A second alternative would have been to adopt the next standard pressure, either at 850 or 1400 psi, with a temperature of 900 to 940 F, and the elimination of reheat. The third possibility was to explore higher pressures with a view to reducing the combined capital and operating costs. In considering these alternatives, the thought was kept in mind that though the proper time for superposition at Twin Branch had not yet been reached, nevertheless it was unthinkable that the time would not arrive within the next 10 years. In other words, somewhere around that time it would be found uneconomical to continue to operate a 25-year-old plant at any high load factor. Therefore, in exploring possibilities for better economies it was felt it would be well to keep in mind the adaptability of some of the present equipment for future rejuvenation of the still-economical but gradually obsolescing 650-psi generating plants.

Preliminary investigation having shown that considerable economic gain was possible by going to higher pressures, a series of all possible cycles was explored. The table on page 356 abstracts the results of that study with reference to four cycles given serious consideration; some have been used at plants generally well known. Cycle No. 3 is substantially the Windsor and Logan cycle.

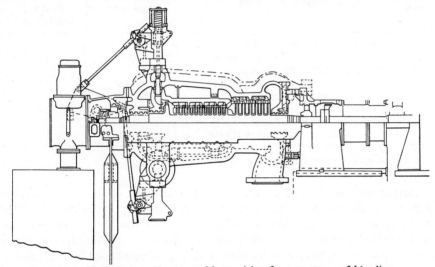

FIG. 3. Turbine employs 18 stages with provision for two stages of bleeding.

Although the thermal gain between cycle No. 3 and cycle No. 1 is of the order of only 7%, there is another more vital benefit—the gain in high-pressure capacity of 32%. This not only affects the economics of higher pressure but also has a very vital bearing on what pressure and what blocks of high-pressure capacity will be available for superposition on the series of 650-psi plants which went into service in the 1920–30 decade; Twin Branch was one.

The next consideration involved an investigation of the technical possibilities and limitation of 2400 psi.

Boiler more Critical than Turbine

It was felt at the outset that the turbine would set no particular limit on the pressure, but that the boiler would require careful investigation. A thorough study was given to the circulation problems that would be involved at pressures up to 3000 psi, and the final consensus indicated that 2400 psi (2500 psi drum pressure and 2400 psi outlet pressure at the superheater) would be the limit at which a boiler with natural circulation should be built. Above this pressure there seemed to be no question but that forced circulation would be required, as in the case of European boilers. As expected, the pressure of 2400 psi selected as a limit for the boiler proved to be no stumbling block in the turbine design. Investigation of boiler feed-pump design, suitable for pressures up to 3000 psi, has been progressing at the same time. These studies indicated that existing designs could be adapted to the proposed pressure, if intelligent use were made of available metals so that pump maintenance would not be materially increased.

SUMMARY OF RELATIVE EFFICIENCIES AND HIGH-PRESSURE CAPACITIES FOR FOUR HEAT CYCLES

Cycle No.	Description of cycle	Net station heat rate in %	† High-pressure capacity for fixed L-P capacity in %
1	2400 psi abs., 940 F, 900 F reheat, 1 in Hg abs.	100	178
2	1200 psi abs., 940 F, 850 F reheat, 1 in Hg abs.	102.9	135
3	1200 psi abs., 940 F, no reheat, 1 in Hg abs.	107.3	135
4	850 psi abs., 940 F, no reheat, 1 in Hg abs.	108.4	100

† All low-pressure units have 400 psi abs. throttle pressure.

Having become satisfied on the technical soundness of the problem, there remained to be considered the more difficult part, namely, the economics of the proposed development, which, of course, are determined by two factors—the additional cost and the counter-balancing benefits of going to the higher pressure and temperature. The economics were materially strengthened by three facts:

1. The new boiler code does not require materially thicker boiler tubes for the pressures under discussion than the present standard for 1400 psi;

2. The development of various ideas of boiler design which make utmost simplicity highly desirable and almost mandatory in drums, tubes, surfaces and circulation circuits, thus reducing expense;

3. A co-operative spirit among the leading manufacturers who participated in the new development because it definitely promised a more economical method of generating power.

After the additional cost was evaluated against the benefits from capacity and thermal standpoints it was found that there was adequate justification for proceeding with the higher pressure. More than two decades had shown an unfailing eventual reduction of the cost differential between the more-economical and less-economical power-generation cycles as experience and standardization were accumulated. Further, such a step offered help toward solving the problem of eventually modernizing the existing 650-psi plants. Hence it was finally decided to proceed with the higher pressure.

Straight Regenerative Reheat

The proposed cycle is a straight regenerative-reheat type. This is still in its preliminary form and will doubtless have to be modified in details as the design progresses. Its outstanding features are the direct-line, simple-flow characteristics, so typical of the Logan and Windsor cycles, with a variable crossover pressure at a maximum of 415 psi.

Another unusual feature of the heat cycle is the idea of exhausting the boiler feed-pump turbines into a separate condenser instead of a heater using main condensate. It is believed that this will yield a higher financial efficiency and ease of operation, but the question has not yet been definitely settled.

Circulation Problem

The main boiler problem—proper circulation—is caused by the relatively small difference in weight of the water and steam at 2400 psi, which emphasizes the need for simple circulation passages and large down-comers to obtain proper hydraulic head. The design finally evolved a boiler with very little convection surface except a few rows of tubes immediately in front of the superheater. The boiler will be of Babcock & Wilcox manufacture with a rating of 550,000 lb of steam per hour at a superheater outlet pressure of 2400 psi and temperature of 940 F. The integral reheater will be designed to handle 450,0000 lb of steam per hour at a pressure as high as 450 psi and will reheat to a temperature range of 850 to 900 F. As might be expected, practically all of the heat-transfer surface in the furnace and boiler is vertical, the only exception being the tubes underneath the wet furnace bottom and the tubes on the top of the primary furnace; the superheater and reheater are vertical. Only convection (no radiant) surface will be used for the superheater; a by-pass provides for temperature control. It is now expected that the reheat surface will be confined to the two side walls of the furnace.

The large portion of the heat input concentrated in radiation to the furnace walls will, it is believed, definitely preclude the slagging of the upper region of the furnace. The economizer will be of the return bent-tube type and the air heaters will be of the Ljungstrom type.

The wet furnace bottom will provide for continuous discharge of slag that will drop over into the ash hopper and water-sluicing system.

Switched as Single Unit

The plan contemplates switching of the high- and low-pressure elements as a single unit, and tying the 132-kw winding of the three-winding transformer bank to either of two buses by a duplicate set of breakers. This will permit balanced operation of the three units. Although a low-tension breaker is called for between the generators and step-up transformers, the sole purpose is to permit sectionalizing for the generators and allow the transformers to be used as 132-kv to 27-kv banks. No electrical transmission lines are called for, the existing network being more than adequate to absorb the increased power generated by the new unit.

5. AIR BREAKERS FOR LOGAN'S AUXILIARIES†

INSTALLATION of a 50,000-kva topping high-pressure turbo-generator at the Logan, W. Va., plant of the Appalachian Electric Power Company called for a considerable increase in auxiliary power capacity and for switching equipment of the utmost reliability. While oil circuit breakers have been giving good performance for many years on both indoor and outdoor circuits, some rather serious oil fires at about the time the Logan project was started demonstrated that they were not altogether an unmixed blessing inside a plant.

From any point of view the quantity of oil stored in a power plant should be kept at a minimum. Consequently, when a considerable amount of 2300-v switching equipment had to be purchased for the Logan auxiliaries, the field of available apparatus was thoroughly canvassed to ascertain if oil-less equipment were available. Discussion dating from 1929 on the desirability of developing a dry-type 2300-v breaker, but which did not fructify during the boom 1929 days, were resumed with the I-T-E Circuit Breaker Company. Its officers indicated a desire to undertake the necessary development.

In the Logan plant 2300 v and 220 v have been used for station auxiliary service. The latter voltage did not have much to recommend it, but it had been there for many years and could not conveniently be discarded. The 2300-v motors were controlled from truck-type oil circuit breakers of fairly modern design, but some branch circuits were still controlled by small, obsolete oil breakers. In general, oil-immersed compensators had been used for starting squirrel-cage motors.

The I-T-E switchboard consists of 17 circuit breakers, three rated 1200 amp and the remainder 600 amp. The 1200-amp breakers are used for the transformer supply and bus-tie connections, the 600-amp breakers for the feeders. All motors, including a 1250-hp boiler-feed pump drive, are started at full voltage by closing the feeder breakers remotely either at the motors or at a control switchboard.

The outstanding features of the new switching structure are separate compartments for the power and control connections, tubular busbars, liberal channels for the control wiring, and main female disconnects clamped directly onto the tubular busbars and enclosed in a composition casing. Safety to operating personnel has been obtained by insulating exposed live parts and by the usual mechanical interlocks which prevent operation unless the truck is in the correct position.

Silver main contacts and tungsten-alloy auxiliary contacts are used to conform to modern practice. Considerable thought has been given by the manufacturer to the design of the auxiliary contact mechanism and the feature tending to increase contact pressure due to short-circuit stresses.

Manual operating levers and mechanical devices to assist in moving the breaker into and out of position are features of the design.

† *Electrical World*, October 22, 1938.

Use of a comparatively large auxiliary transformer bank (6000 kva), and the electrical proximity of large generators and a high-voltage transmission line dictated a circuit breaker having an interrupting rating of 100,000 kva. To be conservative the specifications called for a duty of 125,000 kva. Although that much capacity was not quite available during short-circuit tests, the performance of the breaker left no doubt that it could handle the slight additional increment to meet the specifications if called upon to do so.

Contact Burning Slight

Only twice during tests were the arc chutes removed and the contacts examined. A moderate amount of dressing of the main contacts was required, but burning of the arcing contacts was so slight as to require no attention.

The tests demonstrated that this particular breaker had the ability to interrupt the maximum kva it could encounter at the plant, and there was sufficient indication that it would in all probability be able to interrupt successfully the full rated 125,000 kva. It was felt, therefore, that as far as this particular application is concerned, a breaker having all the advantages of an oil-less interrupter had been obtained without any economic handicap. Whether that can be done in general cannot be predicted as a special study is required in each case.

6. LOGAN OPERATING EXPERIENCE†

SINCE the starting date of Logan station of Appalachian Electric Power Company until August, 1939—15,638 hours of elapsed time—the high-pressure section was available 8965 hours or 57.3% of the time, while the service hours were 8750 or 56% of the time. With a gross generation of 255,527,000 kwhr, the capacity factor for this period has been 40.8%. The availability of the high-pressure turbine alone was 86.4% and its service hours 56% of the elapsed time. The single high-pressure boiler, which is used for the low-pressure section of the plant through the pressure-reducing and desuperheating system when not supplying the topping turbine, had service hours equal to availability amounting to 71.8% of the period.

Since July, 1938, when the last major turbine difficulty was remedied, the record is more pleasant to report. The availability of the high-pressure section of the plant increased to 81.5%, the service hours to 80.2%, and the capacity factor to 61.8%. For the turbine only, the availability at 97.2% and the service hours at 80.2% have been quite satisfactory. The high-pressure boiler performance is encouraging with availability and service hours increasing to 84.7%.

During a recent 13-month period, the station heat rate of 14,000 Btu per kwhr compares favorably with the design expectations, remembering that a serious shortage of circulating water in the summer months, necessitating spray nozzles and recirculation, is an uncontrollable influence on overall plant economy. Under such conditions the condenser pressure rises as high as 3.4 in Hg, and the additional auxiliary power to operate the spray-pond pumps amounts to 8% of full-load generation.

Although the high-pressure turbine is rated 40,000 kw and the five low-pressure turbines aggregate 50,400 kw, making a station rating of 90,400 kw, much higher loads have been consistently possible because of hydrogen cooling of the topping unit and nozzle changes on the low-pressure units. Under favorable winter conditions, with ample circulating water, a gross station load of 103,000 kw has been attained representing a reserve capacity of 12,600 kw, or 14% over the name-plate rating.

Starting and Stopping

Basically, the system of starting and stopping is one in which the low-pressure boilers, the low-pressure turbines, the high-pressure boiler, and the topping turbine are treated as entities, each of which is operated separately but in sequence. Whenever the high-pressure boiler is out of service, at least two of the low-pressure boilers and two of the five low-pressure turbines are operated. The desired load on these units is maintained by backing down on the low-pressure boilers while steam flow is increased through the reducing valve as the high-pressure boiler is started and loaded. The high-pressure boiler could not be started without at least one of the low-pressure boilers in service because the

† ASME Annual Meeting, Philadelphia, Pa., December 4, 1939.

induced-draft fans are turbine-driven. Then the topping turbine is warmed through a manually operated atmospheric exhaust. When ready, the topping turbine under control of the speed governor is synchronized, then put under control of the back-pressure governor, and loaded while flow through the reducing station decreases.

The reverse order is followed when taking the high-pressure equipment off the line.

Under running conditions, the reducing valves are closed, the load is varied on the low-pressure turbines which, in turn, controls the steam flow and load on the topping unit through the medium of the back-pressure regulator on the topping turbine.

A most interesting incident occurred once when the entire station load was lost. The pressure in the 200-psi steam header, supplying the low-pressure units, dropped to as low as 60 psi, but this was sufficient to keep in operation the turbine-driven induced-draft fans on the high-pressure boiler without firing the low-pressure boilers.

High-pressure Turbo-generator

Major outages on the high-pressure turbine have resulted from failure on two occasions of the first-stage shroud band. The only other serious difficulty occurred when the thrust bearing let go, but since its repair there has been no recurrence.

Overheating of blades in the topping turbine, caused by motoring of its generator with insufficient steam flow when low-pressure units have tripped out, is prevented at Logan by a somewhat special reverse-power relay.

The scheme consists of a sensitive reverse-power relay calibrated to operate on the power component only and to close its contacts to trip the generator off the line in case the power flow is from the busbars into the generator with a magnitude of 0.5% of its rating for more than one minute, limited by a timing relay. A single-phase, reverse-power lockout relay makes the above sensitive reverse-power relay inoperative provided the generator is supplying 40% of its rated power to the busbars. This lockout relay insures that the generator will not trip off the line unnecessarily due to improper operation of the sensitive reverse-power relay on surges or otherwise. An alarm sounds the instant that the generator starts motoring so the operator can correct conditions before tripping occurs, if possible.

During the last year there have been two occasions when the low-pressure units have tripped out from failure of the group excitation for these units. On both occasions the high-pressure unit was prevented from motoring by the successful functioning of the special reverse-power relay.

When the topping turbine is being shut down, the rise in exhaust steam temperature is negligible before the exhaust is opened to the atmosphere. But in bringing the turbine on the line after a short shutdown period, the stored heat raises the exhaust steam approximately 175 F until the unit is loaded and under control of the governor.

There have been no vibration or noise problems with the topping turbine but the use of vibration and eccentricity recorders has proved helpful during starting periods.

The hydrogen-cooled generator has functioned most gratifyingly even under capacity load with adverse high cooling-water temperatures. Although this generator was designed for a gas pressure of 15 psi and has had test runs at this pressure with satisfactory gas consumption and lower winding temperatures, day-by-day operation has been confined to a pressure of 0.5 psi. Even with cooling water at 90 F, a higher gas pressure was not needed to effect proper cooling of the windings. Hydrogen purity is easily maintained at 99 to 100%, and gas consumption, averaging less than 400 cu ft per day, has been costing less than $300 per year.

Except for the original difficulty with removal of trapped air from the system satisfactory performance has been obtained from the pad-cooling arrangement for direct water cooling of the stator. There have been no signs of water leakage in the pads or in the piping, nor has it been necessary at any time to replace the condensate used as make-up for this pad-liquid system. Two heat exchanges are provided for the cooling of the pad liquid; both are used during the warmer months but only one is needed during most of the year.

High-pressure Boiler

The one million-lb-per-hr, dry-bottom, pulverized-fuel boiler, which supplies 1250 psi 925 F steam, has been subject to more outage than the turbine. Ash-screen tube failure resulting from impaired circulation and excessive heat input required replacement. To remedy this difficulty, smaller-diameter tubes having spiral circulating strips and an increased slope were substituted and burner angles adjusted with satisfactory results.

West Virginia bituminous coal with ash-fusion temperatures ranging from 2350 to 2700 F has been burned without difficulty. Slagging has not been serious and cleaning of boiler tubes by soot blowers and lancing has been easily accomplished.

Load swings of 10,000 to 15,000 lb per hr are readily handled; on occasion, it has been possible to drop load from 900,000 to 540,000 lb per hr momentarily. With tangential firing, a minimum boiler load of 300,000 lb per hr can be carried without loss of flame stability. With a combination of gas and coal, a load of 175,000 lb per hr has been carried successfully for a 2-hour period. Four auxiliary natural-gas burners are used for starting, and together are capable of carrying a boiler load of 50,000 lb per hr.

The four Hardinge mills, each rated at 15 tons per hr, have been rendering satisfactory service with a power input of 16 kw each at full capacity. To date these four pulverizers have handled a total of over 350,000 tons of coal. Experience indicates that the exhauster blades, made of boiler plate, require replacement after a mill has handled 35,000 tons; the wear on the ball charge amounts to 0.086 lb per ton of coal.

The two Ljungstrom air heaters, when cleaned once each shift with steam soot blowers, have maintained their capacity and have exhibited no tendency to fouling. However, the exit flue-gas temperature leaving the air heaters remains about 30 or 40 F higher than the manufacturer's performance guarantee.

Feedwater make-up is supplied from an evaporator whose feed is coagulated river water in which temporary hardness usually predominates, forming a scale in the evaporator not readily susceptible to cracking. To offset this, approximately 0.2 ppm of tannin is added to the evaporator feed for 1.0 ppm of hardness. Total dissolved solids in the vapor leaving the evaporator is confined to 2 ppm or less. A pressure-type deaerator, located midway in the feedwater-heating circuit and using exhaust steam from the turbine-driven induced-draft fans, has consistently maintained the dissolved oxygen of the feedwater at 0.005 cc per liter or less. Scavenging of final oxygen is obtained by continuous feeding of ferrous sulphate and caustic soda at the deaerator outlet in proportion to the dissolved oxygen in the feedwater. At no time does the boiler water show the presence of ferrous iron. Ortho-phosphates and caustic are fed in shots into the feedwater at the discharge side of the regulator for maintaining a pH value of 9.6 and a phosphate concentration of 10 to 25 ppm as PO_4. Boiler water total solids range between 100 and 250 ppm.

7. HYDROGEN COOLING†

UNTIL about 10 years ago the only cooling medium utilized for electrical rotating machinery was air, not that air has any particularly suitable characteristics, but it is the common surrounding medium which, of necessity, must be used for open-type machinery. The need for machines of constantly increasing size brought about the development of closed systems of ventilating, which facilities were provided for conditioning the cooling air, but primarily for removing dust particles. Elimination of the particularly objectionable oxygen had not been attempted although inert gas had been used for some time in transformers to prevent oxidation of oil or insulation.

TABLE 1. AVAILABILITY AND OUTAGE DATA, HYDROGEN-COOLED SYNCHRONOUS CONDENSERS
SEPTEMBER 30, 1939

Machine location	Total hours from time put into service	Total hours down for inspection and maintenance	Total hours available for service	Availability factor	Total hours out of service for any reason	Total running hours in hydrogen	Total running hours in air	Total outage hours due to H_2 construction	Availability factor on basis of H_2 construction
Atlantic City	78,888	3723	75,165	95.28	4301	70,694	3893	1621	97.95
Benton Harbor	75,552	1450	74,102	98,08	1834	71,840	1878	1142	98.49
Fort Wayne	75,936	4230	71,706	94.43	4685	69,186	2065	1116	98.53
Fostoria	24,408	3969	20,439	83,75	4094	18,544	1770	301	98.77
Kenova	34,584	1006	33,578	97.10	16,863	17,673	48	58	99.83
Scarbro	75,120	3250	71,870	95.67	4309	70,811		846	98.87
Scranton	52,824	442	52,382	99.16	11,055	41,769		165	99.69
Turner	94,440	1890	92,550	98.00	2856	91,584		720	99.24
Total	511,752	19,960	491,792	96.10	49,997	452,101	9654	5969	98.83
Total—machine years	58.44	2.28	56.16	96.10	5.70	51.63	1.11	0.68	98.83

Since the use of any medium other than air necessitates a gast-tight enclosure, it is desirable to use that particular gas or mixture of gases which offers the most advantages to outweigh the more expensive construction required. The desirability of using hydrogen for this purpose was first suggested by Dr. W. R. Whitney of the General Electric Company about 18 years ago. Consider-

† AIEE Winter Convention (with F. M. Porter), New York, N.Y., January 26, 1940.

TABLE 2. SEGREGATED OUTAGE DATA, HYDROGEN-COOLED SYNCHRONOUS CONDENSERS SEPTEMBER 30, 1939

	Atlantic City	Benton Harbor	Fort Wayne	Fostoria	Kenova	Scarbro	Scranton	Turner	Total	Per cent of forced outage
Total years in service	9.02	8.62	8.67	2.79	3.95	8.58	6.03	10.78	58.44	
Routine inspection { Collector compartment	112	107	136	25	16	34	31	63	524	2.63
Routine inspection { Main shell	95	180	229	†	†	160	28	414	1106	5.54
Bearings		247	304	157	60	20			788	3.95
Collector-compartment seal		334					83		417	2.09
Collector rings and brushes	406	300	166	79		492			1443	7.24
Excitation equipment	73	11	21	181	2	23	290	960	1561	7.83
H_2 piping and accessories				2				160	162	0.81
Lubricating system	127		42	50		10		90	319	1.50
Rotor			2942	3346		1464			7752	38.82
Temperature equipment				12			10		22	0.11
Vibration			92			998			1090	5.46
Water cooling equipment	2910	241	247	44	408	49		108	4007	20.07
Miscellaneous		30	51	73	520			95	769	3.85
Total	3723	1450	4230	3969	1006	3250	442	1890	19,690	100.00
Outage because of construction for H_2	1621	1142	1116	301	58	846	165	720	5,969	29.90
Shutdown for miscellaneous construction or not required	578	384	455	125	15,857	1059	10,613	966	30,037	
Total hours out of service	4301	1834	4685	4094	16,863	4309	11,055	2856	49,997	
In air	3893	1878	2065	1770	48				9654	
In hydrogen	70,694	71,840	69,186	18,544	17,673	70,811	41,769	91,584	452,101	
Total	74,587	73,718	71,251	20,314	17,721	70,811	41,769	91,584	461,755	
Total hours in service	78,888	75,552	75,936	24,408	34,584	75,120	52,824	94,440	511,752	

Left-side row groupings: "Forced outage hours" (Collector compartment through Total), "Hours out of service" (Outage because of construction through Total hours out of service), "Running hours" (In air, In hydrogen, Total).

† Indicates interior of machine was inspected at time of other major maintenance work.

able study and numerous tests made to determine the advantages of its employment indicated the superiority of hydrogen especially in the case of large high-speed alternators. The difficulty of providing a simple and reliable shaft seal, and the ability of designers to utilize air cooling satisfactorily delayed the use of hydrogen in generators until several years ago.

In the meanwhile attention was directed to the hydrogen-ventilated synchronous condenser. In 1928 two such machines were placed in service; one was the 20,000-kva unit at the Turner station of the Appalachian Electric Power Company. This was followed at intervals, by seven others on the systems with which the authors are associated, comprising in all a combined capacity of 161,000 kva. Successful operation of these machines led to the installation in 1937 of a 50,000-kva hydrogen-cooled generator at Logan, W. Va. This was followed by a 66,667-kva generator placed in service in the early part of 1939. An additional condenser was placed in operation in July 1939; still another in November 1939, and two more generators are in process of installation and will go into service early in 1941, making a total of 432,000 kva of hydrogen-cooled equipment on the American Gas and Electric Company system.

Experience with hydrogen has thus been far greater on the system with which the authors are associated than anywhere else.

The basic advantages of hydrogen are: low density, high thermal conductivity, high forced heat convection, prevention of detrimental effect from corona, no combustion or oxidation, cleanliness of atmosphere, ease of adaptation to outdoor service, and quiet operation.

Problems

The basic problems involved in the design, application, and operation of hydrogen-cooled rotating machinery, particularly synchronous condensers, rank in the following order of importance:

1. *Gas-tight and Explosion-proof Shell.*—While the possibility of an explosion is extremely remote, it is nevertheless necessary to take every reasonable precaution to minimize the effect of such an occurrence. Therefore an explosion-proof shell consisting of boiler plate has been employed.

2. *Water Cooling.*—The development of internal surface coolers of utmost reliability was essential, since the coolers are ordinarily inaccessible. It was desirable to install the coolers in such a manner as to permit periodic cleaning without removal of gas from the machine.

3. *Collector Compartment, Rings, and Brushes.*—For all synchronous condensers, excitation has been obtained externally from motor-driven or electronic exciters. The problem of finding the proper design and material for brush rigging, brushes, and collector rings suitable for operation in hydrogen was one that has to be determined largely by experience. Ventilating of the collector compartment, and the accumulation of brush dust brought up the questions of how often to clean the compartment and how to prevent any appreciable amount of this dust from being carried into the stator and rotor.

Operating Experience

Availability, determined for each machine, is above 94%. The total running time in hydrogen for all machines was 51.6 machine-years, or approximately 88% of the time in service. With trouble, which could be repaired immediately, it was usually found desirable, if possible, to run the unit in air at reduced capacity.

Collector-ring and brush trouble is all too frequent on standard synchronous condensers. Any difficulties with oil piping, water connections, or coolers inside the shell have been attributed to hydrogen construction. On this basis an availability figure, in so far as construction for operation in hydrogen is concerned, has been determined; the lowest availability is 98% and the average for all machines is 98.8%.

Inspections of main shells and collector compartments represent only 8.2% of the total forced outage time. The majority of the hours lost was because of rotor difficulties (38.8%) and troubles with the water cooling systems (20%). Of outage importance also were excitation equipment (7.8%), collector rings and brushes (7.2%), and vibration (5%). Of the forced outage only 30% can be attributed to construction for operation in hydrogen. Hydrogen piping and accessories accounted for 0.8% of the forced-outage time.

Rotors and Stators.—On most of the machines no difficulties occurred in connection with the rotating fields or the stator windings.

Water-cooling Systems.—Outage time of appreciable amounts in connection with the water-cooling systems occurred on all but one machine. The utilization of spray ponds and cooling towers necessarily requires periodic cleaning of the apparatus involved. Experience with these external water cooling equipments has been satisfactory for both summer and winter operation. Some difficulty has occurred because of strong winds blowing away the spray and thus causing excessive water consumption adversely affecting the efficiency of this equipment.

Experience with the internal surface coolers likewise has been very satisfactory; in only one case has leakage occurred when a slow leak developed at one of the vertical tube connections to its horizontal header plate. A slow but steady accumulation of water in the oil-and-water-leakage detector at the bottom of the machine gave ample warning of this difficulty, which was readily rectified by expanding the bell-shaped end of the tube into the socket of the cooler header-plate.

Outage Due to Hydrogen

Bearings and Lubrication.—Although the bearings are normally inaccessible, their reliability has been found to be such that maintenance on these parts of the machine is relatively small.

Due to the operation in hydrogen, there can be no oxidation of the oil and therefore aside from dissolving a limited amount of hydrogen, the oil remains in such condition that it is unnecessary to renew or recondition it periodically. On most of the machines under discussion the oil which was originally placed in the bearings is still being used.

Hydrogen Equipment.—Experience with the miscellaneous equipment necessary for operation in hydrogen has been most satisfactory. The automatic charging equipment for admitting gas to the machine has been entirely reliable and no failure to maintain the desired machine pressure has occurred. In the case of synchronous condensers, since the pressure is maintained above atmospheric, the gas leakage is always outward and therefore it is impossible for the purity to drop. The purity indicator is therefore of real service only when the machine is being filled with gas.

Hydrogen–Air Mixtures

It has been reported that 4.1 % is the minimum percentage mixture and 74 % the maximum percentage mixture of hydrogen in air which will burn under the most favorable conditions. This is true for the ranges of temperature and pressure ordinarily encountered but it depends upon the size, material, and shape of the container. In the case of mixtures of low-hydrogen content complete combustion under the best possible conditions will not occur until a mixture of approximately 10 % is reached. In igniting various mixtures of hydrogen in air, tests indicate that no appreciable or rapid increase in pressure is obtained in a sealed chamber for mixtures below 10 % or above 70 %. These values are therefore generally used as the lower and upper explosive limits for hydrogen–air mixtures.

The above figures are based on uniformly diffused mixtures of hydrogen in air and do not take into account the cooling effect on surrounding objects, such as stator, rotor, cooling coils, etc., as they exist inside a synchronous machine. Incomplete diffusion and absorption of heat by adjacent material considerably decrease the range in which the mixture is flammable. However, it is absolutely certain that a thoroughly diffused hydrogen–air mixture at temperatures and pressures ordinarily encountered, will not be flammable below 4.1 % or above 74 %, with the practical upper and lower limits closer together than this. This lower limit is the one which has been used in maintenance work on the authors' systems as a basis for determining when it is safe to enter the machine or to start it in air after removing the hydrogen. For normal operation of the condensers in hydrogen a lower limit of 90 % hydrogen in air is used thus allowing an ample safety margin.

Inert Gas Mixtures

The inert gases ordinarily used for the scavenging of hydrogen-cooled machines are nitrogen and carbon dioxide. It is necessary to know definitely the flammable limits of mixtures of these gases with air and hydrogen in order to determine when the proper amount of either gas has been used in changing from air to hydrogen and vice versa. If these scavenging gases are used in insufficient amounts or in a haphazard manner, the effect is worse than if no inert gas were employed, because there is not only the possibility that an explosive mixture may exist at some location in the machine, but also a false sense of security may be given to the operator.

On the basis of perfect diffusion and theoretically, 1.4 machine-volumes of nitrogen or 1.3 machine-volumes of carbon dioxide are required in changing from air to hydrogen in order to eliminate the possibility of having a flammable mixture present in the machine at any time. On the other hand, in changing from hydrogen to air, 3.2 machine-volumes of nitrogen or carbon dioxide are required.

For the filling operation, after approximately 0.8 machine-volume of carbon dioxide has been introduced, a 75% concentration of carbon dioxide is obtained at the top of the machine, with approximately 100% carbon dioxide at the bottom, which means an average of about 90%. In removing hydrogen, a concentration of 95% carbon dioxide is obtained at the top of the unit, after approximately 1.2 machine-volumes of scavenging gas have been used. Everything considered, it is believed best to obtain as a minimum a 75% concentration of carbon dioxide at the top of the unit when changing from air to hydrogen, and a 90% concentration when changing from hydrogen to air.

Higher Gas Pressures

The maximum benefits from hydrogen cooling are not obtained until gas pressures considerably higher than atmospheric, of the order of 15 psi above atmospheric, or even higher, are employed. Increasing the hydrogen pressure enables the gas to remove the heat from the windings more efficiently. The rate of surface heat dissipation by forced convection increases almost directly with the absolute gas pressure, thus more readily transferring the heat to the coolers. As a result, for the same cooling-water temperature, lower field and stator temperatures are obtained as the gas pressure is increased. This is true up to some limit not as yet determined. At the time of the purchase of the Turner machine, the advantage of high-pressure operation was fully realized, and test runs were made at 15 psi. However, the gas leakage at this pressure was excessive, primarily because the condenser was not especially designed for operation at pressures above 2 psi. In planning and building the Fostoria unit, high-pressure operation was taken into consideration, and its rating of 36,000 kva is based upon a gas pressure of 15 psi.

The effect of high-pressure operation can be seen readily. The temperature rise of the stator is 11 degrees less, and that of the rotor 19.5 degrees less with 15-psi hydrogen than with 0.3-psi hydrogen. Bearing temperature rises were also several degrees less. This indicates that the higher pressure gas is more effective in transferring the heat from the bearings and windings to the cooling water. The gas temperature drop through the coolers was 18.5 degrees for operation at a pressure of 0.3 psi and only 12.5 degrees at 15 psi or a reduction of 33%.

The machine carried 25,000 kva at 15 psi gas pressure, or 25% above its ratings at 0.3 psi with the same stator temperature rise and with a somewhat higher temperature rise of the rotor. It is evident that approximately 24,000 kva could have been carried at the higher pressure with a field temperature rise corresponding to 20,000 kva at low gas pressure, or a 20% increase.

Early in 1935 studies indicated the desirability of installing at Logan a 40,000-kw 3600-rpm high-pressure turbo-generator. Cooling of the rotor for

TABLE 3. TEMPERATURE RUNS ON 11.5-kv 36,000-kva SYNCHRONOUS CONDENSER AT FOSTORIA, OHIO, SEPTEMBER 4–5, 1937

H_2 gas pressure, psi	1	5	10	15	20
Stator					
Kva—actual	28,800	30,500	32,800	34,400	37,300
Corrected for normal voltage ..	30,500	32,500	34,500	36,300	39,000
Voltage ..	12,080	12,210	12,170	12,150	12,350
Amperes	1379	1445	1556	1621	1742
Temperature, resistance temperature detector (C)	57	57	57.5	60	61
Temperature rise (C)	35	32.5	33.0	31.5	33.5
Rotor					
Volts	134	142	150	158	174
Amperes	650	685	720	750	807
Temperature by resistance (C)	100	101.5	104	106	115
Temperature rise (C)	78.0	77.0	79.5	77.5	78.5

	CE†	OE†	CE	OE	CE	OE	CE	OE	CE	OE
Bearing temperature (C)	56.5	57.0	57.0	57.0	57.5	57.5	58.0	58.0	62.0	61.5
Temperature rise (C)	34.5	35.0	32.5	32.5	33.0	33.0	29.5	29.5	34.5	34.0

	1	5	10	15	20
Gas temperature (C):					
From coolers	22.0	24.5	24.5	28.5	27.5
To coolers	38.0	38.0	37.0	39.0	39.0
Temperature drop	16.0	13.5	12.5	10.5	11.5
Water temperature (C):					
To coolers	20.5	21.0	21.7	24.5	24.2
From coolers	25.0	26.3	27.5	30.5	30.8
Temperature rise	4.5	5.3	5.8	6.0	6.6
Outside air temperature (C)	15.5	15.3	15.0	16.0	15.8
Temperature cooling water to field thyratron tubes	34.0	36.0	37.0	39.5	40.0

† CE—collector end; OE—opposite end.

a generator of this size and speed could be accomplished only by using hydrogen ventilation. Following the extensive and successful experience with the seven hydrogen-cooled synchronous condensers then in service, the authors unhesitatingly accepted hydrogen cooling for this new generating unit.

Conclusions

Operating experience with eight synchronous condensers, dating as far back as 1928, and with two hydrogen-cooled alternators, permits some rather definite conclusions:

1. Construction for operation in hydrogen is one of the minor factors adversely affecting machine availability. The fact that two machines have each been operating for over two years with no direct supervision indicates that the reliability is as great or greater than that of air-cooled units.

2. From an economic standpoint gas consumption is an almost negligible item. The cost of hydrogen used is generally less than 5% of the savings obtained by reduction in windage losses.

3. Accessories required for operation in hydrogen have introduced no complications of any significance, and the reliability of this equipment has been fully demonstrated.

4. From a safety standpoint there has appeared no reason for concern when the machine is normally operating in hydrogen. As to the necessary handling of flammable gas, if reasonable precautions are employed there is considerably less danger involved than is ordinarily taken for granted in connection with operating and maintaining electrical equipment.

5. More than 45 machine-years of outdoor operation involving five machines during all possible weather conditions, demonstrate the adaptability of this apparatus for such service.

6. General inspections every 5 years have been found sufficient. Internal difficulties occurring between regular inspections have been reliably and quickly indicated by the various protective devices.

7. Field tests have demonstrated that in general the air-cooled capacity of a given synchronous condenser can be increased 25% by the utilization of hydrogen, at slightly above atmospheric pressure, and augmented an additional 20% by raising the gas pressure to 15 psi.

8. When designed for gas pressures up to 15 psi, reliable performance can be obtained and advantage taken of the additional cooling effects of a hydrogen atmosphere.

9. Utilization of hydrogen, particularly at pressures above atmospheric, permits not only the carrying of additional continuous loads, but also allows heavy swings of short duration with entire safety because of the ability of hydrogen to remove heat rapidly and efficiently from possible hot spots in the windings.

10. The condition of the windings, particularly in the machines which have been in service for more than seven years, affords every reason for anticipating insulation life considerably greater than is now obtained from units exposed to the injurious effects of oxidation and corona.

11. It seems clear, to the authors at least, that the hydrogen-ventilated synchronous condenser is preferable and more economical for units above 10,000 kva than the equivalent air-cooled machines. The future offers possibilities of higher voltage and further development at 1800 or 3600 rpm to take maximum advantage of hydrogen cooling.

12. Although tests seem to indicate that increases above 15 psi do not produce correspondingly proportional gains in cooling or capacity, operation at gas pressures above this level should, and undoubtedly will be, explored.

13. The comparatively limited experience with the hydrogen-ventilated turbogenerators at the Logan and Windsor plants is insufficient to draw any conclusions safely. There is definite reason to believe that results which have been obtained with the condensers will be duplicated in the operation of generators, which are substantially the same in construction.

13a VEP

14. Hydrogen cooling permits the construction of large high-speed turbo-generators, which could not be cooled by air properly. By going to hydrogen a more compact design results which, combined with savings from reduced windage losses, more than balances the cost of explosion-proof construction and accessories necessary for operation in hydrogen.

15. Judging by the action of the power industry and its engineers in the past three years it is increasingly apparent that hydrogen cooling has opened up a new field for the development and design of large capacity and high-speed machines. Hydrogen gas has been accepted by most of the profession as being the ideal cooling medium for obtaining greater capacity, greater efficiency, and longer life for synchronous condensers, turbo-generators, and for similar large rotative equipment.

8. INITIAL EXPERIENCE WITH 2400 psi AT TWIN BRANCH†

WHEN the electric light and power industry was born, in 1882, the era of expansion of the physical frontiers of the United States had barely more than begun. The next half century was the great period of frontier expansion; it also was the period of development of the electric power industry.

The termination of that half century of progress unfortunately was marked by a period of depression, deeper than any previously encountered in the country's history. And with the depression came prophets of disaster. Persons in high and low places preached the doctrine of the closed frontier—the idea that not only had physical frontiers been pushed out to their limits, but industrial frontiers had reached a permanent stopping point also. Under this doctrine new plants or better plants were out of the question; the country had all the physical plant it needed and all it was necessary to do was to divide more equitably what could be produced with a mere portion of existing facilities. These ideas were expounded regardless of the obviously calamitous implications to the country's future they carried with them. If the frontiers of the country were truly closed, what else lay in store for this and, more particularly, coming generations except poverty and stagnation?

Fortunately, not every one in the United States accepted the doctrine of the finished plant—of the closed frontier. With the passing of the deepest part of the depression new projects involving new techniques were undertaken by almost all progressive industrial and utility groups. New horizons in power production were exposed. Higher pressures, higher temperatures, larger boilers, higher-speed turbines, hydrogen cooling—all were mighty contributions to sounder economics and to the advance of the technology of electric power production.

Among these developments, one of the most outstanding is the Twin Branch 2500-psi project. To have brought this about, with a natural circulation boiler, in the face of so much that had to be done without precedent, is truly a technical achievement of the first order. The background of experience, the know-how, the courage to undertake so bold a project that all this called for, is a great tribute to the vitality and progressiveness of the organization which sponsored it, and of the industry of which it is a part. To place on the line, at a time like the present, a successful record-breaking, high-pressure, steam-electric power development, which in its more than six months of commercial operation has already created new thermal efficiency records, is a major contribution to national defense.

But striking as these phases of the Twin Branch development are, they are not nearly as significant as its contribution to the expansion of our economic horizons. If the private enterprise system, if democracy and freedom are to survive, then expansion and growth of the entire economic system must continue. No compromise can be made with the idea of a closed frontier.

† *Electrical World*, October 18, 1941.

13a*

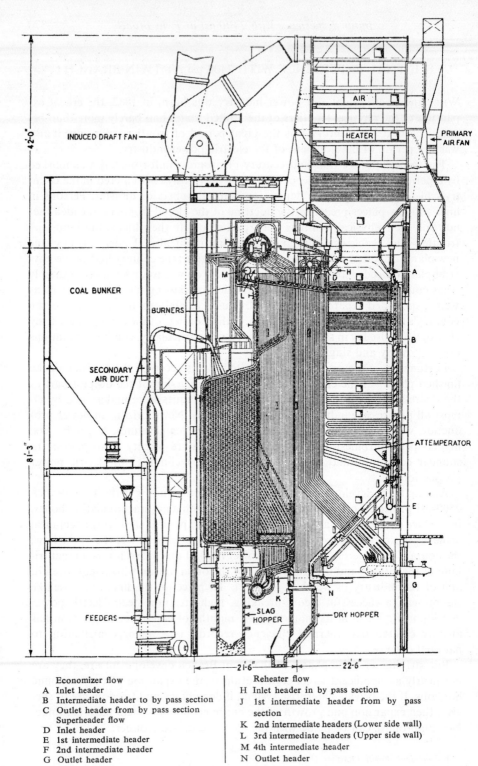

Economizer flow	Reheater flow
A Inlet header	H Inlet header in by pass section
B Intermediate header to by pass section	J 1st intermediate header from by pass section
C Outlet header from by pass section	
Superheader flow	K 2nd intermediate headers (Lower side wall)
D Inlet header	L 3rd intermediate headers (Upper side wall)
E 1st intermediate header	
F 2nd intermediate header	M 4th intermediate header
G Outlet header	N Outlet header

Fig. 1 Single-unit assembly for boiler for 2500 psi steam generation.

From the very beginning the aim of the Twin Branch 2500-psi project was to make a forward step in the economics of generation of central-station power. It was believed that the time had arrived when the technological limitations to high-pressure and high-temperature steam-electric generation could be advanced and thus contribute to both the capital cost and thermal economies of steam-generated power. This idea was reinforced through extensive discussions and negotiations with the group of forward-looking manufacturers of the principal items of equipment without whose help and co-operation the project could not have been undertaken.

No electric utility system can afford to rest long on its technical-economic oars. To be in a position to sell electric service economically it is necessary for a utility to be in the forefront of advanced sound economic practice.

Twin Branch, located near South Bend, Ind., is the main generating station of Indiana and Michigan Electric Company, which is part of the central system of American Gas and Electric Company.

Being the most expensive fuel point on the central system, Twin Branch is the logical place for the thermally most efficient units. The local load, initially estimated to reach a demand of some 165,000 kw by the time the 2500-psi unit went on the line, has actually reached a value of some 225,000 kw thus far in 1941, with a now-expected peak in 1944 of some 290,000 kw.

Changes from Original Plan

Technically, the project which was rated initially at 67,500 kw, was outstanding for the decision to resort to a single natural-circulation boiler, a cross-compound generating unit, of which the high-pressure end would be rated at 22,500 kw at 3600 rpm with a hydrogen-cooled generator. At the low-pressure end was a 40,000-kw unit, at 1800 rpm with an air-cooled generator, and turbine-driven high-pressure boiler-feed pumps, low-shell-pressure heaters throughout, solid switching of high- and low-pressure turbo-generators as a unit, stepping up through a single transformer bank.

First among the changes was the increase of the size from 67,500 kw to 76,500 kw. As the heat cycle was developed, it was found that not only could greater capacity be developed from the low-pressure machine than was originally contemplated, but that the additional steam flow to the high-pressure turbine made possible better economies. Another significant change was from turbine to motor drive for the boiler-feed pumps. Here a thorough analysis indicated that, on account of lower first cost and improved cycle efficiency, motor drive was more logical.

A third change was the use of high-pressure heaters for the 14th stage high-pressure and the crossover extraction points made necessary by the changed position of the high-pressure boiler-feed pump. The determining factors here were the fear of difficulties with 490 F water in the pump and the additional pumping load this would have involved.

The boiler was built as a natural circulation boiler, with mounting confidence in the soundness of the design from a circulation viewpoint.

Not only was the boiler plant itself built on the compact lines that 2500-psi makes possible, but the same ideas were followed in the layout and arrangement of heaters, pumps and related equipment and in the arrangement of turbine and condenser. The new section of the plant is a simplified masonry block. The design ended up with a cubage per kw of capacity of less than 25 cu ft.

The new plant was thus a radical departure from older practice in many of its aspects, but the departure was at no point either revolutionary or of such a nature that it was not possible to harmonize the physical, thermo-dynamic and electrical features of the plant with what had previously been erected there.

Operation

First placed in service late in March of this year the new extensions, except for several simple though time-consuming corrective steps, has been in continuous commercial operation at the design pressure.

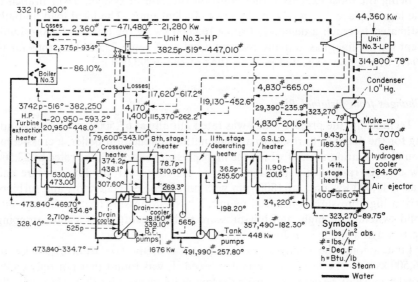

Fig. 2. Heat balance.

Availability.—For the entire first six-month period of operation combined availability of the unit was 72.5%. Of the total outage time 43% due to the high-pressure turbine and 57% due to the high pressure boiler. The combined availability for the last three months, however, is up to 88%, with every indication that the unit will continue in service until such time as we are ready to make modifications in the arrangement of boiler heating surfaces. Except for short periods of time when loads were limited by bearing vibration of the high-pressure unit, the plant has operated at a load ranging from 50,000 to 75,000 kw with the average for the entire period of operation being about 60,000 kw, or close to 80% of the design capacity of the unit.

Temperature Distribution.—Examination of the cross-section of the boiler unit gives an impressive picture of the difficult problem that confronted the designers. To incorporate in this unit the necessary superheat and reheat surfaces giving a 940 F final steam temperature and 900 F reheat steam temperature required a unique and bold arrangement of surfaces. To attain these final steam temperatures under the estimated variations in heat transfer rates, that would be caused by slag formation, radiant surfaces were installed in the first and second open passes.

Slag formation and distribution and, therefore, heat absorption and distribution between various surfaces, did not, however, follow the anticipated course and this brought about what proved to be the only serious difficulty with the boiler.

Slagging Trouble.—Trouble occurred first at a load of about 50,000 kw. At this point the reheat steam temperature reached 900 F and the thermocouples on the carbon-steel reheater tubes of the first and second open passes indicated excessive temperatures. In addition, the total heat absorption of the reheater was such that its correction was beyond the available control range. Expected accumulation of slag on these reheater walls did not take place. Furthermore, load could not be increased to speed up the slagging due to excessive tube temperatures. The unit, therefore, was shut down and the reheater walls were covered with a thin shell of refractory material by Guniting these surfaces with "baffle mix". This was done to give the slag a better chance of adhesion and also to permit increased boiler output and thus to speed up the rate of slag accumulation on the reheater tubes forming the side walls of first and second open passes.

Abnormal accumulation of slag was experienced in the second open pass. Reheater tubes forming the hopper were damaged externally by large masses of slag falling into the hopper. The unit was shut down for the third time for further corrective measures. To eliminate the steaming in the economizer, further insulation of reheater tubes in the convection pass was resorted to in order to permit greater gas flow through the reheater pass and thus better apportion the flow of gas through the divided boiler paths. Steel armor plates were placed on the reheater hopper sections as a protection against falling slag and to reduce the heat input into this reheater surface. To provide a better means for the control of and to prevent slag accumulations, telescopic soot blowers and a slag drip were installed in the second open pass. These modifications permitted the unit to operate up to its full design load.

Control of Blade Deposit.—Control of boiler feedwater and steam purity, of carryover and turbine-blade deposit are matters of experimental study. Whether the original accumulations were caused entirely by boiler-drum disturbances following steaming in the economizer or were caused in some part by carryover attributable to actual steam solution (as advanced by Spillner), they were sufficient to increase the first stage turbine pressure approximately 140 psi above the normal value with clean blades. When the boiler unit was shut down for modifications these deposits were washed out by filling the shell half full with cold condensate and rotating the turbine rotor with the turning gear. The increases in concentrations of the water during the first washing were as follows:

Constituent	Parts (ppm) per million
OH	29
PO_4	9
C_1	1
SO_4	113
SiO	106
Dissolved solids	376
Suspended solids	14

Bearing Instability.—Excessive vibration, caused by instability of the high-pressure turbine rotor in its bearings, first took place within two days of the unit's initial operation and at approximately half load. This condition was corrected by altering the oil flow in the bearings and the trouble seemed to have been solved. However, after several weeks of continuous operation at loads above 70,000 kw, the bearing vibration recurred, taking place only at these high loads. This proved to be the phenomenon of "oil whip" encountered on lightly loaded high-speed bearings. Bearing alterations were also made to assure a more positive oil pressure and the unit has since been operating satisfactorily under all loads.

Starting and Shutting Down.—Operating flexibility in bringing the high-pressure turbine on the line and in shutting down was the source of some early anxiety. Due to the lightness of the turbine rotor and to close inner clearances, the allowable difference in expansion between the rotor and the shell needed to be held within closer limits than first anticipated or desired. However, after careful studies of several subsequent starting and shutting-down cycles, a thoroughly well-defined procedure was established.

Starting from cold, the unit is brought up to a load of 60,000 kw at 2300-psi in about eighteen hours with ease. This starting cycle has proven conservative and can probably be shortened materially with full safety.

Pumps and Heaters.—In the design period a major concern over the auxiliary equipment centered on the high-pressure boiler-feed pumps and the 3200 psi high-pressure heaters. The heaters have given no trouble and the carefully developed procedure for placing them in and out of service, necessary to maintain tightness of tubes in the tube sheet, is proving entirely practicable. It is further being refined to simplify hot starting.

Some minor trouble was experience with the high-pressure boiler-feed pumps, none, however, resulting in plant outages nor being attributable to high-pressure design. In order to give a compact layout, the lubricating system was incorporated in the design of the base plate and the oil reservoir and channels for returning oil from bearings were welded integrally with the base plate.

It has been definitely proved that 2500-psi natural circulation boilers and 2300-psi steam turbines are technically sound and commercially workable.

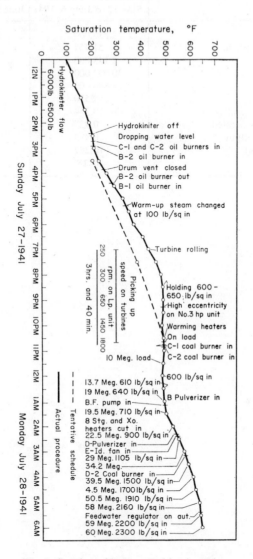

FIG. 3. Starting of log to rated load—18 hr.

INITIAL ECONOMIC PERFORMANCE

	Accumulative 622.83 hr May 12 to June 6, 1941	Accumulative 168 hr June 24 to June 30, 1941
Generated kwhr	32,236,000	11,092,000
Avg. kw generated per hr	51,757	66,024
Auxiliary kwhr	2,143,000	706,100
Output	30,093,000	10,385,900
Coal used (lb)	29,886,000	9,722,000
Oil used (gal)	2570	0
Heat value of coal	10,354	10,886
Lb coal per kwhr generated	0.927	0.876
Lb coal per kwhr output	0.993	0.936
Btu per kwhr output	10,282	10,189
Thermal efficiency	33.18	33.49

9. CO-ORDINATION OF MAJOR UNIT OVERHAUL†

ELECTRIC power systems with wartime loads are operating with plants loaded closer to maximum capabilities and with less reserve than ever before. While it is a definite duty under war conditions to operate with reserves as close to zero as possible, reserves may from time to time slip below the zero line and become negative. Under such conditions, knowing the stress to which equipment will be put, it is more important than ever to maintain equipment properly, and particularly to set up carefully considered scheduled programs of major maintenance and overhaul.

Scheduled Outages

In peacetime, each system can perform and operate on the basis of its requirements and its own experience, but in wartime it is necessary to go further and incorporate not only other available experience, but as many new ideas as possible and, particularly, broad major area co-ordination. Obviously the object of all this is to obtain maximum utilization of existing installed capacity. It is necessary to re-examine previous ideas as to the frequency of scheduled outage periods, the time when such outage is to take place, and the duration of each period in order to accomplish the necessary and indispensable maintenance and overhaul.

Of the three main pieces of major equipment in a steam-electric power plant, namely, the boiler, turbo-generator and condenser, the turbine was commonly considered most critical from the standpoint of the effect of periodic, or scheduled, maintenance on plant or system capability. This may have been the case when it was not uncommon practice to have a group of boilers which included some spare capacity, so that boiler maintenance generally did not reduce plant capability as long as the turbine was ready to run. But that practice, to a considerable extent, has been discontinued in most modern plants, and total steam-generating capacity and turbine capability are so closely balanced that the loss of any boiler, or a reduction in steam-generating capacity, will invariably result in a reduction in unit or plant capability.

This has brought about the condition where the boiler, in general, is more critical than the turbine from the standpoint of plant or system capability. The surface condenser, in any well-designed plant, has usually been the source of minimum disturbance to the continuity of plant or system capability. However, regardless of the relative influence of the three major pieces of equipment on continuity, the object of scheduled outage for overhaul and maintenance is, or should be, to replace all parts that have been exhausted by wear and tear since the last general or extensive maintenance, or to carry out those preventive measures which are likely to assure uninterrupted operation of the unit until the next scheduled outage can be arranged.

† *Electrical World*, October 31, 1942.

Boilers.—To attain these objectives, how frequently should a boiler be scheduled for outage and what are the determining, or governing conditions? The answers depend to a considerable extent on the design of the boiler, the type of fuel, the performance of the boiler with that type of fuel, and to a great degree on the way the boiler is handled.

A period of planned or scheduled maintenance, with every phase of the work properly organized and each item of material on the spot, ready to be installed in its particular place, is indispensable for high reliability or availability of equipment.

Turbine.—The steam turbine, of course, plays a major part in any program for scheduled overhauling. Not so far back, it was considered desirable to open and overhaul a steam turbine once very twelve months, or after approximately 8,000 hr of operating time. The unreasonableness of that, however, has been recognized for some time and most power plant operators have not only lengthened that time, but are satisfied that they are able to do so without detriment to the turbine or its continuity of service. This has been carried to the point where, with intelligent watching, many operators today believe it entirely safe to run turbines, unopened as long as 25,000 to 35,000 hr.

If such aids to sound operation as continuous eccentricity and vibration recorders are used, if stage pressures, and particularly stage-pressure differentials, are watched carefully, and if power capability of the unit or its steady decrease is kept under eye and remedial measures, such as washing, are taken, then the period between necessary openings of turbines can be extended to the 25,000 to 35,000 hr band mentioned.

It is necessary to distinguish between the loss of power due to accumulation of deposits and plugging of nozzle and blade passages, and the gradual loss due to wear of packing and diaphragms and partial or even entire loss of blading. The latter cannot be offset except by replacement of the lost or worn parts. Watersoluble deposits or plugging, however, can be removed by periodic washing with saturated steam.

Nevertheless, even with the most intelligent observation of turbine performance, and such remedial measures as can be effected without opening a turbine a period is reached for any unit where it becomes necessary to take it down if it is to be run with confidence and assurance for another 25,000 to 35,000 hr.

Condenser.—Intelligent operation with the help of proper screening and reversal of flow where there is trash and automatic and continuous chlorination where algae formation is encountered, is particularly helpful in getting optimum availability from condensers.

Co-ordination for Minimized Outage

The object of any program of co-ordination of major unit outage is to maintain the maximum margin feasible between demand on a system and load capability of the various plants serving the system. For an individual system this means careful study and evaluation of the shapes of the annual load and capability curves. The latter involves taking into account not only seasonal variations in hydro capability but seasonal variations in steam-plant capability. However, in wartime, with rapidly growing loads, three other factors have to be taken into consideration. These are the rate of growth of new load, because such growth can overbalance the seasonal trend factor; the rate of bringing in new capacity on the system; and the broad integrated, regional-area picture. While the last is desirable in peacetime, it is a "must" in wartime.

Reliance on Instruments

Prophylactic measures include the utilization of well-known instruments developed in comparatively recent years, such as shaft-eccentricity and vibration recorders. Value of the latter was shown when the No. 2 bearing of turbine "A" which had been operating with a vibration of approximately $\frac{1}{2}$ mil, that suddenly increased to about $1\frac{3}{16}$ mils. Inspection revealed that a tooth of the jaw coupling had broken off completely because of a long-existing fatigue crack. The instrument indication prevented what might have resulted in much greater damage.

Careful comparison of the progress of wear from one inspection to the next will make it possible to predict with rather good accuracy when a part needs to be replaced. While some parts, in spite of their appearance, will last much longer than their predicted life period, the risk of damage to other parts and the consequent forced outage of the machine will in many cases make it a good investment to replace parts in advance of the failure point.

Conclusion

The problem of co-ordination of major overhauls of the generating units on a large system, or in a large co-ordinated area, is more involved than can be covered here. However, what has been said should indicate that co-ordination and advanced planning of such a program can bring excellent results. Eveun so

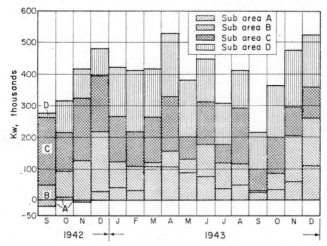

Comparatively steady level of total reserve resulting from a well-coordinated wartime program of planned major outage for maintenance.

a word of precaution is necessary. Advanced planning has definite advantages, but the problem is so important that it is necessary to keep working on it continuously, particularly watching equipment. Loads may come up faster, or not as fast as contemplated. New capacity additions may behave equally erratically. Changes in both of these factors will have a material influence on the reserve available on the system, and therefore on the overhaul program that the system can stand. Where hydroelectric capacity is a factor, its vagaries also have to be taken into consideration.

10. TWO YEARS' EXPERIENCE WITH 2500 psi †

MORE than 6 years ago a 2500-psi steam electric generating unit was projected for installation at the Twin Branch plant of the Indiana and Michigan Electric Company. It was placed in operation in March 1941. The story of that design and construction was fully covered by a series of articles in the October 18, 1941 issue of *Electrical World*. It is not proposed to repeat that story but rather to supplement it by giving an account of our experiences gathered over the subsequent 2-year period of operation.

During that time many lessons have been learned and solutions found to many old and new problems which, of course, were often delayed in execution because of wartime conditions. However, this did not cause any diversion from the main objective. The work went on, alterations were completed, and the unit has been operating at the design values of capacity, pressure and temperatures since April 1943. The boiler, however, has been operating at, or has been in a condition to operate at, full design values since December 1942.

Two main outage intervals were experienced during which substantial changes were made in one or more parts of the boiler. Both of these periods, however, were scheduled months in advance of the time when the unit was actually taken out. In other words, the difficulties encountered were such that they could have been endured. There was a period of some three months when the unit was operated at a pressure of approximately 1750 psi. But since the changes, completed on October 1, 1942, the unit has had no major difficulty. For one year the unit, in so far as the boiler is concerned, has been in a condition to operate at full design values of capacity, pressure, and temperatures, and has actually operated at these values for a period of eight months. This record speaks for itself.

The faith of the authors in the soundness of natural circulation for this design pressure was completely substantiated by measurement of circulation when the boiler was first brought to operating pressure. Circulation values exceeded calculated performance.

With 940 F steam temperature and 900 F reheat temperature, 43% of the Btus absorbed per pound of steam at the maximum design load of 550,000 lb per hr is used for superheat and reheat. On the other hand with 1300 psi and 950 F, only 27% is used for superheat; and with 400 psi and 700 F, only 14% is so used. Because of this, it was necessary to install superheater and reheater surface in zones of higher gas temperature than is usual on boilers operating at lower pressures.

Two tubes in the first division wall failed on two occasions when the load was raised to 480,000 lb per hr and the seventh and eighth burners were placed in service.

The first failure was a crack in a tube, the leakage being such that an orderly unloading of the unit was permissible. The second failure, however, was due to

† ASME Annual Meeting (with E. G. Bailey), New York, N.Y., November 29, 1943.

a sudden opening up of tube resulting in an instantaneous shutdown of the boiler. The automatic protective devices provided, coupled with the thorough training program in which the operating staff had been schooled, and the detailed manner in which preparations were made for such an eventuality, made possible a perfect handling of the situation. Turbines and all auxiliaries where shut down in proper sequence and the boiler was bottled up so that the drum was not subjected to any undue thermal stress.

Feedwater Chemistry

The problems in the boiler-feedwater chemistry, anticipated by some, did not materialize. The feed water treatment and boiler-water conditions that are maintained are comparable with those on our 1500-psi boilers. Regarding steam purity, the conductivity of steam from the boiler averages 0.8 microhm, equivalent to solids carryover of about 0.5 ppm. Tests on evaporated-steam samples substantiate this value.

A slow increase of stage pressures takes place on the high-pressure turbine. Whenever these pressures approach the established limitations, the condition is corrected by dropping the primary-steam temperature by attemperation for a period of several hours. Turbine-supervisory instruments with the wide range and complete control of steam temperature makes it possible to carry out this operation within a period of 2 to 4 hr while maintaining an output in excess of 65,000 kw.

Boiler Feed Pumps

Operation of the high-pressure boiler feed pumps has reached a point of reliability comparing favorably with that of pumps in plants of much lower pressure. However, a number of corrections had to be carried out before this availability was achieved. There are three 5-in. Ingersoll-Rand 8-stage boiler feed pumps direct-connected to 2500-hp 3600-rpm motors. Normally one pump is used to carry the full load of the plant. These pumps have forged-steel barrel-type casings, and the inner-casing assembly can be removed as a unit.

The first and most difficult problem encountered was vibration of the outboard, discharge end of the pump shaft and bearing which caused disturbances of packing, excessive shaft-sleeve wear, scoring of bearings, and wear of main oil-gear drive. This vibration was eliminated by adjustments to balancing-drum clearances, better shaft alignment, and rotor balance.

Operation and Performance

Operating and maintenance problems resulting from the high pressure have been relatively few, in fact less than on any of the 1500-psi plants of most recent construction on the American Gas and Electric Company system. There has been no difficulty with any of the high-pressure piping or valves. Some slight leakage is being experienced with the boiler-drum safety valves due to warping of seats and disks. Corrective measures are still in the experimental stage. This is true also of the boiler-drum gage glasses. The high pressure apparently causes some assembly distortion which, coupled with other factors, is resulting in considerable leakage and shattering of glass, thus requiring frequent and specialized maintenance attention.

The figures of 95% as a desirable and necessary availability factor and 10,000 Btu per kwhr as the thermal performance for a modern high-grade steam plant have been held up for years as ideals. The performance figure has been very closely attained and the availability figure has been exceeded.

The authors believe that the operating experience of the past two years has verified the judgment of those responsible for undertaking this distinctive development in power generation. Without going into the merits or demerits of other thermal cycles or equipment, Twin Branch experience has shown that 2500-psi natural-circulation boilers and 2300-psi steam turbines with their related equipment can be made to effect a power plant that is technically and commercially sound.

11. GLEN LYN'S 100,000-kw EXTENSION†

ALL major equipment for the 100,000-kw extension to the Glen Lyn station was ordered by the end of October 1941, just five weeks before Pearl Harbor.

By the time the decision to proceed with the extension had been reached, American Gas and Electric's central system peak was in the neighborhood of 1,200,000 kw; so a 100,000-kw unit was possible and sound from both a system operating and an economic point of view. Actually the unit chosen was an 85,000-kw nominal (95,000-kw capability) cross-compound unit. Five such units had already been placed in service or were on order: two for Philo, two for Cabin Creek, and one for Twin Branch. However, several months after the order was placed and war-time turbine orders began to pyramid, the War Production Board insisted that the unit be changed to a single-cylinder, single-flow design for two reasons: 1. Save use of large machine tools and space in the turbine manufacturer's shop; 2. Save critical materials, as the cross-compound unit weighed more. This was done in spite of the fact that plant performance would be adversely affected.

The Glen Lyn extension consists, therefore, of a 100,000-kw single-cylinder, single-flow turbine and two 475,000-lb-per-hr Combustion Engineering steam-generating units operating at 1300 psig pressure and 925 F temperature. The turbine constitutes the largest power generating unit, steam or hydro, installed south of the Mason and Dixon Line.

Coal Handling

All coal consumed by the plant is delivered by rail and whenever possible deliveries are arranged so that the cars are unloaded into a track hopper provided with crushing equipment. Conveyors take the coal over the bunkers. Coal for storage is dumped from a trestle and transported to the storage yard by tractor-bulldozer equipment.

Steam Generating Plant.—The two steam generators are designed for 1375 psi, 925 F at the superheater outlet.

The furnace has a "dry bottom" and utilizes pulverized fuel having a ash-fusion temperature range of 2200 F to 3000 F. The front, side, and rear furnace walls are constructed of bifurcated tubes; the tubes forming the roof of the furnace are finned. Total waterwall heating surface is 8088 sq ft. Furnace dimensions are: width 22 ft $4\frac{1}{2}$ in., depth 21 ft 0 in.; gross volume 26,900 cu ft. Boiler heating surface (generating surface) is 9042 sq ft.

The superheater is a two-stage interbank type with controlled by-pass, having a heating surface of approximately 14,900 sq ft. The economizer is 18 tubes high, 36 ft wide, and 22 ft long; it has fin-tube construction and 16,080 sq ft of heating surface.

One 44,300 sq ft Ljungstrom regenerative air heater is provided per boiler.

† *Southern Power & Industry* (with E. H. Krieg), December 1945.

The ash pits are designed to store the furnace ash in a semi-wet state. Soot and slag removal at critical areas is accomplished by hand-operated soot blowers of the type required—wall, retractable, or standard.

The four upper coal burners are served by one pulverizing system and the four lower ones by another. A manually-operated butterfly damper is provided at each exhauster outlet to close when the exhauster is idle.

Temperature Control

To regulate steam temperatures to the turbine at 925 F, plus or minus 10 F, the unit is equipped with a superheater by-pass damper and a spray-type de-superheater. The by-pass damper is located in a "dead" pass at the rear of the generator section of the boiler and is controlled automatically in parallel with the spray-type desuperheater located in the steam header after the non-return valve. The desuperheater consists of five sprays or nozzles in a section of pipe equipped with an alloy inner sleeve; each spray has an automatically-controlled inlet valve and one or more sprays go into service as required to maintain the desired temperature. A total of 20,000 lb of condensate can be supplied per hour if required.

Turbine and Equipment

The new 100,000-kw unit is about the same capacity as five other units installed on the central system during the last few years. Therefore, the requirements for system reserve, cold, hot, or spinning, were not disturbed by this selection.

The turbo-generator is an 1800-rpm, single-cylinder, single-generator type. Throttle steam conditions are 1300 psi, 925 F. This is the highest temperature single-casing turbine built so far by the General Electric Company, the previous limit having been 900 F.

Past practice has been to supply oil to bearings and control system by gear-type oil pumps driven directly from the turbine shaft. That type of oil supply has caused misalignment of the long shaft when the oil tank and pump were placed in the basement, some 30 ft below the driving gears on the main turbine shaft. These difficulties prompted the development of an independently driven oil supply system. The Glen Lyn unit is the first land-type turbine to have its bearing and control oil supply entirely divorced from the turbine shaft.

The turbine is equipped with an automatic low-vacuum trip so that loss of vacuum operates the emergency-stop valve just the same as the emergency governor does and closes off the main steam supply. Although the chance of the vacuum trip and the emergency-stop valve failing to operate is very slight, a further safeguard against excessive pressure and resultant condenser failure is provided in the form of a relief diaphragm mounted on the turbine exhaust.

Feedwater Heating

Five-stage regenerative heating is used; the drains from the high-pressure heaters are cascaded to the deaerator (Heater No. 3). Even though a deaerating hotwell is provided on the main condenser, experience has demonstrated the desirability of deaeration at a positive pressure, to insure that only completely degasified water will be fed to the boiler. The deaerator needs no air-removal apparatus because the operating pressure never falls below atmospheric. Under starting conditions, its pressure is "pegged" by boiler steam supplied through a small pressure-reducing system. This not only simplifies the equipment, but it also insures a fool-proof and continuous supply of completely degasified water to the boiler under all conditions of load swings.

The deaerator was placed at a high elevation so that there is a positive submergence head of 60 ft on the tank pump suction. This assures an adequate supply of water to the tank pumps with minimum likelihood of failure due to flashing or cavitation.

Feedwater is pumped from the deaerator to the boiler in two steps by constant-speed, motor-driven pumps. This permits the use of heaters designed for a maximum pressure of 800 psi, but entails handling hot water (at 400 F) in the boiler-feed pumps. Some small power saving might be effected by doing all the pumping at the lower temperature of the deaerator storage, but this would introduce the disadvantage of higher investment cost for the heaters (2000-psi design), piping and valves.

The low-pressure heaters are equipped with non-ferrous tubes, but it was necessary to use steel tubes temporarily for the two high-pressures heaters because of the unavailability of copper alloys under wartime limitations. These two heaters were re-tubed as soon as non-ferrous tubes became available.

Instruments and Controls

Through the use of remote hand control, automatic controls and complete instrumentation, it was possible to place the heaters, pumps, fans, piping and valves, and all other auxiliary equipment in locations that would give the best performance and economical installation without sacrificing operating facility. The control and instrumentation equipment were engineered to check operating results and to perform automatically many regulating functions in order to give the operators greater freedom in supervising performance.

To control boiler output, an automatic combustion-control system of an all-electric type was provided. This control maintains boiler output by regulating the rate of combustion in response to turbine load changes as reflected by steam pressure variations. The control power units are driven from the station battery by reversible motors equipped with dynamic braking to minimize coasting. Each power unit is provided with a separate push button, hand auto-transfer switch, and a position indicator located on the control board. Thus, remote manual control is always available for every control element.

Electrical Equipment

The generator has a rating of 100,000 kw at 0.9 power factor when operating with hydrogen at 0.5 psig. With hydrogen at 15 psig the rating is 128,000 kw at 0.87 power factor.

The winding is of the transposed bar type and is made in two sections which are independent of each other electrically. Each of the sections is connected in wye for operation at 13,800 v between phases.

The generator frame is built up of steel members and plates welded together. The four surface coolers are built into the stator frame inside of the housing. The housing is so constructed as to be satisfactory for operation with hydrogen pressure from zero to 15 psig.

Performance

The unit went on the line May 6, 1944, and since then has had a splendid service record. It was in service 96% of the first year of operation, with an average hourly gross generation of 97,168 kwhr. Peak gross generation for one hour was 113,000 kwhr.

Boiler availability was especially good. During the first year, one boiler was in service 96% of the time—only forty hours less than the turbine. It burned 168,987 tons of coal, generating 3706 million lb of steam. The other boiler burned 161,283 tons of coal generating 3536 million lb of steam. Pounds of water evaporated per pound of coal averaged 10.97 for both boilers over the entire year and the station water rate was 9.57 lb per net kwhr. Pounds of coal per kwhr gross generation were 0.8202 and per net generation 0.8727. Monthly Btu per net kwhr has been down to 11,009 with a yearly average of 11,219 Btu per net kwhr.

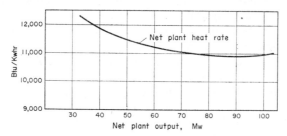

Net plant heat rate.

Low-grade Coals

One of the outstanding benefits from the new installation is the ability to burn poorer grade coals that could not be handled by underfeed stokers. This coal is considerably higher in ash content than normal coal, but under wartime exigencies had to be used.

12. CONSIDERATION OF ECONOMY AT TIDD†

FUNDAMENTAL bases for economical generation can well bear restatement. Briefly they are:

1. Proper site location from the standpoint of area available, water, coal, tie-in to the transmission system, and economical transmission to load centers.

2. Size of units, particularly turbines and boilers, terminal pressure and temperature for the steam turbine, and cycle of operation.

3. Operating arrangements and techniques.

An important consideration in the choice of the Tidd site was that that particular point on the Ohio River is not only the shortest distance from any point on the river to Canton which was available for development but that it makes possible the most economical tie-in to the 66-kv transmission system serving the Ohio Valley.

Thus major economies were brought about by minimizing the cost of transmission, and effecting optimum tie arrangements between Tidd and Windsor with the least increase of short-circuit requirements in the buses.

The turbine is a 100,000-kw single-shaft, 1800-rpm unit operating at 1350 psi and 925 F. While the temperature is lower than originally intended it was determined by the limitations of a single-cylinder machine of that size, and by wartime conditions which made mandatory the consideration of designs that used a minimum amount of material, particularly alloys. The cycle is basically straight regenerative but carried to the point of maximum economy attainable within the range of pressures and temperatures adopted.

Simplicity and Automation

It was recognized from the very beginning of the design that cost of labor had greater significance. Hence, ultimate manning requirements were given full consideration and no step in design was taken whether it involved location of equipment, choice of equipment, installation of such facilities as elevators or communication equipment, without considering the effect it would have on the operating setup. This eventually led to the adoption of a number of novel ideas, all of which have proven highly effective. Among them are:

(a) Centralization of controls for boilers, turbine and condenser at a single operating center in the hands of a single skillful and well-trained operator.

(b) By confining operation to generation transmission control would be handled elsewhere and would be put on an automatic basis.

(c) To the maximum extent possible other operating features were automatized.

The essential thing sought in the entire design was simplicity. This was helped materially by the adoption of the philosophy that the plant itself is part of a system and that it did not have to be designed to be self-contained and self-

† *Electrical World*, August 2, 1947.

sustaining like a fort in a desert, so to speak. In turn this made possible the elimination of a vast amount of duplication of facilities, by-passes, and all the other ramifications which enter a design once the idea of continuity is adopted.

It made possible the adoption of the large size turbine-generator. It played a vital part in the simple auxiliary service supply layout. Although not fully taken advantage of in choosing boiler size in the initial unit, it was exploited to the limit with unit No. 2, now in construction, with the adoption of a boiler of 120,000-kw output. Again, a considerable number of economies, both initial and subsequent, were obtained by adhering strictly to the idea of designing for maximum capability of the turbine chosen without adding to nor subtracting from its capability by lavishness or skimping on the sizes or ratings of other items of equipment forming part of the series.

Planning on such a basis is not only rational but produces results attested by the fact that on a system noted for its economy of generation, the Tidd plant is not only thermally the most efficient station but also the most economical one. This despite the fact it is a single-unit station, with provision for future expansion, and thereby labors under an ineluctable handicap in almost all its costs elements when compared with stations more fully developed in terms of their ultimate capacity.

13. REHEAT CYCLE AT SPORN†

DURING the war the need to get capacity on the line in the minimum possible period made substantial duplication almost mandatory, but with the greater freedom to plan new facilities in the post-war period it became feasible to try again to move forward on the front of economical energy generation. This has been at least partially accomplished by the 2000-psi, 1050 F, 1000 F reheat installations at Philip Sporn (Nos. 1 and 2) and at Twin Branch (No. 5). In each instance effort was directed to combine in optimum fashion the elements of site, cycle, fuel, and equipment, which warrants further discussion:

Site Considerations

Offhand, it might appear that the selection of an adequate power plant would not be an involved affair and that adequacy from the standpoint of area, condensing water, and location above flood plane are perhaps all one has to be concerned about. But this is a superficial view. The following also should be considered:

> Location relative to present and future load centers and power flow on transmission lines; foundation conditions; amount of reasonably flat area available without moving roads, railroads, or large quantities of earth; railroad connections; transmission requirements; availability of economical coal supply; and ash-disposal facilities.

Perhaps these are merely subsidiary to the main theme—location—but they are so important that they cannot be treated independently. There are lesser requirements for an ideal site, such as facilities for operating personnel, and tax costs. But the foregoing factors indicate clearly the nature of the problem and show why an exploration of a hundred miles or so along a typical river frequently will disclose, instead of the expected score or more favorable locations, not more than perhaps two or three first-rate plant sites.

Fuel.—Fuel is a most essential element to consider in site analysis. A fundamental for developing the most economical energy supply is the need to exploit the lowest over-all fuel cost. This does not mean necessarily the lowest-cost fuel in every case, and certainly not the lowest-cost fuel at the mine. It might definitely indicate the necessity for exploiting a high-quality fuel at considerable distances from the mine. Conversely, it might indicate the need, if the plant were located close to the mining regions, of exploiting a fuel of the very lowest quality—perhaps one that would have no commercial value if freight had to be added to its mine cost. This raises the inevitable problem of determining whether to locate a power plant as close to the source of fuel as possible, entailing added transmission cost, or at some distance from the mine with a higher fuel transportation cost but lower electricity transmission cost.

There has been considerable questioning of the economies to be obtained from higher pressures and temperatures. Thermal efficiency alone can never be a proper objective in a successful commercial operation. But the developments

† *ASME Transactions*, May 1948.

in power generation over the last quarter of a century at least have clearly shown the economic benefit of more efficient generation. It is now clear that many of those who have questioned the economics of more efficient generation have merely been slow in acknowledging the economic benefits that have flowed from such new developments. While others, by holding back, have been able to take advantage of the development work carried out and the concomitant headaches borne by pioneers.

The next major step in economical development of generation on the system was left for Twin Branch Unit No. 5 and for the Philip Sporn station.

Twin Branch Addition

Of all of the major stations on the central system, Twin Branch is the most poorly located for fuel, it is farthest from the center of mining and therefore has the highest freight rate. Nevertheless, the decision to install another unit at Twin Branch was sound for the following principal reasons: 1. The large saving in transmission investment that was possible; 2. It represented an incremental installation—perhaps the last increment that could be made at Twin Branch within the limits imposed by the circulating-water facilities; 3. The present addition of Unit No. 5 fitted into a future program for developing an additional power-plant site in Indiana nearer local coal supplies, and the interconnection facilities between Ohio and Indiana.

Unit No. 5 at Twin Branch is similar in terms of equipment, cycle and basic arrangement to that at Philip Sporn station. The present discussion will be devoted to the latter plant.

Sporn Station

Site Features.—The site consists of approximately 275 acres with 6000-ft frontage on the Ohio River. Contiguous coal reserves in Ohio and West Virginia of approximately 22,340 acres have an estimated 132 million tons of coal. This contiguity extends to the coal on the Ohio side because of the ownership of a sufficiently large block of connecting coal under the river bed of the Ohio River. The river at that point is not only navigable with a 9-ft slack-water navigation channel, but has a minimum flow of about 2200 cfs so that there is no question about the adequacy of the site for a plant of at least three times the size of the present project. The coal reserves give ample assurance of an adequate and economical coal supply with a minimum of transportation cost throughout the depreciable life of the plant. The extent of that coal supply and its contiguity to the plant give further assurance that developments in underground gasification, if they should materialize, can be exploited in the future for such savings as may become possible.

Heat Cycle.—In adopting the 2000-psi, 1050 F initial temperature, 1000 F cycle the major objective was optimum financial return, taking into consideration not merely economy of fuel but also reliability, for fuel economy means little to a plant that is shut down.

The heat rate at a net plant capacity of 141,000 kw is estimated at 9270 Btu per net kwhr output with an expected 90% boiler efficiency. This is a better heat rate than has been incorporated into any steam-electric station up to the present time.

In studying the economic possibilities of the various cycles, there was no question that a regenerative-reheat cycle offered the best heat performance. This led to comparative cost estimaties of reheat versus nonreheat, with the balance in favor of reheat. As finally adopted, reheat gave roughly 5% better heat rate, with about 2% higher first cost. This makes the reheat cycle a most attractive investment for this station.

Only six stages of bleed heating are used to achieve a final feedwater temperature of 441 F. Because of the large size of the installation, gland steam leak-off condensers and a separate evaporator condenser were justified. Placing the heater drain pump on the 18th-stage heater was found more economical than on the 20th-stage heater which cascades its drips to the condenser.

The means used to cool the main generator is another example of the effort made to incorporate high reliability in the cycle. After leaving the condenser, a portion of the condensate is passed through the generator hydrogen coolers. The purpose is to preclude the frequent cleaning of the hydrogen coolers which would result if river water were used.

Although condensate temperatures are sufficiently low during winter months to cool the hydrogen and the generator, they are too high during the summer. A condensate cooler therefore is provided in which river water cools the condensate to the degree required in the summer. Even after passing through the hydrogen coolers, the water is still cooler than in the main condensate stream. Therefore the cool condensate is sent to the suction of the hydrogen-cooler pump to pass again through the hydrogen cooler—to condensate cooler—to pump cycle. The cool condensate in this cycle "floats" on the main condensate stream but does not mix with it. Thus not only is the objective of reliability attained, but also fuel economy, since the main condensate stream is not cooled off by the cooler condensate in this subcycle.

Turbine Features.—The new cross-compound turbine is a recently worked out combination of high-pressure and low-pressure turbines. The 3600-rpm high-pressure element has a rating of 35,000 kw and a capability of 42,000 kw. Its throttle conditions are 2000 psig, 1050 F. The two emergency stop valves as well as the turbine inner shell will be chrome-nickel-molybdenum, columbium stabilized, and the outer shell will be molybdenum-vanadium.

The 1800-rpm low-pressure element has an intermediate-pressure turbine in tandem with a double-flow low-pressure section. The whole is rated at 95,000 kw with a capability of 108,000 kw, and its generator will be almost identical with those at Glen Lyn No. 5 and Tidd No. 1 and No. 2. The intermediate-pressure section receives steam from the reheater at 375 psig, 1000 F, and exhausts at about 8 psia to the double-flow low-pressure section. Its high-temperature parts will be of molybdenum-vanadium.

Among the features that improve the reliability and safety of these turbines are the following: Protection against sudden temperature changes by an initial-pressure regulator; protection against excessive starting speeds by a two-speed, motor-operated, synchronizing device on the main operating governor; automatic steam-seal regulator and unloading valve; and automatic low-vacuum trip.

Boiler.—The total heat input of 1.3 billion Btu per hr to produce 150,000 kw gross generation, makes the boiler one of the largest, if not the largest for net input that has been projected anywhere. It is designed for maximum continuous steam output of 935,000 lb per hr at 2035 psig, 1050 F at the superheater outlet. A temperature of 1050 F will be maintained between 850,000 to 925,000 lb per hr by an attemperator between the primary and secondary portions of the superheater.

After passing through the high-pressure turbine, the steam will be exhausted at about 400 psi, and some 85,000 lb per hr will be returned at 650 F to the reheater where it will be raised to 1000 F. This temperature will be maintained at 1000 F between flows of 750,000 and 850,000 lb per hr by an attemperator just before the inlet header to the reheater.

Flue gases discharged from the Ljungström air heaters go into the stack at about 235 F. The 12-in-long cold-end layer of the air heater will be made of Cor Ten steel, which, together with recovery of air heated by boiler-room radiation losses, will resist corrosion caused by the low flue-gas temperature.

Another outstanding feature will be the attempt to run without induced-draft fans. To do this the entire casing will be made tight for a furnace pressure of 15 in. of water. However, induced-draft fans will be installed in case this forward step in the art of coal burning is not successful.

No coal scales will be installed for individual boilers, but all incoming coal for the entire station will be weighed while on the belt.

A major source of dust and dirt within a plant is the expansion joints in the hot-air ducts between Ljungström air heaters and the burners. In the Philip Sporn plant all hot-air ducts under pressure will be welded, and bellows-type expansion joints will replace the usual slip-type joints.

Initially, outside air will be brought in on the windward river side of the plant at a point where the cleanest air should be obtained by taking advantage of prevailing winds. Such air will be distributed throughout the boiler room and condenser pit by two large ventilating ducts on each side of the elevator shaft. This clean air will be distributed throughout the boiler room to pick up heat and push dirty dust-laden air to the forced-draft fans, this action being aided by natural thermal air currents. Provision is being made to install air filters in the future, should these prove necessary.

Electrical Arrangement.—The outstanding feature of the whole is its extreme simplicity: One 52,500-kva and two 62,500-kva 3-phase transformers step up the voltage of the high-pressure generator, and the two windings of the low-pressure generator, to 132,000 v through a single switch to each of the two 132,000-v buses. No low-voltage switches or buses are employed. Auxiliary power is furnished by two 6000/8000-kva transformers each connected to one of the two low-pressure generator windings and stepping down to 2300 v. Seven feeders at 132 kv take the output of the station to the system. Each of these is switched through a single oil switch operable on either the main or the reserve bus.

14*

14. NEW CRITERIA FOR ECONOMY AT SPORN†

A NEW frontier has been opened on the Ohio River within the past few months with the successful opening of the Philip Sporn plant—a new milestone in the development of large-scale economical generation of electric energy by steam.

Impressive as are the size of the boiler, the throttle temperature and pressure adopted, and the further development of the reheat cycle, other features of the plant are even more noteworthy. In these days of monumental and ever-growing federal projects, the completion of this undertaking effectively demonstrates the ability of private capital to conceive, undertake, and carry through large-scale developments. Here is a power plant conceived as an entity to develop 900,000 kw, first step of which carries it two-thirds along the way to completion. When the project is completed, it will have a capability of generating 6 to $6\frac{1}{2}$ billion kwhr per year—almost a billion kilowatt-hours in excess of the generation at Hoover Dam in an average year, and equal to 40% of that produced in 1949 by the Tennessee Valley Authority in its score or so of plants.

The performance bogey of 9300 Btu per kwhr to feeders—0.715 lb of coal per kwhr—is another striking feature of this development. This new yardstick of efficiency in the conversion of fuel to electric energy has implications of enormous scope from the standpoint of conservation of resources and national defense.

The fundamental concept behind the Philip Sporn plant design was to set new criteria of economy—both capital and operating—in large-scale steam-electric generation of electric energy. Everything about the project has been bent to that end: Its location, design, construction and operating organization, and the operation of the plant itself.

At the present writing Unit 1 is in full commercial operation; Unit 2 is just about ready to go; Unit 3 is scheduled to go on the line on July 1, 1951; and Unit 4 on May 1, 1952.

Each of these units, originally rated at 137,500-kw capability, has been re-rated to 150,000-kw as a result of actual operation of the similarly designed Twin Branch Unit 5 and of Sporn Unit 1. The design of the plant permits economical and ready expansion by two additional units—Units 5 and 6. Reasonably early construction of these units can be expected because of the rate of system growth.

The prospect of a large-scale generating center where 6 to $6\frac{1}{2}$ billion kwhr of energy will be generated annually is sufficiently close to realization to make a discussion and appraisal of the design details of interest.

Joint Ownership

The Philip Sporn plant, located on the Ohio River at the boundary between the two largest operating groups in the American Gas and Electric system—

† *Electrical World*, June 5, 1950.

Ohio Power and Appalachian Electric—has been developed logically as a joint station of these two companies.

Even though ownership is divided it is pertinent to inquire if it is reasonable to plan a 900,000-kw station and concentrate so much capacity at one point. A rational answer to such a question could not be given except on the basis of system growth—past and expected. Shown in Fig. 1 is the history of the system

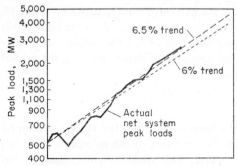

FIG. 1. Peak loads versus long-term trends.

annual peak demand, plotted logarithmically from 1928 to 1950, with a projection of expected growth in demand through 1953. For comparison there are shown 6% and 6.5% trend lines, based on the interval 1928 to 1953. Thus, on the basis of an expected peak of approximately 2 million kw in 1953 (with an actual recorded demand of over 2 million kw as of March 31, 1950) the plant, as constituted in 1953 with a rating of 600,000 kw, will have a capability equal to 24% of the then peak. Assuming that the plant will have reached its ultimate size of 900,000 kw in 1958, its capability will represent slightly over 25% of the then peak. On the other hand this will be only 18% of the peak expected in 1965 on the basis of a long-term 6% trend in growth. In view of the economic advantages to be gained from exploiting the site to 900,000 kw, the development of a plant of this size was considered feasible and, in fact, highly desirable.

Though each company will have a plant at that location of only 450,000 kw the joint operating arrangement gives each the capital benefits of a high-grade site, much lower cost of site per kw of capacity, and very substantial savings resulting from the development of general facilities—such as coal-unloading and ash-handling, coal-storage, trackage, hydraulic system and machine-shop. It also enables each company to get the operating benefits of lower fuel costs, lower supply costs, and lower costs of supervision, operation and maintenance.

Fuel Supply

At its ultimate capacity of 900,000 kw, the Sporn plant will burn close to 200,000 tons of coal a month. No construction of a plant of this magnitude could have been initiated on a sound basis unless reasonable assurance of fuel adequacy and economy first had been obtained. This assurance was provided by an extensive contiguous coal reserve with the plant roughly in the center of the area; the reserve encompasses some 22,348 acres and an estimated 90 million tons of recoverable coal. An initial underground mining installation will provide a capacity of 600,000 tons of washed coal per year subject to no freight costs. The remainder of the requirements for the first four units is being obtained from local commercial mines at short river-transport distances from the plant. This coal will be delivered by barge to a dock equipped with ample facilities for unloading economically at the rate of one million tons per year, or for loading coal from the extensive storage provided at the plant for delivery to other plants of the two companies located on the Ohio River or its tributaries. Thus by intimate co-ordination of the two problems of site selection and coal supply, adequacy and economy of supply running the probable life of the present plant have been assured.

Heat Cycle

Once the cycle has been chosen and the design developed there is substantially nothing which the operators of the plant can do to improve the thermal performance of the plant. This commitment is final as long as the plant is operated on the basis of the original design. It can be changed only if the plant is entirely rebuilt, or if a major modernization or improvement program is carried out. In many respects the economic soundness of the plant to continue as the center for production of economical energy is influenced by the choice of the cycle. Yet the judgment necessary for determining a cycle can very seldom be given if the heat balance is considered independently and out of context with the other elements in a power plant, and if what has gone before and what might be in prospect are overlooked.

Progressive Economy

The essential operation in a steam-electric plant is a conversion of thermal energy to electric energy, and all equipment and functions are subsidiary to that main operation. Hence when it came to justification of capital expenditures it was far more desirable to doubly scrutinize non-productive expenditures, such as those involving factors of space, safety, and decoration, than the very essential expenditures which might contribute to plant efficiency, provided these latter are fully self-supporting. It was most helpful within the organization's thinking to realize that progress in economical steam generation has only been brought about by a willingness to explore and exploit relatively small individual gains, which collectively pushed efficiency to levels beyond any that were obtainable previously, and that no progress would result if full advantage was not taken of the steps that were immediately available.

Progress and performance in nine steam-electric power generation developments on the American Gas and Electric system are presented in Fig. 2. Each of these was distinctive and unique, and each represented a contribution toward economical generation in its own particular sphere as against previously established standards or limits. The cycle at the Sporn Plant—2000 psi, 1050 F and 1000 F reheat—takes its proper place as the culmination of a long line of forward-marching strides in thermal economy. It represents the highest practical thermal efficiency obtainable today by the use of the regenerative reheat steam cycle.

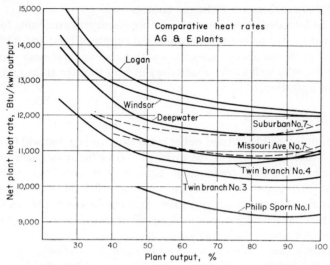

FIG. 2. Culmination of long line of progress in thermal efficiency.

Turbine and Boiler Units

From studies of turbo-generator units, within the limits of temperatures and pressures contemplated, it was concluded that the most economical results, both from a capital cost and an operating standpoint, would be given by a unit having a nominal rating of 125,000 kw and a maximum gross capability of 160,000 kw, thus giving a feeder capability rating of 150,000 kw.

The design adopted calls for a single pulverized-fuel boiler. Although the boiler was designed for full forced-pressure operation without induced draft such equipment was installed initially for ease in working out the details of the new step. No header connections are provided on the steam end. Probability methods for determining plant reliability have often failed to take into consideration system size and growth, functioning of major interconnections, arrangements for emergency supply, and the effect of various schemes of generation on both capital and operating costs. In the final analysis what one is concerned with is the greatest reliability per dollar of investment and the lowest overall cost of service. This result was given by the basic scheme chosen.

A cross-compound turbine was chosen to take maximum advantage of the high internal efficiencies possible by going to 3600 rpm on the 2000-psi, 1050 F high end and to 1800 rpm on the 1000 F reheat low end. These introduced very little additional cost beyond the additional space required because the high- and low-pressure units were treated as a single unit on both the steam and electrical ends.

Even with the cross-compound idea to contend with, the design yielded a unit figure of 23.3 cu ft of building space per kw of gross capability on a 4-unit basis, and an expected figure of 22.9 cu ft when the plant is fully developed.

On the boiler end no insurmountable problems of size seemed to present themselves, due to the experience with a 120,000-kw (gross) generating unit that was brought on the line very successfully in 1948 at the Tidd station.

The turbo-alternators are hydrogen cooled.

For the boiler a dry bottom was chosen because it was attainable for the full range of coal needed to assure an economical supply of fuel; and it eliminated some of the problems always associated with handling large-scale molten-slag hearths or floors as well as the uncertainty of heat transfer always present in a wet-bottom operation.

Electrical System

No distribution will be carried out at generating voltage but will be at 138 kv for the first three units, and of the order of 300 kv for subsequent units. This made possible simplified connections between generators and transformers without the intervention of low-voltage switching. Adoption of 3-phase transformers for the step-up service, utilizing three transformers per machine, one for the high-voltage machine and two for each of the two windings of the larger low-voltage machines, eliminated low-voltage buses.

The auxiliary supply, except for emergency-tie connections, was developed strictly on the unit basis, utilizing a 2300-v bus and 2300-v service for motors from about 150 hp and upward, and 550-v service for motors from approximately 125 hp and below.

Based upon extensive experience all generators, principal transformers, and all main buses have been fully covered by differential relay protection, with emphasis on providing maximum safety with minimum expense. The use of full protection on equipment and buses made possible the adoption of a single switching scheme on all high-voltage feeders, with a net saving in all-around cost and an improvement in the degree of reliability.

Control and Operation.—Previous experience with a completely centralized control system at Tidd station had indicated the soundness of the basis idea. Further development of that idea on Missouri Avenue No. 7 and Tidd No. 2 gave additional reinforcement to on original concept. It was therefore carried out to the full at Sporn.

It has been demonstrated that the operation of a plant from a centralized control point made possible the utilization of a much higher grade of personnel which could be given greater scope and responsibility, while the existence of such personnel gave the plant an additional assurance of safety and continuity of operation, not subject to hazards encountered in industrial relationships. New devices, such as the electrically transmitted boiler-gage-glass image, furnace-flame image, and the improvements in remotely operated soot blowers contributed to the further development of a basically sound and relatively new idea.

Control of the electrical end was developed as part of the basic concept. First the control of turbines was made part of the centralized plant control and the plant discharges its responsibility by placing the unit on the high-voltage buses. Distribution of the energy is placed in system hands. Control and operation of high-voltage feeders are not only placed within the high-voltage yard, and at some distance from the plant, but are made automatic. Thus economy of operation and lesser probability of service disturbance are promoted at the same time.

Personnel.—All the key personnel and substantially all the top supervisory personnel came from other points on the system. Selections were determined primarily by considerations of knowledge and skill required in a plant of this kind, by the desire to promote every talent and ability that had shown itself capable of being promoted and by a decision to accept the temporary weakening of organization at existing plants that would inevitably result from a loss of capable men to the new plant.

It was felt that such weakening not only would be temporary but that, in compensation, opportunities would be presented for developing and promoting still younger men into the positions vacated.

Substantiating Operating Results

Unit No. 1 of Sporn plant started supplying power to the system on November 24, 1949. A load of 20,000 kw was maintained for approximately 8 hours and during the next 24 hours the load was gradually increased to 60,000 kw, with the generators operating in air. This operation was maintained for a trial period of four or five days.

During this time all the feedwater heaters were placed in service. The evaporators were set up for maximum production of distilled makeup in order to accumulate a reserve for the commercial starting of the unit. Also during this period the boiler, which was started with treated river water, was operated at 1000 psi to reduce the possibility of carrying over silica to the turbine until the water conditions in the boiler were improved.

The unit was removed from service after the peak load on November 28, as planned for completion of necessary construction work, balancing of the turbo-generator sets, and testing and filling of generators with hydrogen.

The unit was returned to service on Saturday, December 10. Load was increased over the week end to 140,000 kw on Monday morning and to a gross generation of 150,000 kw on Tuesday morning, which is the system peak day. This load increase was accompanied by an increase in steam pressure and leveling off of steam temperatures so that at 150,000 kw, output, the design conditions of 2000 psi and 1050 F at turbine inlet were realized, and 10,000 F reheat temperature to the low-pressure unit was obtained.

A number of minor corrections in construction and design are indicated by operating experience as necessary and will be carried out at the convenience of the system load schedule.

In particular, a most satisfactory performance has been turned in by the centralized control facilities. A number of plant shutdowns, including one or two under emergency conditions, have proved the anticipated simplicity and the value of the remotely operated devices for quick restarting of the plant. In the matter of special provisions for protection of those features of the reheat cycle which are supposedly "endangered" during trip-outs, recent experiences have again demonstrated a complete absence of situations that would require special attention.

Performance of the unit is furnishing ample grounds for the belief that full realization of the theoretical bogeys will be achieved. The weighted average heat rate for the period from December 15, 1949, to April 12, 1950, gives 9411 Btu per net kwhr generated. During this inverval the plant took in its stride a number of shutdowns, and a period of operation with poor-quality coal during the coal strike that prevailed in the first quarter of the year and also a period of operation with some or all high-pressure feedwater heaters out of service for repairs. One check run at constant gross generator output of 151,000 kw gave the following results:

Auxiliary power: 8000 kw, making net generator hourly output 143,000 kw.
Coal rate: 0.8344 lb per kwhr; 11,129 Btu per lb of coal
Heat rate: 9286 Btu per kwhr
Boiler efficiency: 89.85%
Thermal efficiency: 36.77%

To measure the maximum peak capability the unit was operated for three consecutive hours with a gross hourly generation of 161,000 kw and auxiliary power consumption of 8000 kw. Under this condition the generators were operating under 25 lb of hydrogen pressure.

15. ECONOMICS OF 200-Mw, 2000-psi UNITS†

THE Kanawha and Muskingum plants of the American Gas and Electric system are the direct result of studies which had as their objective the definition of a plant cycle that would be the logical successor to the seven large units of 150,000 kw capability installed or under construction at Philip Sporn, Tanners Creek and Twin Branch plants. Ever-increasing prices for fuel, labor and equipment have necessitated a design to improve heat rates and reduce operating labor costs, without prohibitive increments in investment.

It was recognized at the outset that increased size of units would be of help. The long pioneering experience of the company with high pressures, high temperatures, high speeds, reheat, large sizes of units, and lower grades of fuels gave confidence in the efforts to reach the objective.

In addition to system requirements, the boiler design had great influence on the size of the unit. The Philip Sporn boilers had already been operated for substantial periods of time with an hourly furnace heat input of 1.43 billion Btu. Successful operating experience indicated that an appreciable increase in this level of input could be achieved without introducing any particular difficulties.

With this assurance, serious consideration was given to a number of other possible avenues of improvement: 1. Cycles with two reheats; 2. cycles using single-shaft high-speed units; 3. cycles using pressures higher than 2000 psi, up to a maximum of 5000 psi; and 4. cycles using temperatures, primary and reheat, as high as 1200 F.

Each of these possibilities was analyzed, evaluated, and finally eliminated from consideration at this time. Two reheats were not favored because of radical changes in boiler, turbine, and piping designs which would be required. Single-shaft turbines could not give the assurance of best overall economies which could otherwise be obtained. Cycles with ultra-high pressures would introduce new and unexplored boiler problems which could not be accepted for solution in the short time available for development. Temperatures from 1050 F to 1200 F, while offering substantial thermal savings, introduce metallurgical problems which in turn introduce problems of procuring materials in critical supply. It was felt that such requirements could not be justified under the present defense mobilization program.

Consequently it was resolved to use the largest units which could be adapted to the system without disturbing reserve requirements or operating and maintenance schedules, and which would make possible consolidation of the many gains already experienced: Thus, the decision was made to use 200,000-kw (net) cross-compound units, to limit the throttle pressure to 2000 psi, and to utilize ferritic steels in the construction of piping and turbines at 1050 F, both for primary and reheat steam temperatures. This combination, as will be brought out in the description of the heat cycle, will be capable of developing a net continuous output of 200,000 kw at a heat rate of slightly over 9000 Btu per net

† ASME Annual Meeting (with S. N. Fiala), Atlantic City, N.J., November 25, 1951.

14 a*

kwhr. It is the logical culmination of the numerous forward steps taken previously at many points in the development of an optimum power supply on the
AGE system.

The Kanawha and Muskingum plants are being designed with the minimum possible variation between the two plants; they are thus very nearly
alike except for differences in the boilers. The Kanawha boilers will be of dry
ash-pit design and the Muskingum boilers will be of fluid-bottom design.

Equipment was arranged to give the most compact and economical layout
attainable. That this has been accomplished is demonstrated by the low volume
of building structure per unit of capacity—19.3 cu ft per kw including all service
and office facilities. The satisfactory development of remote operation through
centralized control rooms made possible the location of the control room on
the roof of the turbine room without detrimental effect on plant operation. It
had a material effect on the total building volume and volume economy.

The Cycle

Essential details of the cycle on which the plant is projected to operate are
shown. Overall plant thermal economy is estimated at slightly above 9000 Btu
per kwhr. These calculations are based on full-load conditions, and with average
allowances for circulating water temperature, steam losses, feed make-up,
pressure drops, terminal temperature differences, blow-down losses, and auxiliary power requirements. Seven stages of extraction feed heating are planned
with an extreme full-load feedwater temperature of 461 F.

The Turbine

The turbo-generators are arranged cross-compound. Gross capability of
each complete unit is expected to reach 217,260 kw, operating with $1\frac{1}{2}$ in. Hg
back pressure.

The use of an external valve chest with turbines of this size has several decided advantages. It makes possible the concentration of high temperatures on
a relatively small and reasonably symmetrical portion of the high-pressure
turbine shell and also makes feasible the use of a forged valve chest, fabricated
in this case from a ferritic forging. Aside from internal valve parts of austenitic
material in the emergency stop valves and control valves, all high-temperature
elements operating in the range of 1050 F will be made of ferritic chrome-moly-
vanadium material.

Returning from the reheater at 1050 F, the steam again enters the high-
pressure turbine-shell midsection at a point adjacent to the initial steam inlet,
whence it flows toward the coupling end through four stages. The steam then
leaves the high-pressure shell through two 24-in. lines and enters the single-flow
or intermediate section of the low-pressure turbine through two massive intercept valves. Steam pressure at these valves when operating at maximum load
will be about 260 psi. Passing through the intercept valves, the steam expands
through sixteen single-flow stages followed by four similar double-flow stages.
A large overhead duct will form the crossover between the single- and double-
flow sections of the low-pressure turbine.

The four reheat-steam stages in the high-pressure shell were introduced to reduce reheat-steam temperature prior to its entering the intercept-valve bodies and low-pressure turbine shell, and to avoid serious thermal-stress problems associated with 1050 F reheat steam. However, the specific volume of the steam is nearly doubled and this factor necessitated the unusually large physical size of the intercept valves.

The two intercept valves will be mounted at the top front of the low-pressure machine. Each will be operated by independent hydraulic cylinders and reach-rods. In action, the valves will operate as shut-off or throttling elements. A speed governor activating the intercept valves will regulate overspeed of the low-pressure unit. Like the high-pressure emergency stop valves, the intercept valves are also being furnished with a steam-sealing device which eliminates stem-bushing steam leakage when the valves are in their normal or fully open position.

Overheating of the low-pressure turbine exhaust hood is expected to be more critical than on previous machines of similar design. For this reason automatically controlled water sprays are located in the low-pressure crossover duct and used for attemperating the steam leaving the single-flow low-pressure turbine casing. Water will be injected during all no-load operations and for all loads up to 5% of full load.

The lower half of the low-pressure turbine exhaust hood is fabricated from steel plate, rather than built up from iron castings as has been past practice. A considerable saving in weight is expected from this change and the final installation should be more rugged and less sensitive to temperature shocks.

Adoption of steam seals in lieu of water seals is an interesting change from usual design standards. These seals will provide a number of operating and maintenance advantages. Greater operating flexibility is expected. The usual difficulty experienced in adjusting water seals while bringing units to synchronous speed or while changing load will be eliminated. Rapid starting of turbine units will be simplified. Serious thermal shocks to the turbine shafts can be avoided with ease. By the elimination of motor-operated valves and the water runner parts required with water-seal operation, the system is simplified and less subject to mechanical trouble and maintenance.

Record Size Boiler

In view of previous successful experience with natural-circulation boilers operating above 2000 psi at Twin Branch and at Philip Sporn plants, there was no concern regarding the adequacy of natural circulation for the conditions of this plant. Extensive circulation tests, which were conducted on the 2300-psi Twin Branch unit No. 3, demonstrated that the separation of steam and water provided by the cyclone steam separators assured a more than adequate differential head for proper circulation. These tests also demonstrated that circulation in individual tubes, as well as the overall circulation rate, increases with increases in the rate of heat absorption.

The Kanawha plant is equipped with a dry-bottom type furnace in order to handle high ash-fusion coals. Each boiler is designed for a furnace heat input of 1.84 billion Btu per hr almost one-third greater than the heat input requirements of the Philip Sporn units. In terms of heat input, these boilers may be classified among the largest, if not actually the largest, steam boilers ever projected. Primary steam flow is 1,335,000 lb per hr, and reheater flow is 1,216,000 lb per hr. The furnace will be 56 ft, wide by 27 ft deep, with an average height of 90 ft.

Each boiler will consume approximately 80 tons of coal per hr. This high firing rate will be sustained by the use of eight pulverizers on each boiler, sized so that seven pulverizers will carry full load on the boiler even when burning the poorest quality of coal contemplated—15% ash, 10% moisture, and 40 grindability. One pulverizer will thus be available at all times for maintenance, which can be accomplished during regularly scheduled working hours, without reducing output capability.

Each boiler will be furnished with three Ljungström regenerative air preheaters which will bring the combustion air to a temperature of 550 F. There will be three induced-draft fans and two forced-draft fans, the latter being sized and motored for full-pressurized furnace operation. The induced-draft fans are being equipped with tight shut-off dampers and bulkheading arrangements to permit maintenance on any one of the three fans while the boiler is in operation.

The boiler casing is designed for pressurized furnace operation. Such operation will provide better control over furnace combustion conditions, improved temperature control, and increased boiler efficiency through reduction of stack losses. An appreciable operating saving through reduction in induced-draft fan-power requirements is also anticipated. The induced-draft fans are provided to avoid any possibility of boiler outages during peak-load operation as a result of casting leakage and will furnish a very desirable degree of flexibility in operation, although it is expected that these boilers will operate without induced-draft fans approximately 95% of the time.

Superheaters and reheaters are entirely in the form of convection surface; experience with radiant surface in various earlier installations has not been as favorable. The problem of slag formation and removal, resulting from the higher gas temperature required for the all-convection surface arrangement, has been found controllable and the difficulties so introduced are more than offset by the advantages accruing from the complete elimination of radiant superheater and reheater surface.

Superheater and reheater surface has been divided and so arranged as to give the necessary temperature control with minimum use of reheat attemperation and without danger of overheating reheater tubes during periods of trip-out, starting and shutting down of the boiler. Superheat and reheat control below full load will be obtained primarily by recirculation of flue gas from the economizer outlet to the furnace, with water-spray attemperation at the reheater inlet and between stages of the superheater for critical temperature control.

Condensers

The condenser is of the single-pass divided-circulation type, with 90,000 sq ft of cooling surface consisting of $\frac{7}{8}$ in. OD × 18 BWG arsenical admiralty tubes,

30 ft overall length. A storage-type hotwell is provided with 1100-cu ft storage capacity. Condenser design provides for condensate deaeration, with the standard residual oxygen guarantee of 0.03 cc per liter, although operating experience with similar condensers indicates that oxygen removal will be considerably better than this.

The steam-jet air ejector is of the two stage triple-element type with separate inter and aftercoolers, each element having sufficient capacity to handle normally expected air leakage. Steam for operating the air ejector is normally obtained from the high-pressure turbine exhaust through a pressure regulator. During starting up and at loads below approximately one-half load, steam is obtained from the boiler intermediate superheater header at 2150 psig, 900 F and reduced to 150 psig at the nozzles. Steam and water space priming ejectors also use steam for the intermediate superheater header, reduced to 400 psig at the nozzles by means of a pressure-reducing orifice.

Two circulating-water pumps, each 70,000-gpm capacity, provide cooling water for the condenser of each unit as well as general service water for washing traveling screens. Condensers are installed low enough to permit siphon operation. To reduce basement depth, circulating-water pumps are installed in the screen house at the river, some distance from the plant.

Water flows through the two halves of the condenser in opposite directions, and provision is made to backwash the condenser at full load by means of two special butterfly-type reversing valves, operated simultaneously by a single switch on the control panel. Electrical protective features insure that the two valves at opposite ends of the condenser will operate together.

Electrical Features

The Kanawha generators, although of larger size, will be quite similar to those previously installed at the Philip Sporn and Tanners Creek plants, with relatively minor differences in design details. The 1800-rpm double-winding low-pressure unit will be rated 140,000 kva at 0.8 power factor and 0.5-lb hydrogen pressure. Under these conditions the short-circuit ratiowill be 0.8. The 3600-rpm high-pressure generator will be rated 75,000 kva at 0.85 power factor and 0.5-lb hydrogen pressure, with a short-circuit ratio of 0.7. At 30-lb hydrogen pressure the low-pressure generator will be rated 175,000 kva at 0.8 power factor, and the high-pressure machine 93,000 kva at 0.85 power factor.

Shaft-driven exciters on each generator are supplemented by a motor-generator set for starting up the combined generator unit. A second such motor-generator set will provide emergency excitation in the event of failure of any of the main exciters. Voltage control for each unit will be provided by amplidyne regulators.

The large size of these generating units necessitated the selection of 18-kv terminal voltage, the economic upper limit in this case. Conventional aluminum bus structures will connect the 18-kv terminals of the high- and low-pressure generators directly to three-phase power transformers, stepping up to 132 kv. Disconnect switches in these leads permit convenient maintenance and testing of the units.

Three main power transformers will be provided for each unit, one for the high-pressure generator and one for each of the two windings of the low-pressure generator. These transformers will be bussed on the 132-kv side, and connected to the double-bus 132-kv yard immediately adjacent through two oil circuit breakers which can also serve as a bus tie.

Initially, five 132-kv feeders will be employed, two to Cabin Creek, two to Glen Lyn and one to Philip Sporn. Each feeder will have a single breaker with selector switches to the two buses and a bypass switch to one of them so that two 132-kv buses may be used as main and transfer bus for maintenance and for emergencies.

These three transformers which are the forced-air, forced-oil cooled type, are located immediately adjacent to the station to give the shortest possible leads from generators to transformers. The transformers will be protected by the usual differential relay setups and by low-pressure carbon-dioxide fire-fighting equipment.

Provision will be made in the switchyard for installing a 330-kv double-bus system to take the output of future generating units, with a large transformer bank for tying the 132-kv and 330-kv buses.

16. 4500 psi AND DUAL REHEAT AT PHILO†

HISTORY is being made at Philo with a throttle pressure of 4500 psi, which is above critical steam pressure, a new high in superheat—1150 F—and the first double reheat in any commercial installation. Startling as these technical achievements are, other and more fundamental aspects of this project are even more significant.

For one thing the Ohio Power project reflects the dynamic character of the electric power industry. Here, in the very important field of energy generation, new frontiers are being pushed forward and new avenues of approach to improvement in capital costs, in efficiency, and in operating costs, are being opened.

The fact that a dynamic spirit is being maintained in this industry should have an important bearing on maintaining the indispensable good opinion of the American public, of all regulatory and governmental bodies, and of investors. We know that government steps in when a vacuum is created as a result of lack of performance by private industry—by failure to meet a challenge. Here, however, is another demonstration of challenge again being met, as it has been met, with some minor exceptions, by the private power industry throughout its almost 75 years of existence. Investors in the industry are bound to be affected and influenced by this renewed demonstration of the resourcefulness and dynamism of the industry; of its ability to cope with unfavorable factors like materially higher costs of equipment, labor, and supplies—particularly fuel.

Developments like Philo also have an important influence on the morale of the people within the industry and those outside whom the industry is trying to attract. No people of ability, imagination, and enterprise can be expected to be attracted to any venture that is static or declining. An industry that cannot attract its share of the young and the able cannot, over a long period of time, do anything but decline.

There has been some indication of an effort to interpret this challenging development in the economics of power generation as an aswer to the challenge of atomic power. This is wholly without foundation and does not make very much sense. The electric power industry long ago indicated its awareness of the potentialities of atomic energy and its determination to explore and develop that resource whenever or just as soon as it can be done with benefit to its consumers. But while waiting and helping to bring this about, it would be an act of folly to neglect opportunities for further development of technologies that have done so much to bring the industry to its present flourishing state.

The electric power industry, far from challenging any particular method of generating energy, demonstrates by this Philo project its readiness to adopt and bring forward every new development that can advance its technical and economic well-being.

While the use of higher steam pressures, higher temperatures, reheat and higher reheat, larger units, and single-boiler-single-turbine unit system designs

† *Electrical World*, June 29, 1953.

all date back to the thirties, the principal steps forward were all taken within the past 12 years.

It has been evident for some time that a flattening out in this curve of progress would rapidly set in unless new avenues of approach and progress were opened up. The more the problem was studied the clearer it became that the way for further progress lay in breaking away from present conventional design limits in pressures, temperatures and reheat, and thus from the limitations on further reductions in capital costs.

This approach has brought forth a 120,000-kw unit embodying many new and forward-reaching ideas in the technology of economical power generation. Background work for this unit included more than 30 years of experience within the American Gas & Electric system and a continuous program of co-operation and research with General Electric and Babcock & Wilcox.

Managements of these companies decided that it would be advantageous to install a commercial-scale pilot unit as moderately sized as could be built expediently. This would embody the many new developmental ideas and the most advanced technology in order to obtain experience in the design, manufacture, construction, and operation of all the equipment involved.

It is fitting that this new step toward more economical power generation should be made at Philo because it was there that one of the earliest moves in this direction was made—the installation of a 600-psi reheat unit in 1924.

Between 1924 and 1953 there was installed the first 1200-psi reheat unit at Deepwater (1930); the 2300-psi reheat unit at Twin Branch; the 2000-psi, 1050 F/1000 F reheat units at Philip Sporn and Tanners Creek; and the 2000-psi, 1050 F/1050 F units at Kanawha and Muskingum. All of these resulted from the collaboration of the three companies which have now brought together their experience to make possible this latest forward step in power generation.

New Concepts

While many new concepts will be included in this developmental unit, the following will be particularly outstanding:

1. Feedwater inlet pressure of 5500 psi and turbine throttle pressure of 4500 psi constitute not only new highs in their respective classes but are approximately double any pressures previously developed commercially.

2. The pressure employed is well above the critical pressure of 3206 psi, thereby requiring the unit to be of the once-through type, eliminating the need for the heavy and costly drums found in conventional evaporative boilers.

3. Mainsteam temperature of 1150 F, is 50 deg higher than anything previously built.

4. Double reheat to further increase operating efficiency.

5. Cyclone furnaces of the type developed by Babcock & Wilcox.

6. A new design of feed pumps to deliver feedwater at 5500 psi.

7. A new high in overall plant thermal efficiency of more than 40% has been achieved. This compares with the 1951 average for the United States of 25% and

the best plant performance in 1952 of 36.68 % at the Tanners Creek plant on the American Gas and Electric system.

Because of the high efficiency of this unit it will be possible to replace a 40,000-kw turbine-generator installed in 1924–25 without appreciably increasing the cooling-water requirements at this location, where its availability is somewhat limited.

Steam Cycle Progress

Increases in thermal efficiency, in part, have been obtained over the past years by continually increasing initial temperature and pressures. Fig. 1 shows that increasing temperatures permit increasing gains in efficiency by going to higher pressures, but at the 1050 F level no appreciable gain in thermal efficiency can

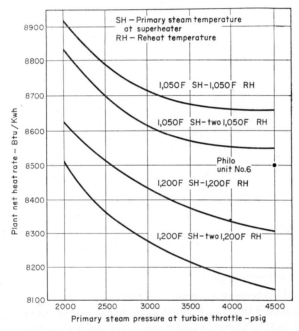

FIG. 1. Effect on heat rate of increasing temperature and pressure.

be made by going higher than approximately 4000 psi. However, as temperatures approach the 1200 F level, considerable gain can still be obtained in going to 5000, 6000 or even 7000 psi with units of large capacity. The cycle chosen thus will not only exploit a favorable combination of higher pressures and temperatures but will also establish an experience mark in anticipation of the higher pressures and higher initial temperatures which are inevitably coming.

The proposed cycle is shown in Fig. 2. It is still in preliminary form and possibly may be somewhat altered in the final design shakedown. Aside from

the high temperature and pressure involved, some of the salient features are
these: Double reheat, seven stages of feedwater heating, and low and high-
pressure feed pumps. Main steam flow will be in the order of 675,000 lb per hr.

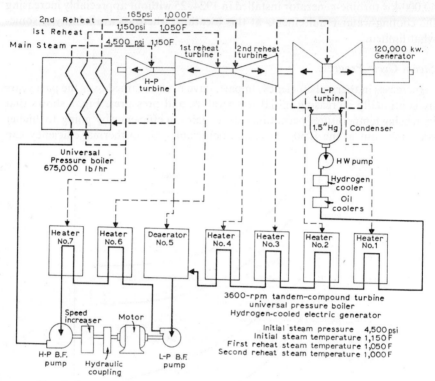

FIG. 2. Supercritical-pressure steam cycle for 4500 psi, 1150 F superheat and
double reheat, 1050–1000 F.

New Type Steam Generator

A Universal Pressure steam generator will be furnished by Babcock & Wilcox
to supply steam for this new high-pressure high-temperature cycle. This will be
the first time that this type of steam generator will be used in a commercial installation in this country. It is of the once-through type into which water is
pumped at one end and superheated steam taken out at the other end. Although
the steam generator acts like a single continuous tube from one end to the other,
it is made up of many sets of tubes in parallel. To assure the uniform and predictable distribution of water between these many passes, a high-pressure
drop of 1000 psi is taken between the feedwater inlet header and the super-
heater-outlet header. At full load feed, therefore, pumps must supply water to
the feed header at 5500 psi. To withstand these tremendous pressures, the entire
high-pressure side of the steam generator except the headers is made up of
small diameter tubing of considerable thickness, varying in composition from
carbon steel at the inlet to stainless steel at the outlet.

As this is a once-through unit type, there is no need for the usual high static head to provide circulation and separation of the steam and water. The steam generator can thus be arranged in a very compact manner, resulting in an appreciable reduction in building volume. This represents one of the major advantages inherent in this type of equipment.

The steam generator is rated to deliver 675,000 lb per hr of steam at 4500 psi and 1150 F, as well as reheat approximately 615,000 lb per hr of steam at 1150 psi and 800 F to 1050 F, and 525,000 lb per hr of steam at 165 psi from 630 F to 1000 F. The high figure of 81 % of the total heat input to fluid in this apparatus will be utilized in the superheating of steam in the main super-heater or in the reheaters. It is apparent why the term "boiler" is no longer applicable to this equipment.

Cyclone Furnaces.—Crushed coal will be fired to this unit through three cyclone furnaces of the type recently developed by Babcock & Wilcox. Liquid slag will be tapped out of these cyclones into the small primary furnace into which the cyclones fire and thence to a water-filled slag tank. Gas leaving the primary furnace will be mixed with cool recirculated gas to give a gas temperature of 1900 F before entering the convection surfaces. This will insure clean surfaces in the convection pass. The gas leaving the convection surface of the reheaters and the superheater passes partly to the recirculation fans and mostly to a tabular type air heater and thence to the induced-draft fans and the stack. No separate flyash-removal equipment is provided because the cyclone furnaces will remove all but a small amount of exceedingly fine ash which will be easily dispersed by high stack velocities.

Control Provisions.—Pressure in the steam generator will be controlled solely by the discharge pressure of the feed pumps. A pressure regulator at the turbine throttle will therefore control pump speed rather than firing rate as in other in-stallations. Heat input to the unit will be controlled primarily by main steam flow and temperature. Temperatures at the two reheat points will be control-led by the interaction of dampers in the gas passes leaving the reheater and superheater surfaces.

Special provisions must be made for starting and transient conditions ex-pected with this steam generator. To do this recirculation valves with pressure-breakdown orifices will be provided to permit fluid circulation through the unit during these times. Flow will be to the deaerator which in turn will pass any excess heat in the form of saturated steam to the condenser for removal from the cycle. Carefully controlled desuperheating arrangements will be required to prevent excessively high-temperature steam from reaching the deaerator or the condenser. Operation of these recirculation valves will be automatic to provide the protection necessary during transient conditions. No provision will be made for circulation of steam through the reheaters during startup due to their location.

Turbine-generator Unusual

A General Electric turbine-generator will operate at 3600 rpm with a special, high-pressure turbine in tandem with a machine of the reverse-flow type. Steam at 4500 psi, 1150 F will flow through four steam leads from the steam generator to four pairs of stop and control valves located under the front of the machine. The control valves will be throttling valves to vary load on the machine by varying steam pressure to the first stage.

At full load steam will be exhausted from the high-pressure turbine at 1225 psig and will return to the steam generator where it will be reheated to 1050 F. It will then re-enter the midsection of the high-pressure rotor of the reverse-flow type section and turn forward through several stages before exhausting to the second reheater at 185 psig. The steam will then return at 1000 F to the mid-section of the turbine adjacent to the inlet from the first reheater and flow through several stages before crossing over to the double-flow low-pressure section. Intercept valves will be provided in each of the two hot reheat lines to protect the machine from overspeeding in the event of a tripout.

Because of the high temperature the stop valves, control valves, steam leads, flanges, and part of the inner shell of the high pressure machine will be constructed of austenitic material. Studies are now being made by General Electric to determine whether the rotor of the high-pressure machine can safely be made of ferritic material with provision for steam cooling or whether it will be necessary to include some type of austenitic material in the highest temperature zones. The high-pressure machine will utilize the now common double-shell construction which allows the lower temperature part of the inner shell and all of the outer shell and its fittings to be made of the more available ferritic materials. The turbine from the first reheat point to the condenser will be fabricated entirely of ferritic material.

Controls.—Controls will be more complicated than on a conventional machine to the extent that two sets of reheat stop and intercept valves must be provided. Another special feature required will be an initial temperature regulator. This device will sense the steam temperature at the turbine throttle, or some indication of that temperature, and will close the control valves in the event the temperature drops either too low or too fast. This will protect both the steam generator and the turbine by preventing conditions such as the sudden carryover of water in the event of loss of fires. The protection that this equipment will provide is similar to the protection that an initial-pressure regulator provides on a conventional boiler.

The generator for the unit will be hydrogen cooled, with facilities for operation at 30-psi gas pressure to obtain maximum rating. A tentative maximum rating of 125,000 kw, 0.85 power factor has been selected.

The machine will be connected directly to a single, 3-phase step-up transformer with forced-oil, forced-air cooling. It will be connected on its high side to the 132-kv station bus. The normal station auxiliary supply will be taken directly from the generator leads.

The larger and more important station auxiliaries will be operated at 4000 v and the smaller equipment connected to a 550-v auxiliary bus.

Pumps.—At present two alternate arrangements of pumps are under consideration. Both will provide two sets of pumps, each set having a capacity of

60% of full-load flow and having adjustable-speed hydraulic couplings for the regulation of turbine throttle pressure. The two arrangements being studied are as follows:

1. Low (LP) and high (HP) pressure feed pumps are shown schematically in the cycle diagram. Feed water is taken from the deaerator by two constant-speed, barrel-type feed pumps of conventional design and pumped to approximately 2000 psi. The water is then fed through the two top feedwater heaters and heated up to approximately 515 F. It next enters the two newly-designed 6500-rpm, adjustable-speed, barrel-type pumps which boost pressure to 5500 psi.

Each set of pumps is driven by a common motor, that is, one end of a motor shaft is coupled to one of the constant-speed LP pumps, and the other end is coupled to the adjustable speed coupling and step-up gearing driving one of the HP pumps. This arrangement has the advantages of eliminating 5500-psi heaters and their piping and valves. It has the disadvantage, however, of requiring the pumping of hotter water, with its slight thermodynamic loss, and of requiring in effect twice the number of pumps, with the attendant increase in first cost and maintenance.

2. The other arrangement under consideration calls for single, high-speed feed pumps located under the deaerator and pumping the feedwater through two high-pressure heaters at 5500 psi. This makes for a simpler arrangement but calls for the development of feedwater heaters to stand the very high water pressure as well as for the use of additional valves and piping good for this pressure.

Feedwater Heater Choice.—The feedwater heaters in this cycle will be more or less conventional if the arrangement of feed pumps described as alternate 1 above is utilized. On the other hand, two high-pressure heaters of new design will be necessary if the second arrangement is selected. European as well as B & W experience clearly indicate that the pH of the feedwater should be kept high, around 9.2 to 9.5. This means that all feedwater heaters must be provided with tubes which stand up under these conditions. The deaerator, operating at approximately 175 psi, will have to receive special design consideration to assure that it will satisfactorily absorb the slugs of water and steam which will be dumped into it by the recirculation valves which are part of the starting-up procedure.

Two-pass Condenser

Because of limited circulating water availability and the desire to use existing tunnels. a two-pass condenser will be utilized. Similar precautions must be taken in the selection of tubes and tube sheets as in the case of the feedwater heaters to provide adequate life in the face of the relatively high pH of the steam. As any solids which enter with the feedwater must either pass through the steam-generator tubes and turbine and continue to circulate in the cycle or be deposited in these tubes, the turbine, or elsewhere, it is imperative that the cycle be made as hermetically tight as possible. This precaution will be observed throughout the cycle but will be especially stressed in the condenser where the biggest possible source of contamination exists. Every possible means will be utilized in order to make the condenser completely tight.

Problems Anticipated

Every effort will be made to anticipate a number of operating problems during the initial operation of this unit. Water conditioning and the control of solids in the steam generator, turbine, and other parts of the cycles are problems which will have to be watched carefully and which probably will lead to new concepts in the treatment of feedwater.

Control of flow, especially in valves subjected to the tremendous pressure drops contemplated, is a problem which will require not only the best thinking during design stages but may require some design changes after initial operation.

Control of the cleanliness of the heat-absorbing surfaces and of the steam and reheat temperatures will require careful attention and planning. This problem is being tackled experimentally.

Starting and shutdown procedures will have to be carefully worked out during the first stages of operation to protect all component equipment and yet obtain the high degree of flexibility which is inherent in this Universal Pressure type steam generator.

Behavior of the various alloy steels at these high temperatures and pressures will have to be carefully observed as no operating experience is available under these conditions.

These are some, but by no means all, of the many problems which will have to be faced to a lesser or greater degree. It is expected, however, these will be satisfactorily solved during the design and construction or during the initial operating period of this machine.

Throughout all of the design work ahead sight will not be lost of the two primary aims of this entire project—developing new techniques and design arrangements to yield immediate improvements and pointing the way toward design which will provide greater economical prospects which make this project significant.

17. REASONS FOR 450 Mw and 3500 psi†

SELECTION late last month of the Philip Sporn plant on the Ohio River as the site for the second of two 450-Mw supercritical turbine-generators ordered by American Gas and Electric reveals the pattern by which these units will be integrated into the system.

Only a month or so earlier the way was cleared for Indiana & Michigan Electric to install the first unit at the new Breed plant on the Wabash River. Neither location had been determined when purchase of the largest steam-electric generating units ever undertaken by the electric power industry was announced early last May. This article will give some of the reasons which guided the selection of the units.

Both 450-Mw units will take steam at 3500 psi, 1050 F with double reheat to 1050 F. They represent the latest effort to provide the system with low-cost, highly-efficient generating capacity utilizing all of the best features in everything that has been developed to date in power generation. But they will conform to the necessity of employing materials that are commercially prudent and justifiable. Design of these new units is a natural evolution from pioneering work done in connection with design and construction of the Philo 4500-psi supercritical unit which is about ready to go into service.

The need for pioneering in the supercritical pressure range, and the decision to proceed with the installation at Philo of the smallest practical supercritical pressure generating unit which would give suitable design and operating experience for extrapolation to sizes commercially desirable was recognized several years before reaching that decision. A pressure of 5000 psi was chosen as the highest under which a boiler could be designed within existing boiler code rules. That selection of pressure was accompanied with its mating temperature for best thermal performance, which proved to be 1150 F.

While it was known from the start that it would not be possible to justify 1150 F as an operating temperature on a commercial unit, this temperature was adopted only because of American Gas and Electric experience and judgment that the magnitude of the problem could not be well enough defined without actually going through with an installation somewhat beyond the practical range. And because Philo was planned to serve as a prototype for a new series of machines and to push forward the frontiers in thermal conversion economy it was felt that a step beyond practical limits was essential.

Experience integrated in the course of planning, engineering, designing, and operating more than twenty 200 to 225-Mw units on American Gas and Electric and Ohio Valley systems, gave rise to the conviction that it was necessary to boost unit size, pressure, temperature, and reheat to carry forward the line of progress in thermal efficiency and capital cost. This led to the 450-Mw units.

Space Saving

One of the positive measures that can be taken to counterbalance the rising

† *Electrical World*, December 24, 1956.

cost of equipment, construction, and operation is to go to larger units. Increasing unit size brings reductions in cost by the integration of a great number of savings in the components of the plant.

One of the most significant reductions results from the reduced volume per
kw of capacity. Fig. 1 compares the space required by two 225-Mw units
with that for one 450-Mw unit. The two units require some 8,666,000 cu ft vs.
5,350,000 cu ft required by one 450-Mw unit. On a unit basis the reduction is
from 20 to 12 cu ft per kw. Obviously such compression achieves a large decrease in capital cost, a lower maintenance cost, and an increase in productivity
of operating and maintenance labor.

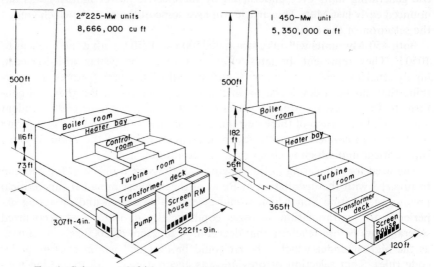

FIG. 1. Cubage cut 38% by single turbine over that required by two turbines
of equal capability.

There is, however, an offsetting factor—addition of the larger capacity unit
calls for an increase in reserve requirements which can be met economically only
by relatively large integrated systems. The American Gas & Electric system will
have a capability of close to 57 million kw by the end of 1959, only a very short
while after the first 450-Mw unit goes into operation. Replacing the capacity
of a single unit of such size during an outage requires adequate transmission
capability and here AGE's 345-kv system comes into play.

There is an important point about size of units that may be worth emphasiz-
ing. As any power system grows, the largest unit size that has been exploited
becomes progressively a smaller and smaller percentage of the total system
capability. Unless the unit size is increased, definite losses are sustained in
opportunities for savings that are thus passed up.

Figure 2 shows the changing relationships of unit sizes to system capability
on the system since 1942. It will be noted that the second 450-Mw unit, when it
comes on the line, will represent only 7% of system capacity. This percentage is
the same as the 90-Mw units represented in 1942. Incidentally experience has

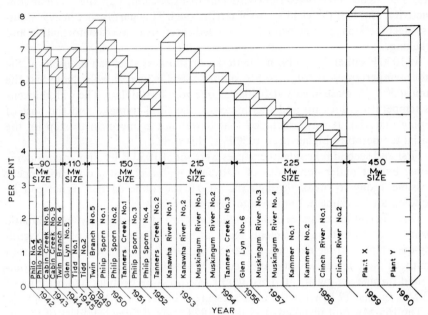

FIG. 2. Steady rise in system peak can more easily accept larger units.

demonstrated that 7 % is a reasonable and sound relationship for the AGE system.

A pertinent question to ask is "Why 3500 psi?" The answer is that higher pressure gives two advantages—improved thermal performance and reduced volume of steam, the latter a significant consideration for a throttle flow of about 3 million lb per hr. But higher pressure requires heavier pressure parts and, therefore, higher capital costs.

Extensive evaluations carried out in design studies leading to the 450-Mw decision indicated that 3500 psi today gives a good balance between these two effects and that it is well within the optimum pressure range of once-through steam generators. Embodiment of that balance in the single-furnace once-through steam generator which is being installed has been made possible by an evolution of designs during a period of almost two years.

Another pertinent question is "Why 1050 F for primary steam and reheat?" The first AGE system post World War II machines designed for 1050 F operation utilized stainless steels for inner shells, steam-chest piping, stop valves, and similar parts because of inadequacy in the metallurgical know-how for low-chrome materials. Subsequently this deficiency was remedied and all more recent 1050 F machines have been built with low-chrome-moly materials.

1050 F *Avoids Steel Problems*

Service experience with the stainless steels has brought to the foreground a number of operational difficulties, not all of which have as yet been cleared up.

Corrective measures for many of these difficulties are still in the laboratory and research stages. Stainless steels have difficult, bothersome heating characteristics. The steels have twice the coefficient of expansion, half the thermal conductivity, and half the yield stress of ferritic steel. This means that heavy, stainless-steel sections must be heated and cooled carefully to avoid deformation and cracking.

At 1050 F either $2\frac{1}{4}\%$ chrome material or stainless steel can be used; at the other temperatures use of Type 316 stainless steel is unavoidable. Note first the almost 50% increase in cost at 1050 F as one goes to stainless. The wall thickens as temperature goes up and this increases the total tonnage and the cost of piping material. It was concluded that it was not economical to step up the initial temperature higher than 1050 F at this time.

First and second reheat temperatures were set at 1050 F by similar considerations. Except under conditions of unusually high fuel cost it does not appear possible today to break through the 1050 F economic barrier. What can be done on an experimental unit is another matter.

This efficient use of space is the result of a series of evolutionary and dramatic design changes and improvements made progressively in system plants extending over almost three decades. It started with the 1200-psi Deepwater plant in southern New Jersey which combined high pressure, reheat, and a cross-compound turbine arrangement. This plant's low-pressure cylinder had a rating of 42 Mw against 12 Mw for the high-pressure cylinder for a 3.5:1 ratio.

The 165-Mw unit installed at Philo in 1929 utilized three cylinders of practically equal rating with a total LP-to-HP ratio of 2:1. Hydrogen cooling and removal of previous limits on 3600-rpm machine sizes enabled the Sporn and Tanners Creek 150-Mw machines to have the more economical ratio of 55:45 between their 1800 and 3600-rpm cylinders. Further advances in the Ohio Valley and Indiana–Michigan machines reversed the rating ratio to 45:55.

In all these arrangements, however, grave disadvantages existed in that not only were the two cylinders of different size but all turbine wheels were different except those in parallel exhaust sections. The alternators also had different ratings. What was worse, difference in size of the two machines required a large excess in space and cubage. It remained for the double-reheat design and adoption of 3600 rpm as the speed for both cylinders to make possible the almost complete duplication of cylinders, wheels, alternators, and maximum and most efficient utilization of space and cubage.

A cyclone-fire, once-through steam generator with a single furnace serves the cross-compound turbine-generator set, both elements of which are designed for 3600 rpm and will have indentical 225,000-kw generators and triple-flow-exhaust steam ends.

Except for the first wheels in each cylinder, all four other wheels are completely interchangeable between the two cylinders. The turbine-driven feed pump located on the main turbine floor and exhausting to the main condensers supplies the one boiler during normal operation.

The generators will be rated at 265,000 kva each at 16 kv. They will feed through individual step-up transformers to the outgoing 345-kv transmission system.

18. SUPERCRITICAL STEAM PRESSURE†

THE technical and commercial success of higher pressures, higher temperatures, reheat, larger units, and single-boiler, single-turbine unit designs has been one of the main reasons behind the ability of power systems in the United States to maintain stable and even lower rate levels in the face of large rises in capital and operating costs. Since the trends which have resulted in the large cost rises have not been arrested, it has been imperative to continue the development of more economical means for mass generation of electric energy.

In 1952 and 1953, when this problem was intensively studied, thermodynamic, economic and related studies, balanced by experienced judgment, indicated that the path of super critical pressures, high temperatures and double reheat was the correct route to follow. At that time we had in operation many large-size (215 Mw) high-pressure, high-temperature (2000 psi, 1050/1050 F) reheat units. In Europe once-through boilers at high pressures and temperatures had been operated for many years.

Further study and evaluation of this problem indicated clearly that super-critical-pressure units in large sizes represented the most practicable answer for continued progress in the economical generation of large blocks of electric energy. But since there were far too many unknowns to embark on full-scale production units, it was decided to build a prototype to prove the soundness of the basic concepts in supercritical pressure, higher steam temperatures, double reheat, and particularly the adequate control of feedwater chemistry.

The unit finally chosen for Philo is rated at 110,000 kw net output, with initial conditions of 4500 psi and 1150 F. Two stages of reheat are employed, the first at 1200 psi, reheating to 1050 F, and the second at 200 psi, reheating to 1000 F. Serving this unit is the first once-through steam generator used in this country. It has a maximum steaming capacity of 725,000 lb per hr at 4570 psi and 1150 F at the superheater outlet with two stages of reheat. Designed feedwater inlet conditions are 525 F and 5500 psi. The steam-turbine-generator of this unit is rated 125,000 kw and utilized a hollow-conductor-cooled generator, rated 156,000 kva at 30-psi hydrogen pressure.

The size of this unit was chosen as being the minimum which would yield valid operating experience and which could be scaled to future production units without dangerous extrapolation of size.

This unit has now been in operation somewhat over 11 months. As of February 1 it had operated 4600 hr and generated some 360,000,000 kwhr. In that period it gave enough operating experience to check every basic design conclusion and to indicate means of correcting inadequacies or errors.

Chemical Control

It was recognized from the very beginning that almost perfect control of feedwater conditions was essential to the successful operation of this unit. Initial

† AIEE–ASME Joint Meeting, San Francisco, Cal., February 27, 1958.

concepts of the means and quality of this control were outlined and then largely substantiated by personal observation of German experience with Benson boilers at subcritical pressure, by comprehensive tests on a small supercritical pressure steam generator at the Babecock & Wilcox Research Center at Alliance, Ohio, and by several test programs at various locations on our system.

From extrapolation of experience in the subcritical region and the specific experience and test data indicated above, the original specifications were set at:

Total dissolved solids	500 parts per billion (ppb) max
Silica (SiO_2)	20 ppb max
Iron (Fe)	10 ppb max
Copper (Cu)	10 ppb max
Dissolved oxygen (O_2)	5 ppb max
pH value	9.5 to 9.6.

Due to a prolonged period between completion of the boiler and delivery of the generator, an opportunity was afforded to test out thoroughly all equipment except the turbine. This was done by recirculation through the steam generator and the startup bay-pass system to the condenser. This activity, carried out between October 1956 and March 1957, would otherwise have been of shorter duration.

From the beginning of the recirculation phase, total dissolved solids were within the originally specified maximum. There has been no problem in maintaining even lower values by demineralizing a portion of the condensate flow. Normal operating values have been in the range of 100–200 ppb.

Following the chemical cleaning, recirculation was initially carried out with cold water and without deaeration. The initial iron content of the feedwater was 2000 ppb and was reduced to 100 ppb in 6 hr by filtration of a portion of the condensate flow. Subsequent increases in iron content to even higher than the initial were traced to cavitation-erosion of the by-pass system.

With establishment of deaeration and controlled-temperature firing of the steam generator, values of iron were rapidly reduced for stable conditions. However, interruptions of flow, flow changes, and temperature changes temporarily resulted in peak values of suspended iron oxide.

During the pre-operating recirculation period, outlet temperature of the steam generator below 600 F prevented excessive deposition of iron oxide. As temperature was increased higher values occurred, which had to be tolerated until equilibrium was reached. Initial operation of the turbine with flow through the reheaters and bleed-steam system resulted in feedwater iron values of two to four times normal for a few days.

Demineralization.—Three important factors determined as the result of subsequent operation are:

(a) Iron is present predominantly as a suspended or colloidal oxide rather than in solution even at the low level of 5 to 10 ppb.

(b) The greater portion of the iron oxide in the feedwater deposits in the steam generator before the 750 F temperature region, the outlet iron concentration being relatively constant for a widely varying input.

(c) Based on examination of tube specimens and measurements of metal temperature, it appears that up to five times as much iron oxide deposition can be tolerated without elevating metal temperature than for natural circulation boilers with comparable heat-transfer rates. This is ascribed to the absence of salts, especially sodium hydroxide, and a concentrating film mechanism.

Normal values of iron at the steam-generator inlet are 2–10 ppb, with 1–2 ppb at the final superheater outlet.

Due to the chemical cleaning operations employed, silica was originally low during the pre-operating recirculation phase. However, with initial operation of the turbine and for some time after, contamination from the heaters and re-heaters was evident. This was controlled by demineralization.

From the beginning, and for conditions of 9.5 to 9.6 pH, corresponding to about 2 ppm ammonia, the copper content of the feedwater has generally been less than the control limit of 10 ppb, with the major portion of the copper coming from the steam side of the 70–30 Cu–Ni heater tubes. It has also been confirmed that almost 100% of the copper in the feedwater is deposited in the steam generator.

Because of the good mechanical deaeration, down to about 2 to 3 ppb of dissolved O_2, and the normal low values of iron resulting from pH control alone, hydrazine has not been employed, except for the preliminary pre-operation recirculation phase. Tests are planned in the near future, however, to evaluate the independent effects of hydrazine and pH on iron pickup.

Filtration capacity is provided so that during start-ups all of the condensate flow can be treated. The demineralizer capacity approaches 50% of startup flow and 20% of full-load flow. By filtration alone, an initial high iron-oxide content following an extended shutdown can be reduced to acceptable limits for full-load operation in a few hours.

The filters are of the tubular-membrane element type, employing a pre-coat of purified cellulose fibre ground to 100 mesh. Duration of runs has ranged from a minimum of one hour for extremely high iron content of the water to as long as one week for normal operating conditions.

The demineralizers are of the mixed-bed type and furnish an effluent quality of about 50 ppb dissolved solids and 3 to 5 ppb SiO_2. Operating runs at full flow are about one week.

This account has been given in such detail because it seems to me that chemical control of feedwater is the heart of the tough technical job that one undertakes with a supercritical unit of this type. Water control was the No. 1 problem and results to date indicate that its solution has been entirely successful.

Pump Problems

A great deal of work on the boiler-feed problem was carried out while the project was under investigation and discussion. As a consequence not much difficulty was expected in that direction. We could not have been more mistaken.

Continued unexpected difficulties with the boiler feed pumps have been a most serious concern. The troubles are only indirectly associated with super-critical pressure and, therefore, cast a false shadow on the otherwise successful unit operation.

The pumps consist of two one-half size assemblies, each having a tandem arrangement of low-pressure pump, direct-connected 4000-hp, 3550-rpm in-duction motor, hydraulic coupling, speed-increasing gear and high-pressure pump of variable speeds up to 6500 rpm. The LP pumps take suction from the deaerator, discharge through the two highest feedwater heaters at 2100 psi to the HP pumps, which discharge at 5450 psi to the inlet of the steam generator. Suction temperature of the HP pumps reaches 560 F.

Pump difficulties centered in two main areas. The LP pumps suffered a loss of "Colmonoy" facing material on their wearing parts, possibly associated with concurrent vibration difficulties at the pump center bearings and hydraulic balancing devices. Operating abnormalities suggested dismantling, which re-vealed the problems and the deposition of the "Colmonoy" at the HP pump suction strainers. These difficulties persisted after several minor corrections, so that the LP rotors were rebuilt in December 1957 to improve shaft stiffness, making them comparable in slenderness ratio to numerous other units which are operating successfully. The "Colmonoy" coatings, not actually required, were omitted. The rebuilt design, which omitted impeller rings, has recently resulted in seizure and wear at the impeller. Hardened rings are now being in-stalled to overcome this.

The more aggravating and serious source of pump failures occurred in the "Grapholoy" breakdown bushings used to seal the HP pump ends against their suction pressure. These bushings reduced a 2100-psi outward flow at each pump end to the deaerator leakoff pressure of 200 psi. Compression of the "Grapho-loy" in the rather long single bushings, plus shaft deflections, resulted in a series of seizures and bushing failures. Aggravating these effects was the severe flashing of the 2100-psi, 567 F water to 200 psi.

In May 1957, a temporary labyrinth-type steel bushing of increased clearance was installed until a more permanent seal was designed. This improved reliabi-lity at the cost of higher leakage and greater bushing wear. The higher leakage required increased condensate injection for adequate quenching. This increased wear of the intermediate bushing. A temporary planned program of periodic replacement of bushings was undertaken to eliminate forced shutdowns.

A new seal design was installed on the HP pumps, along with the rebuilding of the LP pump rotors, during December 1957. The new design incorporates a second or inboard seal injection directly from the LP pumps. This injection line by-passes the intermediate heaters, using their pressure drop to insure positive sealing of the HP pumps even at minimum feedwater flow. The relatively "cool" injection water flows into the HP pumps and outward on the shaft, with no flashing, back to the deaerator. Condensate injection still seals the deaerator leakoff but does not have the former quenching requirements. As a result, leak-age flows are reduced. The thermally undesirable flows into the HP pumps, which by-pass the heaters, are a minimum because of the low differential between first injection and pump-suction pressures.

Temporarily, labyrinth bushings are installed in the new-seal-flow arrangement. We are conducting extensive tests with the pump manufacturer on two alternates to develop the final mechanical seal design. Basically, both alternates consist of multiple segments or rings across each breakdown as opposed to continuous bushings. These, it is expected, will give reduced leakage and, since the segments are free to move radially, they should avoid the possibility of shaft contact due to deflection. But the satisfactory performance of the design still remains to be proven.

Starting up

A once-through type boiler requires an external recirculating system to replace the natural circulation which exists in a conventional boiler even during startup. Cold condensate must be circulated as the fires are started, then progressively hotter water, and later steam must be disposed of until the steam at the superheater outlet is suitable for starting the main turbine. To protect cyclone and furnace-wall tubes there has been adopted the basic operating rule of not firing unless at least 225,000 lb of water per hr is circulating. This was based upon a minimum water velocity of 7 fps through the cyclone tubes. Interlocks have been provided to shut off all fires if flow should at any time drop below this minimum.

The design of a recirculation system represented one of the most challenging problems associated with the design of the unit as a whole. All equipment had to be designed for completely tight shut off against full pressure and temperature conditions. In operation, the piping and valves must withstand pressure drops from 4500 psi down to 150 psi at all temperatures between room temperature and 1150 F. In starting up and in passing through this range of temperatures, the valves must pass first a water-like phase, later steam which flashes wet and then dry again, and finally completely superheated steam. This imposes particularly severe service on these valves.

Two means of recirculation are provided: The first and main one is the turbine by-pass system which is normally used in starting and shutting down the unit; the second is the superheater by-pass system which was provided to maintain flow through the boiler in the event of unit trip-out and to assist in restarting the boiler from a hot, bottled condition.

Shortly after starting the recirculation of cold water through the turbine by-pass system, these valves revealed severe erosion. The eroded sections were redesigned to eliminate the sharp change in contour where erosion had been most severe and more stellite was employed on surfaces subject to impingement.

It was subsequently discovered that as fluid temperature was raised above 250 F and some flashing occurred across the breakdown valves, the cavitation noise associated with high-pressure drop disappeared. Several inspections of the valves since that time indicated that the erosion potential is a function of the noise level. Subsequent operation has been entirely successful.

Erosion difficulties similar to those encountered on the turbine by-pass were also experienced with the superheater by-pass valves. Again, the wider application of stellite and minor modifications in contours and flanged fits have eliminated this problem.

Controls

Development of a successful control system for supercritical pressure once-through steam generators meant the introduction of a number of completely new concepts.

Turbine-inlet pressure on a supercritical unit is a function of feedwater flow and turbine control-valve opening at a given steam temperature. On this unit, flow and pressure are regulated by variable-speed hydraulic couplings on the high-pressure feed pumps and an air-loaded bellows on the turbine initial-pressure regulator which positions the turbine control valves.

Philo 6 was designed to simplify the control system as much as possible. Normally, feedwater flow is maintained constant by feed-pump speed, with turbine-inlet pressure held at 4500 psi by the turbine control-valve opening. Both of these values are set by a pneumatic demand signal, originated manually and maintained through the use of feed-back signals from flow and pressure transmitters.

To provide for an alternate means of automatic control, the turbine control valves can be operated directly by the speed governor as on a standard unit. In this case, the feed-pump speed is varied to hold the turbine-inlet pressure constant, using a pneumatic signal originating from a second pressure transmitter. Although the unit may be readily operated by this means, it does not have as rapid a control response as the former method during certain transient phases of operation.

The once-through characteristic of a steam generator requires positive control of turbine-inlet pressure by the above methods. This allows steam temperature to be maintained directly by the firing rate, eliminating the need of attemperators or other devices required when firing is varied to hold pressure, as on a drum-type unit.

The combustion-control system, using steam temperature as a master signal and feedwater flow as an anticipating signal, calls for equal changes in air flow and coal flow. During initial operation, a second anticipating element, the "transition zone" temperature, was used in the control. It was found that this control did not function as expected and, in fact, went in the wrong direction during steam temperature changes caused by conditions other than varying feedwater inlet temperatures. This element was then removed from the control system.

A gas sample is continuously taken from each cyclone outlet and automatically analyzed. This adjusts the speed of each coal feeder insuring proper excess air and, therefore, proper combustion in each cyclone. Considerable satisfaction is taken from the fact that the O_2 analyzers have been operating very satisfactorily as an integral part of the combustion-control system.

Control of reheat temperatures is accomplished by proportioning the gas flow through the divided convection pass of the steam generator, one-half containing the primary superheater and the other half the first and second reheaters in series. The control selects the higher of the two reheater temperatures and adjusts the proportioning dampers to obtain rated temperature on that reheater. Although this method usually results in one of the reheaters being slightly below its design temperature, it is equally as good from a cycle-efficiency standpoint as controlling the lower and attemperating the higher of the two, and it eliminates the problems of attemperation.

The turbine-cooling steam-temperature control was originally a three-element system, utilizing steam temperature as the primary element, turbine camshaft position as a load index, and attemperating water flow as a feed-back signal. It became apparent during early operation that the response of the temperature element alone was adequate for controlling cooling steam temperature, and the feed-back and load-index signals were removed from the control.

The complete control system has proved to be simple, fast acting, and stable during operation. As an example of its versatility, it can be left on automatic while reducing throttle steam from 1150 F to 850 F with successive temperature set-point changes, when taking the unit off the line.

Performance

Operation of the steam generator has revealed no functional problems. After several months in operation, selected steam-generator tube specimens were removed for inspection. No adverse conditions were found.

The two stages of reheat have been performing satisfactorily.

Operation of the high-pressure turbine to date has been carried out with only the austenitic rotor and using steam cooling of the rotor, inlet steam pipes, first-stage nozzle assemblies, inner shell and high-pressure packing. Continuous monitoring by numerous thermocouples indicates that the cooling stream is maintaining the turbine metal temperatures in these vital areas very close to their design values.

During several simulated hot restarts using the turbine by-pass system, thermocouples located in the main stop valves indicated some large temperature differentials across the valve walls. Resulting calculations indicated very high stresses on the inner surfaces. Inspection of the separate stop and control valves revealed no body cracks. Although the integral stellited seats on both the stop and control valves were cracked extensively, none of the cracks extended into the base metal.

Heat-rate tests were conducted on the unit in July 1957. The results are summarized by the following figures of net plant heat rate for summer operation: Predicted, 8760 Btu per kwhr; actual test 8920 Btu per kwhr.

The difference in considerable measure is due to the feed-pump performance and to the use of the higher speed on the recirculating gas fans on the steam generator. If the redesigned pump seals perform up to expectations, leakage should be reduced to a point where approximately 50 Btu per kwhr will be picked up. A good portion of the remaining difference is due to lower than design feed-pump efficiency which apparently cannot be corrected.

15*

Prototype Lessons

With the confidence gained during the detail design phases of the prototype Philo 6 unit, and with the expectation of the good operating results which have been achieved, studies made in 1955 and 1956 indicated that new capacity additions should comprise very large supercritical-pressure generating units.

In selecting the operating conditions for Philo 6, studies of various pressures and temperatures were made to indicate the cycle gains which could be achieved by the ultimate exploitation of materials then available rather than on the optimum economic performance. When, however, selections were to be made for designs which might ultimately lead to capacity additions of as much as 5 million kw, the utmost consideration of economic performance had to be exercised. Complete studies of various initial pressures, temperatures numbers of reheat, and arrangement of auxiliaries resulted in the decision to build initially two 450,000-kw generating units. The first is at the new Breed plant on the Wabash River in Indiana and the second is Unit 5 at the Philip Sporn plant on the Ohio River in West Virginia.

These units will have steam conditions of 3500 psi, 1050 F, with two stages of reheat, each to 1050 F. Since initial temperature has been reduced, no steam cooling of the high-pressure turbine rotor will be required. The steam generators will be once-through units, each with a capacity of three million pounds-per-hour. The first at Breed will utilize cyclone firing; the other at Sporn will be fired with pulverized coal. Both will be designed for pressurized firing without induced-draft fans. The cross-compound turbine-generators will each operate at 3600 rpm; the high-pressure turbine, a second reheat turbine, and a triple-flow low-pressure turbine will drive the primary shaft; the first reheat turbine, a second reheat turbine and a triple-flow unit will be on the secondary shaft.

Steam pressure of 3500 psi was chosen as the reasonable engineering-economic balance between improved cycle performance with higher pressures and the increasing cost of all pressure parts. Studies of the effect of initial temperature indicated that all operating temperatures above 1050 F required extensive use of stainless steel in steam-generating tubes, piping and turbine parts. With temperatures from 1050 to 1250 F the gains of increased performance are not enough to bear the heavy capital burden of the austenitic materials. If the step to higher temperatures is to be made, it must come either through lowering the cost of stainless steels or through improvement of the steels at their present cost levels so that the required capital investment can be justified by the expected improvement in efficiency.

Feed pumps for these 450,000-kw units will be arranged differently from those at Philo. A single 20,000-hp condensing turbine-driven feed pump will supply each of these steam generators. Earlier studies indicated that single turbine-driven feed pumps were attractive for a series of eight 225,000-kw generating units which the company was designing. The first of these units has been in service for 6 months and has proved that the use of a single boiler-feed pump is as valid a concept as the universally accepted concept of unit boilers, turbines, etc. It was natural, therefore, to consider turbine-driven feed pumps on these supercritical units, particularly with the high auxiliary power requirements resulting from these high pressures. To obtain the highest possible cycle gain, it was determined that a condensing turbine should be used which will in effect increase the exhaust annulus area of the main machine. No pumps of this type and size have been built but evaluation of this problem with manufacturers indicated that efficiency and reliability could be expected to equal the excellent performance received from feed pumps operating at current pressures of about 3000 psi.

Based not only on Philo experience but also on added know-how of manufacturers, cupro-nickel tubes in the condensers will be replaced with less expensive copper tubes, and two high-pressure (4500-psi) feed heaters will be used. As a result of Philo experience, the filter for removing solids, primarily iron-oxide, from the condensate stream will be enlarged to approximately one-third of maximum throttle-flow capacity. Controls of the Philo 6 system were so successful that they will be essentially duplicated on these new units. One refinement will be the introduction of arrangements in the boiler for individual reheat temperature control.

With these specific exceptions, the design of the new 450,000-kw units is based squarely on features proved in operation of Philo 6. This enhances the value of Philo 6 as a true prototype for the 450,000-kw units and is added assurance that the commitment to the two large units is soundly grounded in experience. Each of the 450,000-kw units represents approximately $55 million of investment. Where such large sums may be committed to a new design concept, it is hard to see how the gains flowing from prototype experience (in the form of closer margins in design, and greater assurance of successful operation) can be passed up or short-circuited.

In summary, I believe that the Philo 6 design, construction and initial operating experience have been conspicuously successful in breaking through the supercritical-pressure barriers and in providing a sound foundation for the design and construction of new, more efficient generating units of the largest size required by large-system economics. Further, this experience should find useful application in many other power-generation developments—nuclear power reactors, for example. I also believe that Philo 6 experience, and the new units being designed on that experience, indicate that the present optimum performance, which must improve if the continuing rising costs are to be met, will have to be obtained along some new paths.

Higher temperatures will have to be achieved by improved metallurgy and the reduction in cost of high-temperature materials. Improved cycle arrangements must be achieved by such devices as the combined steam-turbine, gas-turbine cycle, which gives fair promise of becoming possible in the not-too-distant future. Improvements in capital cost will have to be obtained through further raising of sizes, simplification of plant layout and simplification and improvements in design and methods of erecting the principal components such as turbines, boiler, pumps, condensers.

We must always look to the formulation of new concepts that are as yet not distinct or fully understood.

19. THIRTY YEARS OF ADVANCE
IN THERMOELECTRIC GENERATION†

THE American Electric Power Company system serves a 7-state area in the industrial heart of the United States covering 48,600 square miles. Generating capacity at the end of 1959 was 5,560,000 kw. Growth of the system peak load and the generated output over the last 30 years are shown in Fig. 1. As a point of reference for the rapid growth rate, a line showing a doubling output every ten years has been drawn, starting with 1940. The chart shows that the system growth has exceeded this rate considerably.

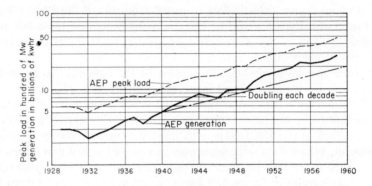

FIG. 1. Generated output more than doubles every ten years.

Overall construction costs for the generation of thermoelectric energy as shown in Fig. 2 have risen in the past 30 years more than 200% while boiler and turbine-generator costs have risen almost 300%.

In the face of these large increases, the low production cost which has been achieved testifies to the success of technological advances resulting from continuous research and development in design, construction, operation and maintenance.

One of the most significant contributions has been the improvement in thermal efficiency of power plants, shown in Table 1. The best AEP plant lowered the heat rate of 13,710 Btu per kwhr (24.9% overall efficiency) to 8700 Btu per kwhr (39.2% overall thermal efficiency), or an improvement of 57.5% in thermal efficiency.

The most important manifestation of this improved efficiency, as shown in Fig. 4, is in fuel cost. Although the price paid for coal has risen 120% in the past 30 years, the fuel cost per kwhr, because of the reduced heat rate, has increased only 26%.

† World Power Conference (with S. N. Fiala), Madrid, Spain, June 2, 1960.

TABLE 1. HEAT RATES—BTU PER NET kwhr

Year	AEP System		USA
	Best plant	Average	National average
1930	13,710	17,740	19,800
1935	13,750†	16,450	17,850
1940	13,370‡	16,270	16,400
1945	12,000††	13,960	15,800
1949	11,200	13,450	15,030
1954	9110	10,720	12,180
1958	9140	9920	11,090
1959 (Est.)	9030	9600	—
1960 (Est.)	8700	9400	—

† Deepwater generating station, a joint plant of Philadelphia Electric Company and Atlantic City Electric Company, the latter at that time a subsidiary of the American Gas and Electric Company (predecessor of American Electric Power Company), established a record for the year 1935 of 11,834 Btu per kwhr.

‡ The Deepwater Btu performance for 1940 was 11,655.

†† The Deepwater Btu performance for 1945 was in excess of 12,000.

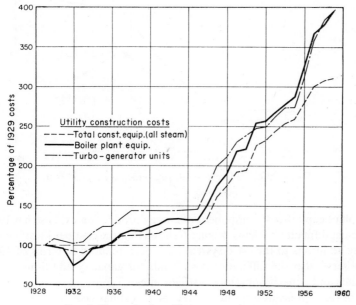

FIG. 2. Sharp rise in utility construction costs.
Published by Whitman, Reguardt Associates, Baltimore, Md.

Other components of production cost have been similarly controlled so that the over-all energy production cost has been held reasonably stable, as shown in Fig. 3.

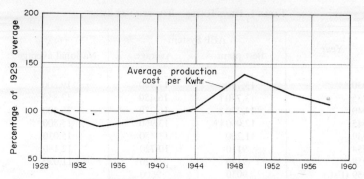

FIG. 3. Stability of production costs on AEP system.

Even through wage rates rose 190% and the wholesale price index was up 92%, though effective cost control maintenance cost per kwhr, Fig. 5, did not increase.

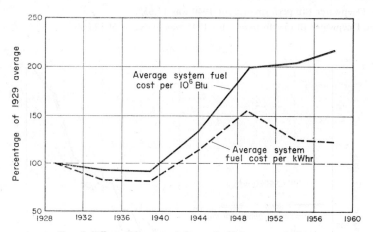

FIG. 4. Effect of improved thermal efficiency on fuel cost.

The plant capital-cost component of energy production expense has been one of the most difficult problems. Figure 6 shows that the cost per kw of generating capacity additions to the AEP system have been reduced in terms of 1929 dollars. Observe also the difference cost trends for capacity additions at existing plant sites and capacity developed at new sites.

One of the significant ways in which capital cost has been controlled has been the reduction in cubic content, Fig. 7, of plants by use of larger unit size, unit-type arrangement of components, the closest attention to actual area and volume needs and the elimination of unnecessary space. Two trends are shown: One for plants on the Ohio River, where deep basement structures are required to accommodate the large rise and fall of river level (73 ft in some locations); and the other for plants off the Ohio River where the rise of level is small.

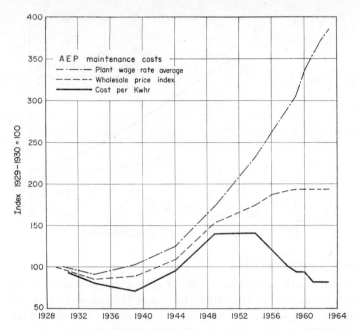

FIG. 5. Control holds maintenance costs per kwhr to 1928 level.

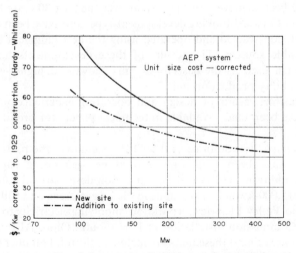

FIG. 6. Increased unit size lower plant capital cost.

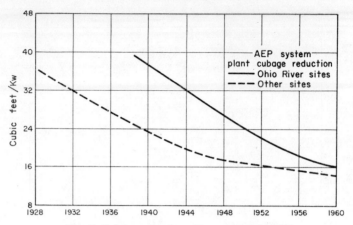

FIG. 7. Cubage reduction of approximately 60%.

Techniques of Development

Improvements in the generation of power came about as new ideas, supported by past experience, were first tested and, when proved, were fully exploited. While considerable laboratory and factory development was almost always going on throughout this period, it was necessary to use the utility system as the proving ground for new designs. Experience, punctuated by hard work and reinforced by numerous perils, created an appreciation of the fact that extrapolation to solidly attainable amplification must be quite conservative, the ratio frequently not being greater than two to one. Developmental installations, therefore, must consist of fairly large generating units representing a large investment and requiring very close co-operation with the participating manufacturers.

The improved efficiencies and the lower relative operating costs and capital costs which have been achieved on the system over the past 30 years have followed this pattern. Forward-looking development-type units have been carefully studied, designed, constructed and operated using the best judgment available from all responsible sources. The results of these developments have been integrated and then used as the basis for design of advanced, higher-efficiency but still high-reliability production units.

As soon as feasible, results of design and operating experience with these developments have been made available in technical papers for broad dissemination.

Table 2 shows capacity additions to the AEP system over the past 30 years. Thirty-five units have been built with a total net generating capacity of 6,179,000 kw. In addition, AEP Service Corporation designed and constructed eleven identical 215,000-kw net output units in the Kyger Creek and Clifty Creek plants of the Ohio Valley Electric Corporation, supplying electricity to the U.S. Atomic Energy Commission installation at Portsmouth, Ohio. AEP is one of the joint owners of OVEC. If these units are included, the total number designed and built is raised to 46 and the total generating capacity to 8544,000 kw. Three

TABLE 2. CAPACITY ADDITIONS TO AEP SYSTEM

	Date	No. of Units	Size Mw	Steam Conditions Pressure psi	Steam Conditions Temperature F	Design Heat Rate Btu/kwhr	Notable Features
Development units							
Logan A	1937	1	90	1250	925	12,000	High-press., high-temp. first 1,000,000/hr, single-boiler, topping installation.
Twin Branch 3	1940	1	72	2400	940/900	10,200	High-press., high-temp., first single-boiler reheat installation; more centralized controls.
Philo 6	1957	1	107	4500	1150/1050/1000	8,650	Supercritical press., once-through boiler, 1150 F temp., double-reheat, cyclone-firing, hollow-cooled generator bars.
Production units							
Philo 3	1929	1	160	600	720/720	12,000	Large triple-compound, reheat 60-Mw topping unit over 3 30-Mw units.
Windsor 7	1939	1	150	1250	925	12,000	
Philo 4 etc.	1941	5	95	1300	950	10,850	Advance in steam temp, large-size, cross-compound.
Glen Lyn 5 etc.	1944	3	100	1300	925	10,900	Single-cylinder 1800 rpm, one installation with single boiler.
Philip Sporn 1 etc.	1949	7	150	2000	1050/1000	9200	Largest-sized units, high-press., high-temp., modern reheat, single-boiler, centralized controls.
Kanawha 1 etc.	1953	5	215	2000	2050/1050	9070	Largest-sized units, high-reheat temp.
Glen Lyn 6 etc.	1957	8	225	2000	1050/1050	8920	Adopted single-turbine driven boiler-feed pump.
Breed and Philip Sporn 5	1960	2	450	3500	1050/1050/1050	8500	Largest-sized, supercritical-press.. once through boiler, double-reahet, single-boiler feed-pump.
Totals		35	6179				

15a*

highly significant developmental units have been listed at the top of the table and the thirty-two production units have been shown in groups of essentially identical designs.

Efficiency Evolution

The evolution of higher thermal efficiencies can be traced in Table 2. Deepwater in 1930 utilized 1200 psi steam pressure and 725 F steam temperature at a time when 600 psi and 650 F were above-average conditions. Operation of Deepwater proved conclusively the economies of higher steam pressure.

In the mid-1930's as load on the system began to increase with improved economic conditions studies dictated topping installations at two older plants. Based on Deepwater experience, 1250 psi was chosen as the steam pressure. Metallurgical progress made available superheater tubing and alloys for turbine parts suitable for raising the steam temperature to 925 F. Using these conditions, Logan A and a similar unit at Windsor were desgined and put in operation, bringing the over-all plant heat rate to 12,000 Btu per kwhr (28.4% efficiency).

The limited application of topping units was recognized at the start and efforts were concentrated on further development of condensing plants. A decision was made to take full advantage of all previous improvements and establish in a single design the most advanced conditions that then could be incorporated with no foreseeable limitations to betterment and refinement in future designs.

The result was a 67,500-kw installation for operation at 2500 psi, 940 F initial steam temperature, 900 F reheat temperature and composed of a single boiler and single turbine-generator. It was projected for the Twin Branch plant and placed in operation in March 1941. This was the forerunner of the postwar single turbine-boiler, reheat-cycle unit development now reaching commercial sizes of 600,000 kw.

That Twin Branch development can be traced back to the large-scale reheat adaptation at Philo Plant in Units 1 and 2 prior to 1929 and in Unit No. 3 in 1929. The heat rate of that Twin Branch unit of 10,200 Btu per kwhr (33.4% over-all efficiency) established a new record of performance.

To maintain this trend, it was necessary to develop higher operating temperatures which was accomplished by the installation of the first steam-electric plant to use 1000 F at 1300 psi, namely the 25,000-kw unit at the Missouri Avenue plant.

During the war-time period (1941–45), no particular advance in technology took place. However, in 1946, as soon as conditions permitted, plans were formulated for the next stage. The result was full exploitation of a single-boiler, single-turbine, high-pressure, high-temperature, reheat-type unit of the maximum size which could be accepted by the system and could be expected to be designed safely by the manufacturer. It was projected for the new Philip Sporn and other plants of the system.

The new units were rated at 150,000-kw net output at 2000 psi, 1050 F, with reheat to 1000 F. Austenitic material was used exclusively in the high-temperature superheater tubes and high-temperature parts of the turbine on the first six of the seven identical units. The last unit of this series benefited from the metallurgical improvements in ferritic material which was used for the inner shell of the high-pressure turbine to check the suitability of the least expensive type of construction at 1050 F. These seven units operated at a heat rate of 9200 Btu per kwhr (37.1% efficiency).

The next series of units advanced the reheat temperature 50 degrees to 1050 F and involved projection in size to 215,000 kw net output.

In 1953 it became quite obvious that unit-size increase alone would not achieve the desired improvement in thermal efficiency, and it was concluded that pressure, temperature and reheat should be extended to the then-appearing practical limits to obtain a base for a series of new unit sizes with improved thermal economy. The result was the placing in service in 1957 of the Philo 6 supercritical unit with steam at 4500 psi, 1150 F and two stages of reheat—one to 1050 F and the second to 1000 F. Successful operation of this installation demonstrated the feasibility of the elevated steam conditions and proved, on a relatively large scale, the once-through boiler operating at supercritical steam pressure.

Metallurgical research has not yet resulted in low-priced, high-temperature materials suitable for a large-scale installation at temperatures above 1050 F. Consequently the units based on the success of the Philo development have been designed for the most economical combination of pressure, temperature, and size factors. In 1960 two units, each rated at 450,000 kw, were installed for operation with 3500-psi, 1050 F initial steam and two stages of reheat, both 1050 F. These units at Breed and Sporn plants will have an overall net neat rate of 8500 Btu per kwhr (40.2% over-all efficiency).

Development of the higher initial steam conditions was accompanied by advances within the heat cycle and supporting auxiliary equipment. Examples are more stages of feedwater heating (nine with the 450,000-kw, units), better distribution of bleed points, use of desuperheating zones in heaters, and use of designs having negative-approach temperatures. Boiler efficiencies have been radically increased through the extensive use of large air heaters, improved combustion techniques, and pressurized firing, which was first introduced on the Philip Sporn series of units in 1949. With pressurized firing forced-draft fans handling cool combustion air could replace the previous practice of using forced-draft and induced-draft fans in series, but the Philip Sporn units and the Kanawha 1 units were provided with induced-draft fans for emergency operation. The reliability of pressurized operating was gradually improved to the point where with the Glen Lyn 6 series of units ID fans were omitted.

Improved overall cycle efficiency has been aided by steam-heating the combustion air, even though steam is extracted from the turbine cycle, because the boiler can be operated with lower outlet-gas temperatures with a given critical air-heater metal temperature. The latter is determined by the sulphur content of the coal burned, and the resulting air-heater pluggage which must be held at a tolerable level even with newly developed in-service air-heater washing techniques.

Auxiliary power requirements have been reduced by improved efficiency of components, largely the result of increased size. For example, two forced-draft fans will serve the 450,000 kw Philip Sporn 5 unit, whereas sixteen FD fans were used with the 160,000-kw Philo 3 unit in 1929. Boiler-feed pumps have also been radically reduced in number. Steam-turbine-driven boiler-feed pumps were used in the 1920's, often three per unit. These were gradually displaced by squirrel-cage, motor-driven feed pumps, generally no less than three one-half capacity pumps being provided per unit.

In 1955, AEP staff studies indicated that substantial efficiency and capital economies could be achieved by reliance on one full-size steam-turbine-driven feed pump. The efficiency of this important auxiliary was not only increased by its larger size but the overall cycle efficiency was improved by the strategic location of the feed-pump turbine in the heat cycle. In the first applications on the eight 225,000-kw units, beginning with Glen Lyn 6, each pump was driven by an extraction turbine which serves three feedwater heaters below the reheat point with only moderately superheated steam rather than the highly super-heated steam that would have been extracted from the main turbine. The 450,000-kw units utilize a single, condensing, turbine-driven pump which improves cycle efficiency by lowered turbine exhaust loss.

Substantial improvements have been made in turbine and generator efficiencies during this period, The major improvement in generator efficiency has come from the development of hydrogen cooling, described later.

Reduced Labor and Maintenance

Lowered unit maintenance and other costs have been achieved by developments in several directions. Foremost has been the use of larger integral blocks of generating capacity. The practice of the 1920's was to use several boilers to serve one or more turbines. For example Philo 3 unit had six evaporating and two reheating boilers to serve three turbine elements—a single high-pressure element and two parallel low-pressure elements. Therefore, the largest amount of capacity that could be forced out of service for an extended period was one-half the unit capacity of 160,000 kw. For reserve requirements this unit could be classified as an 80,000-kw unit. Similarly, Logan A and Windsor 7 could be classed as 40,000 and 60,000-kw units, respectively, since these were the largest blocks of generating capacity which could be forced out of service for an extended period. The units in the Philo 4 series have two boilers and high- and low-pressure turbines so that the largest amount of capacity that can be forced out by a single failure is 50,000 kw. All subsequent units, however, have a single turbine or a single-turbine, single-boiler design so that the full unit can be forced out of service.

Despite the rapidly increasing size of units, Fig. 8 shows the trend toward greater conservatism in the percent of total system capacity exposed to outage if a single turbine, boiler, or significant element in its cycle of operation had a mishap.

Along with these large units have come larger total plant capacities at a given site. Philo plant in the early 1920's was projected to have 80,000-kw capacity. In 1929 this was raised to 240,000 kw and today it has 497,000-kw capacity. Philip Sporn plant was designed in 1947 to have 600,000-kw capacity; today it has 1,050,000 kw, and an ultimate capacity of 1,500,000 kw is projected.

With the increased size of units and the increased concentration of capacity at a given site, more capacity can be serviced effectively by a given size maintenance and operating group.

Of vital importance, not only in reducing labor and maintenance costs but in the safe operation of the large and costly pieces of equipment in a modern power

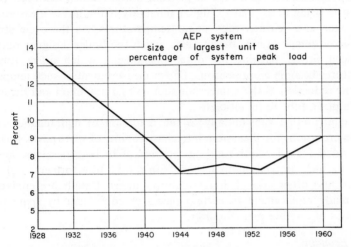

FIG. 8. Despite growth in unit size system availability is more conservative.

plant, has been the development of centralized automatic control. Philo 3 unit in 1929, with eight boilers and three turbines, had control stations located at each pair of boilers, the turbines, the condenser pit and the electrical station's switching dispatcher. In all, eleven operators were required per shift. In 1940 Twin Branch 3, with a single reheat boiler and reheat turbine, was put in operation with control centralized at a boiler panel, a turbine panel and at the dispatcher's panel. Tidd units 1 and 2 of the Glen Lyn 5 series were the first completely centralized control units. All functions of unit control from admission of coal to the boiler to switching of high-voltage output were concentrated in a single control room. Both of the 100,000-kw units were operated with seven men per shift. Today, centralized controls have been brought to the point where on the 450,000-kw Breed unit four operators per shift will control all functions of the generating plant.

Another important feature tending to reduce maintenance and also to increase the safe effective output of units has been the extensive instrumentation of critical areas of major equipment which has been developed over the past 30 years. This instrumentation allows operation of equipment at the true design limits of the components rather than at some artificially created limit which does not bear a real relationship to the properties of the materials in question. Examples of these instruments are: Boiler-drum metal temperatures which determine safe rate of pressure rise in the boiler; turbine-stage pressure measurements which indicate internal condition of the turbine; turbine-metal temperatures measurements in shells and bearings; and, especially important, the turbine supervisory instruments which record vibration, eccentricity and differential expansion to give a continuous record of the condition of the machine. These turbine supervisory instruments were first used on the Windsor unit of 1939 and have come to be standard throughout the industry.

A particularly important type of instrumentation which has been used and is presently being developed to higher reliability is the direct measurement of super-heater and reheater-tube metal temperatures. As the cycle strives for higher steam temperatures, the margins which the boiler designer puts in for unbalanced firing, slagging, or other phenomena which affect tube-metal temperatures become more onerous. Direct measurement of such temperatures allows the reduction of these uncertain factors to a minimum and the exploitation of a given tube to its ultimate.

Also contributing substantially to the reduction of maintenance cost has been the provisions for maximum interchangeability of spare parts. Spare parts are provided for all major rotating sections of turbo-generators, and other equipment such as feed-pump motors, fan rotors, transformers, high-voltage bushings, and other critical, longer-delivery items. Details are engineered to permit ready replacement of certain pieces of equipment by spare units, in some cases of another manufacturer.

Capital Economy

Reductions in capital cost, achieved over the past thirty years, indicated in Fig. 6, have been the result mainly of increased size of unit and development of the unit-type, single-boiler, single-turbine, single auxiliary concept. These cost reductions are all the more significant when consideration is given to the fact that greater use is being made of the more expensive materials needed for operation at the higher steam conditions. Philo 3, for instance, first operated in 1929, utilized 600 psi, 720 F, whereas the Breed-Philip Sporn units of 1959 utilize 3500 psi and 1050 F with their associated high-cost alloys. In other words, the capital cost of new generating plant, when adjusted by a national average cost index, has been brought down about 30% in thirty years without taking credit for the 57.5% increase during that period in thermal efficiency.

Employment of larger units, and particularly the unit system idea, has made possible staggering reductions in building volume. The multiple system of auxiliaries has been abandoned and individual auxiliaries have been greatly increased in size as illustrated by the employment of a single boiler feed-pump serving the 450,000-kw Sporn series of units. The driver is a 20,000-hp condensing turbine.

Switching and transformer arrangements have been simplified. In units designed during the early 1920's, low- and high-voltage switching were required. Low-voltage buses and switching were utilized for plant synchronization and to supply low-voltage feeders and auxiliaries. High-voltage switching was utilized for transmission system switching of transformers and transmission lines. Even when low-voltage buses were abandoned low-voltage switching was still utilized to obtain fast closing time to synchronize units, and high-voltage switching was used for fault protection. Philo 3 was the first unit to utilize high-speed, high-voltage outdoor switching operating at 138 kv without low-voltage switches. This practice became standard. Main step-up transformers have also undergone evolutionary simplification. Practice in 1929 called for three single-phase step-up transformers with a spare single-phase unit. Then multiple, three-phase units were utilized, three per unit, on the 150,000-kw units of the Philip Sporn series; but later, starting with the 225,000-kw Glen Lyn 6 series, a single three-phase step-up transformer was utilized to handle the output of the entire unit.

A most significant generator development over the past thirty years has been the hydrogen-cooling system, which has permitted efficient high-speed machines in sizes which could not otherwise have been designed, shipped, or economically installed. The first hydrogen cooling installation in 1928 on a 20,000-kva synchronous condenser provided the operating and design information for the construction of the first hydrogen-cooled central-station generator—a 50,000-kva unit placed in service in Logan A station and connected to the topping turbine. The efficiency of this machine was 98.4% compared to 97.2% for a similarly rated air-cooled machine. At first, hydrogen was used at 0.5 psi pressure, but this was first raised to 15 psi and more recently to 30 and 45 psi. The 45-psi pressure has been associated with the conductor-type cooling in which the hydrogen is circulated directly through small passages inside the stator and field coils of the generator. This arrangement was first used on the Philo 6,156,000-kva generator in March 1957.

Hollow-cooling of rotor conductors has been demonstrated and will be used on the Breed 450,000-kw unit. Stator windings, however, will be cooled by direct circulation of oil. Because of its improved heat-removing capabilities, water-cooling will be used on the stator bars of the Philip Sporn 5,450,000-kw machine.

The net result of this experience has been that there are now in service on the AEP system 63 hydrogen-cooled generators in sizes ranging from 28,000 to 292,000 kva, totaling 7,400,000-kva capacity.

The universal acceptance of hydrogen-cooling led to simplifications in design, progressive improvements and broadening of the application. Today hydrogen pressures in excess of 45 psi are used in the direct cooling of stator windings of AEP machines; this had made a significant contribution to very much increased outputs by permitting compact designs.

The size reduction achieved through the use of hydrogen and liquid cooling is illustrated by comparing the 160,000-kw generator placed in operation at Philo plant in 1929 and the Philip Sporn plant 450,000-kw unit of 1960 where for the same bearing span a rating of 291,000 kva is being obtained as compared to 62,000 at Philo.

Development for the Future

At this writing the construction of the two largest and most-advanced type units in the AEP system is well along. The first should be in commercial operation in January 1960; the second will follow toward mid-year. These are the

450,000-kw (net output) units for Breed and Sporn plants designed to operate
at 3500 psi and 1050 F with double reheat to 1050 F.

Development of thermoelectric generation will continue along many lines.
Improved efficiency through the use of higher steam conditions awaits a break-
through in metallurgy. Moderately priced materials for service in the 1100 to
1250 F range are badly needed. Studies and research on combined coal-burning gas-turbine steam-turbine cycles continue and development work on a pilot
plant appears to be near realization.

Great effort is expended to improve reliability and lower maintenance costs.
Somewhat larger units will be designed, but it is doubtful if the rapid increases
in size of the last few years will be repeated in the next decade. Improved
efficiency by virtue of higher steam conditions presents a difficult problem
but there are possibilities that there may be a break-through to steam-temperature levels in the 1200 to 1250 F range. If materials are economically
available for these temperatures, there appears to be the possibility of exploiting higher pressures in connection with these higher temperatures. Continued development will be conducted on the low-pressure end of the steam
cycle to minimize turbine exhaust losses and obtain the highest possible turbine
efficiency in the low-pressure range.

As larger and more complicated generating plants are designed, automation
will receive wider acceptance and operators will be relieved of substantially
all routine operations. Their work will be confined largely to operations calling
for most essential judgment decisions.

Effort will also be concentrated on obtaining a series of individually small,
but cumulatively significant, gains in thermal performance, such as reduction
of outlet-gas temperatures from the boiler, improvement in efficiency of auxiliaries and reduction of auxiliary requirements.

Long-range developments will also play their part; perhaps the most important of these may be the development of techniques for direct conversion of
heat into electricity with the possible use of magnetohydrodynamic or similar
devices.

20. CARDINAL—A SUPERCRITICAL SUPERSIZED STEAM PLANT JOINTLY OWNED WITH CO-OPERATIVE†

ON NOVEMBER 4, 1963, construction of the new steam-electric Cardinal plant was begun on the Ohio River adjacent to existing Tidd plant at Brilliant, Ohio, to consist initially of two 615-mw units, and, when completed with a third unit to have a total capacity of almost 2 million kilowatts. Even with only two units, Cardinal will be the largest in Ohio and one of the largest in the world. It will require an investment of about $125 million.

The first two units will burn close to $3\frac{1}{4}$ million tons of coal annually to generate some 9 billion kwhr; the third unit will increase these figures to more than 5 million tons and almost 15 billion kwhr. Most of the coal for the plant will be produced in eastern Ohio with some in contiguous West Virginia. It is expected that Cardinal will be one of the world's most efficient power stations thermally as well as one of the most economical from the standpoint of both capital cost per unit of capacity and operating and maintenance cost per unit of output.

Unique Ownership Contract

More than matching the unusual features of this advanced project in steam-electric generation technology is the unique character of its sponsorship and ownership, and the work of the sponsors and owners in conceiving, planning and carrying through the necessary negotiations. Buckeye Power, an association of Ohio's 30 rural electric co-operatives, and Ohio Power Company, one of Ohio's principal investor-owned electric power companies, will each own one of the two initial Cardinal units.

Transmission facilities of Ohio Power will carry the energy to the sub-transmission systems in the various areas of the state. Ohio Power, five other investor-owned electric utility companies, and Buckeye Power, will own various segments of the sub-transmission and wheeling facilities.

Engineering, design and construction will be handled by American Electric Power Service Corporation. While both units will be similar, neither, by itself, will be quite complete. They will share certain common stations facilities, and even cycle facilities, some of which will be located as part of one unit and some as part of the other. However, the right of each owner to all these facilities, even in the unlikely event of separation of normal joint operation, will be provided for clearly by an exchange of right-to-use cross-easements which will make it possible for each owner to operate its unit independently of the other. The total expenditure for the two units will be divided in two, so that the cost of each unit to its respective owner will be identical.

Wheeling.—Plant output will be fed to the 138- and 345-kw transmission network of Ohio Power which in addition to absorbing its own share, will deliver the full requirement of the thirty Buckeye Co-op members to those

† *Electrical World* (with John A. Tillinghast), March 9, 1964.

investor-owned systems participating in the wheeling operating which will take the power to the tap points or substations from which the co-ops are now being supplied under existing purchase agreements with these utility systems.

Back-up will be provided by Ohio Power to the same extent as is provided for any other plant on its system. For this Buckeye will contribute its share of the reserve found necessary in the operation of Ohio Power's interconnected system, and this will be in the order of 15%.

Interim Capacity Arrangements.—The current total peak of the 30 co-op members of Buckeye is about 200 Mw but by the time Buckeye's unit is completed it is expected that this peak will be close to 300 Mw. Ohio Power will purchase the remaining capacity of the Buckeye unit releasing it in increasing amounts year by year as Buckeye's load grows. When the full capacity of its unit is reached Buckeye will obtain its additional requirements from Ohio Power until such time as it is able to develop sufficient load to make feasible the addition of a second unit as large or larger. For this interim capacity the charge will in effect be a reciprocal charge; it will be based largely upon the history of Ohio Power's use of Buckeye capacity during Buckeye's growth and the charge to Ohio Power for such capacity.

Financing.—An unusual feature of the undertaking is the financing. Since ownership is separate Ohio Power will finance its unit on the basis of its current practice and capital structure. Buckeye will finance by approximately 15% equity contributed in the main by the co-op members and the balance by first mortgage bonds, backed by the plant and the contracts for absorbing its capacity created by Buckeye members and by Ohio Power. It is expected that these mortgage funds will be obtained in the public money market.

The ability of Buckeye to obtain a loan from REA, if justification could be established, at the legal rate of 2% as well as the difficult problems this created for Ohio Power and the other investor-owned co-operating utilities, was fully recognized by both parties. They also recognized the social-economic advance in developing a satisfactory generation program for Buckeye's members that precludes the need for the assumption of any burden for its financing by the federal government.

The discussion of a solution to this problem was helped materially by the imaginative and advanced program for highly economical generation that was started by one of the authors under the aegis of the AEP System Development Committee. Originally begun as Project 200—a general program to improve materially the technics and economics of large scale system thermal-electric generation—it was merged into the development of what came to be known as the Cardinal plant.

Thus, after the project had been completely formulated in a series of negotiations that extended for more than a year, Buckeye was able to adopt a program of financing its mortgage capital in the market place and yet obtain energy costs showing substantial savings over the currently purchased energy costs. This was made possible by the advanced engineering concepts which lowered both capital and operating costs; the very favorable fuel arrangements that it was possible to develop in the eastern Ohio coal region; the ability to exploit to the fullest the economics of large size units more than double Buckeye's initial requirements; and, finally, the arrangements to assure full loading of the unit as soon as it would be put on the line.

Joint Operation and Supervision.—Co-ordinated operation of Cardinal will be carried out through a jointly and equally-owned operating company which in essence will provide a single-plant organization. Supervision will be provided by a jointly staffed board of supervisors to whom the head of the operating organization will report. The board of supervisors will, however, be aided and the talents of the operating organization supplemented by continuous supervisory guidance furnished by a service and engineering organization—most likely American Electric Power Service Corporation.

Since this advanced technical project contributed so much to make the project a reality, it is proposed to discuss the technical phases of the Cardinal plant, not in complete detail, since such detail is still lacking, but sufficiently to provide a clear understanding or what has been undertaken.

Technical Features

As has been indicated, the technical features of Cardinal are the product of the project that was undertaken to integrate the full breadth and depth of the long developmental experience of the AEP system in the exploration of the new ideas in thermal energy conversion, to substantially improve maintenance and availability, to evaluate manufacturers and their capabilities and, most significantly, to check out some important hunches. We urged on ourselves and our associates self-analysis and creative synthesis aimed at distilling the best in our long experience and the experience of others into the design of conversion equipment of the highest excellence.

We started on a solid foundation of supercritical proficiency with advanced steam power generation units—Philo 6, Breed and Sporn 5. These gave design, construction and operating experience. In addition, we had the design and most of the construction experience of Tanners Creek 4, the latest supercritical pressure unit about to be brought into service. Design concepts of Cardinal, therefore, represent a fourth generation of effort. Philo 6, in 1957, was the prototype, the 500-mw Breed and Sporn 5 units in 1960, a little more than three years ago, were the first commercial full-scale applications of the technology of super-criticality. What we did in each of these plants was good at its time but in each were fertile grounds for improvements. Sporn 5, which cost $140 per kw in its first full year of operation had a heat rate of 8782 Btu per kwhr, or an over-all thermal efficiency of 38.9%, Tanners Creek 4 is designed for 8500 Btu per kwhr and the current working estimate of capital cost is $110 per kw. High availability and substantial ease of startup should be demonstrated this month as this unit is put into operation.

The technical design has been developed during a period in which larger sized and higher efficiency units, pioneered for many years on the American Electric Power systems have been receiving broad nation acceptance. It is significant that of the nearly 11,000 Mw of large units committed by US utilities in 1962 and 1963, 7500 Mw, or 70%, are designed for supercritical steam pressure, either single or double reheat.

Capital costs at Cardinal will be low, and will compare favorably with any that are presently being projected anywhere. Supercritical steam pressure with double reheat will be the firm foundation for a most attractive thermal efficiency. Simplified design based on extensive experience, exhaustively re-examined and updated, will give good operational qualities and high availability.

Significant Items.—In the design of Cardinal plant eleven significant items stand out:

1. The first use of a single, supercritical, double-reheat turbine and associated 723-Mva, single-shaft generator.
2. A once-through universal-fuel pulverized-coal dry-bottom boiler.
3. New excitation system for the generator.
4. Grade level location of boiler-room floor and heavy equipment.
5. General cycle simplification and equipment duplication.
6. An 825-ft stack for difficult environmental conditions.
7. Single 725-Mva, 3-phase, main-output transformer bank.
8. Development of strip and deep mined coal, delivered by truck, rail and barge.
9. Modified outdoor construction.
10. Positive quality control in both turbine and boiler manufacture.
11. Elimination of escalation.

In addition the summation of the effects of many other smaller items contributes substantially to the overall excellence of the project.

General Arrangement.—The general arrangement of the plant is shown in cross-section. The two identical units are located directly on the river, each with its own intake structure but with common discharge. The turbines are ranged parallel to the river with a low intermediate heater bay leading into the boiler-room area.

Turbine.—The outstanding feature of Cardinal plant design is the single 3600-rpm tandem-compound steam turbine driving a single 723-Mva water-cooled generator. For many years we have utilized cross-compound machines for two reasons: First, in the large sizes—large relative to the technology of the time—which we have been advancing, tandem machines were unacceptably large extrapolations of then existing experience; and second, turbine manufacturers had represented cross-compound machines as being lower in costs in these larger sizes than the tandem arrangement. However, as experience accumulated, the tandem machine with one generator, fewer parts, and particularly smaller, higher-speed parts, had clearly demonstrated that it would be less costly to build. At Breed and Sporn we have in effect two 3600-rpm tandem machines, each of 250-Mw rating, operating together as a cross-compound set. Favorable experience with these machines, particularly the water-cooled generator at Sporn 5, was the direct basis for the extrapolation of this much larger tandem machine at Cardinal.

To assure high availability and to utilize design, construction and operational experience of Tanners Creek 4, the Cardinal machine, which also will be furnished by the General Electric Company, will utilize the H-P first-reheat and second-reheat sections of Tanners Creek 4 to which are added three double-flow exhaust sections which will have been tested in operation by other utilities, making all 615 Mw flow through a single coupling to the largest generator ever designed. With these five turbine sections and the generator on one shaft, much simplification results in controls, piping, startup and excitation, but the machine grows to a considerable length—200 ft. With this long machine we found it attractive to turn the turbine-generator 90 deg from our usual configuration so that the centerline parallels the river, giving a turbine room only 63 ft wide.

Once-through Dry-bottom Boiler.—Since the initial operation of Philo 6, nearly seven years ago, water-side problems have been virtually nonexistent with our once-through designs. No single tube failure can be charged to water-side deposits and only one tube failure, caused by maldistribution during upset startup conditions, has occured. This highly satisfactory experience reinforced our confidence in the supercritical once-trough boiler from the fluid heat-transfer point of view.

In the face of problems with wet-bottom and cyclone-fired boilers, and their requirement of low-fusion coals, we have adopted the concept of the universal-fuel boiler which can fire any coal available to the AEP system. This fundamental decision was based on extensive studies of availability, maintenance costs and efficiency of wet-bottom and pulverized-fuel boilers. The Cardinal boiler will be designed and manufactured by the Babcock & Wilcox Company and will be the first of such dry-bottom pulverized-coal boilers. In the main, it will resemble the cyclone-fired Tanners Creek 4 boiler but will be fired by five new 50-ton per hour counter-rotational pulverizers developed in a joint program with Babcock & Wilcox at our Twin Branch plant. Fuel will be fed into the dry-bottom furnace by twenty new cell burners.

Steam Cycle.—Thorough and comprehensive review of the economics of steam conditions reaffirmed our Tanners Creek 4 design parameters—3500 psi, 1000/1025/1050 F. In total analysis these conditions are a most attractive balance between high thermal efficiency and reasonable capital costs. One of the satisfying achievements of this work was reducing the cost increment for second reheat to the point where it could be completely justified.

AC Excitation.—When we selected the 723-Mva generator it became obvious that the predicted 3700 kw of dc excitation power would require an inordinately bulky, low speed, geared exciter. Convinced that this concept had been carried beyond practical limits, we insisted on the study which later led to the unanimous selection, both by the manufacturer and ourselves, of an ac system in which a directly coupled 120-cycle ac generator, running at 3600 rpm, supplies main-shaft power for rectification in stationary devices. This new system will be tested exhaustively before startup at Cardinal. Not only is the size of the exiter sharply reduced but an improvement in over-all plant heat-rate of

10 Btu/kwhr results from the higher inherent efficiency of this method. The possible later development of shaft mounting for the rectifiers offers the prospect of a completely self-contained excitation system and removes excitation problems as obstacles to further growth in single-shaft sizes of alternators.

Grade Level Layout.—At Breed and Sporn 5, the boiler was arranged for continuous upflow of gas, placing the full support of the boiler, the heavy regenerative air-heaters, forced-draft fans, recirculating-gas fans and other heavy rotating equipment in the boiler house structure. This led to heavy structural-steel costs and unsatisfactory operating characteristics. As shown, the Cardinal boiler has up, across and downflow of gas, allowing the main operating floor, the fans and the heavy tubular air-heater to be located on a rugged mat at grade elevation. This will ease erection access and result in improved operating characteristics.

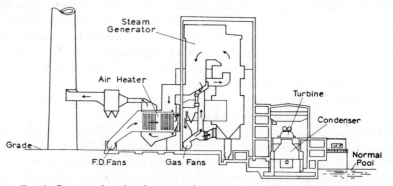

FIG. 1. Cross-section showing general arrangement with grade-level floor.

General Simplification.—We have striven for simplification of auxiliary systems to promote lowered capital cost and favorable operating characteristics. We have reaffirmed our decision, made as long ago as 1956, to use a single turbine-driven feed-water pump to serve the entire unit. The AEP system has 12 units in operation totaling 3184 Mw embodying this concept which, in in the aggregate, have operated over 400,000 hr with remarkably good performance. To simplify design, purchase and construction, we are insisting that the two units at Cardinal be exact duplicates with only minor variations in some auxiliary equipment which need not be duplicated in a two-unit plant.

Modified Outdoor Construction.—Our operating experience has resulted in the evolution of the outdoor-plant idea to our current modified outdoor construction in which all operating equipment on the main operating floors is enclosed, but the upper half of the boiler, the support steel, the air heater and other equipment, are outdoors. This represents a good compromise between first cost and operating in the climate we experience.

Stacks of Record Height.—Due to difficult environmental conditions, particularly that the plant site is in a valley more than 400 ft deep, stack design presented special problems. To assure that products of combustion are properly dispersed and diluted to tolerable levels under all wind and atmospheric conditions, extensive model studies, meteorological surveys and other techniques were used to select proper stack height. On the basis of these analyses, it was decided to build two 825-ft stacks, the tallest in this country. They will be nearly 150 ft taller than stacks built only ten years ago at the Clify Creek plant, then the tallest in the country. They will be of concrete construction with steel liners.

Transformer.—The full output of each generator will be stepped up—on one unit to 138 kv and on the other to 345 kv—by a single three-phase step-up transformer, rated at 725 Mva. These will be the largest used on a single machine. To provide back-up an auto-transformer bank, normally in use between the 138 and 345-kv yd, will be equipped with tertiary winding of generator output-voltage level (24 kv) which can be brought in under emergency conditions to serve either unit.

Coal Supply.—Extensive negotiations have led to arrangements to provide both strip and deep-mined coal, delivered by truck, rail or barge. Flexibility between the two types of coal and the three modes of delivery will assure a reliable fuel supply at attractive prices. Facilities for all three modes of coal delivery will be included in the initial design.

Quality Control.—While quality control has always been a difficult problem, it becomes progressively more so as units become larger and designs more sophisticated. An important part of our work has been the detailed, strenuous review among ourselves and with manufacturers and others of design, manufacture and construction techniques to assure the highest quality in conceptual design and then to transmit these concepts to the final structure. Areas of difficulty on the Breed and Sporn 5 designs, and areas of new design components such as the generator, the three double-flow exhaust ends, and others were reviewed and the adequacy of measures taken to prevent recurrent difficulties detailed. Model testing was indicated in several areas such as the gas and air-flow paths in the boiler, lugs and attachments on boiler tubes, the flow pattern in the steam entering the condenser and other areas.

Escalation.—In one non-technological field—escalation—we have made tremendous strides. Escalation has been literally the bane of our existence since it was introduced in the early 1940's. Several fine jobs, including Breed and Sporn 5, were nearly ruined by escalation, totaling on some pieces of equipment as much as 15% of the contract price. The present achievement has been the commitment of all major equipment for Cardinal, and most of the site labor, on a firm price basis. The net effect has been to reduce escalation on the overall job to an almost negligible point.

Accomplishments

What, then, is the essence of achievement in the creation of the Cardinal project?

We have brought to bear every bit of experience—our own, the manufacturers', other utilities'—and have projected this with all the creativity and imagination at our command into a coal-energy conversion unit which is likely to prove the forerunner of a long series of units on our own system as well as those of other utilities. Each machine, utilizing the supercritical pressure double-reheat steam cycle, will generate 615 Mw net, will have a thermal efficiency and an availability close to the highest, and an operating and maintenance cost close to the lowest. With all this, capital cost below $100 per kw is expected.

Fuel cost will be most attractive at Cardinal. A combined flow of rail, barge and truck coal is expected to result in an average delivered fuel price of under 17 cents per million Btu.

Co-ordinated design, purchase and construction, pulled together by our product engineers, will lead to a smoothly scheduled startup, with the first unit going into operation in July 1966 and the second in January 1967.

Essentially it was the contribution of the economics and integration of the co-operative, and investor-owned facilities that provided the solid basis for the agreement to build.

From the very beginning, this proposal had in it so much that was novel and exciting technologically and organizationally that both parties agreed that it offered the opportunity to evolve new and more solid relationships if subsequent more detailed negotiations could be carried to final agreement.

In developing the contractual arrangements we re-examined almost every concept and every problem involved in the very close kind of juxtaposition and interweaving of electric energyy supply facilities of groups like the co-operatives and the investor-owned companies. These juxtapositions extend over thousands of miles and the problems of service into areas covered, by the distribution lines of either the Ohio utilities or of the Ohio co-operatives, or by both. The problems of rates, industrial expansion, growth of communities, annexation of physical areas to existing communities, responsibility for supporting the investment created by Buckeye and Ohio Power, substantial expansion, and many other difficult problems were analysed and debated. The result was satisfactory understandings or at least resolution of the problems into their components to serve as a basis for future understanding.

Recognition of Mutuality.— It took a special kind of attitude and approach to make this possible. It took, in the first place, a great deal of understanding of each other's respective problems, a very high level of mutual regard, and respect for the other's rights and interests. It took a great deal of what might be called engineering in the negotiations—engineering in the better derivative sense of the terms, which is ingenuity. It took a great deal of adroitness to make possible many of these revelatory and fruitful discussions.

Both groups recognized some of the great contributions that the other could make or could bring to the joint enterprise. Ohio Power and the participating utilities recognized that Buckeye and its 30 rural electric co-operative members were the exemplification of the spread of new forms of social-economic organization for carrying out an important service in modern society. They had come into being originally because of a need that was not being met by others. They had grown and prospered over the years. Buckeye and its people recognized that the utilities, even though they may have been slow to respond to the challenge of rural electric service in the dreadful 30's, nevertheless had come out of the depression with a great deal of vigor and had continued to extend the technology, scope and effectiveness of their social-economic service. By increased mastery of the means of energy production, transmission, utilization, and business organization, they had developed new techniques and new advanced devices for efficiently converting primary energy into electric energy; they had researched, developed, and brought into being new techniques for large-scale transportation of electric energy; and they had given the entire art and business of electric generation, transmission, and distribution a new drive, a new elan, that excited the admiration of the entire world where ever electric power is known.

Each group recognized that these developments created opportunities for co-operation not only for the benefit of their respective interests, but for the benefit of our whole society. The achievement of the co-operatives and the Ohio utilities in developing this program is a prime example of the ability to overcome very difficult, but in some cases merely dogmatic, differences and to work together to achieve new pragmatic solutions to difficult situations created over long periods of time. The great achievement lies in the participants' bringing forth the ability to look to the future for soulution rather than to the past and, while not disregarding their individual interests, in allowing their concern for the interest of the public at large—essentially the people of Ohio—to over-ride all the differences and difficulties in the path of agreement.

CHAPTER 2

HYDRO POWER GENERATION

CONTENTS

1. TWO SMALL AUTOMATIC HYDRO PLANTS†

KANAWHA River is a navigable stream with improvements for navigation dating back as far as 1875 when provisions were made for a 6-ft navigable depth from its outlet to the Ohio River to a point about 90 miles above its mouth by means of eight locks with movable dams and two with fixed dams. The River and Harbor Act of 1930 authorized the replacement of the four upper locks and dams by two of higher lift to increase the depth of the channel to 9 ft. These two new structures, located at Marmet and London, each accommodate a low head automatic hydro plant.

Each dam contains five roller gates slightly more than 100 ft long with a height, including the longitudinal aprons, of 26 ft.

Keynote of the design was simplicity—a simple layout, a small construction organization and a minimum of unproductive refinement. Very little was included that could not pay its way. Only by following this criterion was the cost of the power developed brought into the competitive zone of steam-generated power.

The two plants are essentially identical except for the location of the storage structures for the emergency bulkheads of the dam and the type of bank-retaining constructions at the landward end of the buildings.

Concreting

Because frequent cycles of freezing and thawing are more severe than in areas considerably farther north or south, special attention was given to the specifications for the cement, the type of aggregates and the mix. Standard ASTM specifications for cement were modified to reduce the heat of hydration and resultant tendency to develop cracks during cooling. For the portions of the super structure the cement was purchased under specifications limiting the tri-calcium-silicate compound to 50% of the cement by weight, and the tri-calcium-aluminate to 12%. The tensile strength of mortar after 5 days curing was limited to a range of 225 to 275 lb per sq in.

Two classes of concrete were used. Class A, containing 550 lb of cement per sq yd, was used for portions of the structure exposed to freezing and thawing, for portions in which water-tightness was essential, for sections with heavy reinforcement and for all of the thin wall sections. Class B, containing 440 lb of cement per sq yd, was used for the larger masses of concrete not subject to severe conditions. Hydraulic lime was used as an admixture in both to the amount of 69 lb per cu yd of concrete.

Wherever possible all concrete was deposited in horizontal layers not exceeding 18 in. thickness, the pouring being done as a continuous process. Height of pours did not exceed 18 ft and during hot weather the maximum height was reduced to 8 ft.

† *Power Plant Engineering* (with E. L. Peterson), February 1937.

All concrete was subjected to vibration supplemented by hand spading adjacent to the forms on exposed faces. To prevent any inferior material in the finished concrete at the top of a lift or near its face, a thin layer was removed from the top of every lift and disposed of as waste material. All concrete surfaces were cured for a period of 14 days by continuous spraying.

Hydraulic Equipment

Hydraulic and physical conditions at the two plants were such that it was possible to design virtually duplicate installations. Studies indicated that the most economical combination of units at each plant consist of three of the vertical-shaft propeller type, two to have fixed-blade runners and one an adjustable-blade runner. The turbines for London were purchased on this basis but for Marmet an automatically-adjustable propeller of a new type, later described, was substituted for one of the fixed-blade propellers.

The fixed-blade turbines have a rated capacity of 6600 hp under a net effective head of 23 ft at a speed of 90 rpm. Diameter of the runner is 177 in. The automatically-adjustable-blade runner unit for Marmet has the same general dimensions as the fixed-blade unit, but the rated capacity is 7600 hp. The ad-

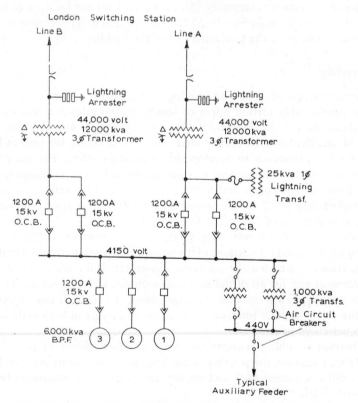

FIG. 1. Simplified electrical layout for two identical automatic low-head hydro plants.

justable-blade, Kaplan-type units, have a rated capacity of 7250 hp under a 23-ft net head at a speed of 90 rpm. Diameter of the runner is 169 in.

Both fixed-blade and the automatically-adjustable-blade turbines employ integral, cast-steel speed rings while the Kaplan-type units have separate stay-vanes, the lower ends of which are provided with feet and foundation bolts which are embedded in the concrete just outside of the throat-ring. Erection of the separate stay-vanes proved much more difficult than that of the integrally-cast speed-ring due to the difficulty of setting the vanes at the exact elevation required to secure a proper fit with the curb-ring.

Main turbine bearings on the fixed-blade and automatically-adjustable-blade turbines are of the adjustable lignum-vitae type lubricated by water from the turbine casing. The bearings on the Kaplan units are of the Cutless rubber type also lubricated by water from the turbine casing.

Electric Equipment

Both plants have the same electrical layout. Although there is a difference in hydraulic output of the three machines in each plant, it was possible to use identical generators at both plants. This permitted interchangeability as well as other obvious economies.

All six generators are rated at 6000 kva, 90 rpm, 4510 v and are designed for a WR^2 of 5 million and a runway speed of 190 rpm. The thrust bearings are the General Electric self-lubricated type consisting of a cast-iron rotating plate and a babbitted stationary plate resting on pre-compressed springs so that the load will be distributed over the entire bearing surface. Guide bearings of the babbitt-lined, sleeve-type are located below the thrust bearings.

Selection of umbrella-type generators, proper proportioning of lengths of shafts between turbine and alternator, a special crane arrangement, and omission of direct-connected exciters contributed to a minimum height of super-structure, which is outstanding in the design.

A single bus serving two lines through separate transformers was employed, each generator being switched by a single modified truck-type low-tension breaker. The low side of the main transformer banks uses two switches in parallel, that arrangement being actually more economical than one single low-voltage breaker of double the rating. Furthermore, the adoption of the two-transformer arrangement has resulted in the total elimination of a high-tension yard.

Special Features

London and Marmet have many new ideas pertinent to hydro-electric plants, especially those of intermediate size. Some of these features follow.

Automatically Adjustable Unit.—The Newport News automatically adjustable runner used at Marmet is a new type with no interconnection between the runner vanes and governor. The runner consists of four vanes with integrally-cast stems supported on roller bearings, two of which are radial and one a thrust bearing. The four vanes which are interconnected by a rack and pinion to a piston in the

16   VEP

upper part of the hub, freely pivoted and so arranged that water impinging on them sets up a mechanical moment, tending at all times to open or increase the pitch of the vanes.

This tendency to open, however, is balanced by two reactive devices, the principal one consisting of the piston previously mentioned, which is urged downward by the differential pressure between head-water and draft-tube. Head-water pressure is admitted to the top of the piston from the scroll-case and draft-tube pressure is admitted to the bottom through ports in the piston stem. An auxiliary or adjustable reactive device consists of a heavy-duty spring located in the hollow turbine shaft and acting downward on the piston. Movement of the runner vanes to correspond with changing load or variations in head is entirely automatic and is controlled by the reactive devices.

Governors.—London and Marmet plants contain one of the first installations of a new Woodward governor known as the "cabinet" type in which the actuator, pumping unit, sump tank, piping and restoring mechanism are consolidated into one compact unit. The governing mechanism and oil-pumping units are mounted on a fabricated-steel sump base as an integral part thereof, and the whole assembly is enclosed in a neat cabinet. The actuator cabinets for the Kaplan units include pilot valves for the operation of the servomotors. A panel attached to the front of each cabinet is equipped with gages showing oil pressures, gate position, and setting of a gate-limit device, and various control switches.

Control and Relaying.—The two stations have been designed and laid out for full automatic operation. However, due to the necessity of the most complete co-ordination between the operation of the plants and the river flow to meet navigation requirements, supervisory control has been superimposed upon the automatic equipment. This makes possible a parallel control from the Cabin Creek steam plant whenever the navigation-pool level requires particularly close control. A scheme of metering permits the Cabin Creek operator to check from hour to hour the total output of the plants, as well as the output of each individual unit. Neither the design nor the plan of operation require operators at the plants beyond the period of initial operation when the usually-encountered minor difficulties must be ironed out. After this initial period, all operating men will be withdrawn from the plants. A daily inspection will, however, be made for general maintenance and cleaning purposes.

2. HYDRO FLOAT CONTROL AT WINFIELD†

THREE hydroelectric plants have been placed in operation within the last three years by the Kanawha Valley Power Co., a subsidiary of the American Gas and Electric Co. All three plants are on the Kanawha River, a tributary of the Ohio. Winfield, the last of the three, is 31 miles above the mouth of the river; Marmet is 37 miles upstream from Winfield; and London is 15 miles above Marmet.

As at London and Marmet, Winfield was constructed at a navigation dam built by the United States. A 50-year license from the government for an annual fee covering the use of water and dam facilities stipulates that navigation requirements are the prime consideration in control of the forebay pool and that power production is a secondary consideration.

Because of this restriction and, more particularly, the low quality of power available due to tail-water elevation at times of increased river discharge as well as the absence of firm-power output, Winfield even more than London and Marmet, would not have been feasible except as part of a large interconnected and integrated power system. Low operating cost was assured by simplification of design and provision of full automatic operation float control.

Care was taken in the placement of concrete to avoid segregation. Chutes were used only where absolutely necessary and the free drop in all cases was under five feet. Concrete was poured in alternate monoliths with horizontal layers not exceeding 18 in. thickness and carried out as a continuous operation until the completion of a block or lift. To prevent voids in the concrete mechanical immersed vibrating equipment was used.

Concrete used at Winfield was of the same type as that at London and Marmet. While the masses of concrete were not extremely large in volume, it was felt desirable, nevertheless, to minimize the heat generation. Among other things, what might be called a moderate-heat cement was used; the specifications limited the heat-producing compounds of tri-calcium-silicate and tri-calcium-aluminate.

Owing to the larger variation in net operating head at Winfield, the larger wheel rating obtainable with an adjustable-blade as against a fixed-blade unit, and to the minimum power requirements under the license, at least two of the three turbines selected had to be the adjustable-blade type if they were to conform to the same dimensions as the units for London and Marmet. Thus, the plant has three adjustable-blade turbines, two of the I. P. Morris Kaplan type and one of the Newport New automatically-adjustable-blade type. The former have a rated capacity of 9200 hp under a net heat of 26 ft at a speed of 90 rpm. The diameter of their runners is 169 in. The automatically adjustable-blade propeller turbine, which has no interconnection between the runner vanes and the governor, has a rated capacity of 9150 hp under the 26-ft head at 90 rpm. The diameter of the runner of this unit is 177 in. The principal change in the Kaplan turbines from those at London and Marmet was in the speed rings.

† *Power Plant Engineering* (with E. L. Peterson), February 1940.

16*

These were integrally cast in three sections instead of having independent stay vanes.

Electric Plant.—Three identical generators are installed at Winfield. Though essentially the same as those at London and Marmet they have 150-kva higher rating—a nominal 6150 kva, 90 rpm, 4150 v.

The electric layout employs a single bus. Each generator is switched by a single modified, low-tension truck-type breaker. Three single-phase three-winding transformers are used to supply 33 kv to the Guyandotte line and 44 kv to two lines to Turner.

Automatic Control.—Several changes from the automatic control at Marmet and London were made for Winfield. One feature, which has greatly simplified the control and speeded the operation of putting the generators on the line, has been the use of pull-in type synchronizing; the machines are brought into synchronism without application of any field. Experience at Marmet and London had shown that for units of this type synchronizing in the normal manner was a somewhat difficult process which caused considerable delay at times in getting the units connected to the system. Therefore the generators were bought with specially-braced windings to permit pull-in synchronization. Thus far this scheme has proved much more satisfactory than the former; the time required to bring a unit into synchronism at Winfield is about 45 sec against 1 to 5 min or more at London and Marmet.

Float Control.—Of special interest at Winfield is the unique operation by float control whereby three large units are controlled by a change in water level of only four inches. Starting, stopping, and loading operations are accomplished by operating the gate-limit control of the governors by a float device, which consists essentially of a Selsyn transmitter attached to the headwater float and three Selsyn receivers, one for each governor. The result is that the gate-limit device assumes a position at all times corresponding to the position of the float and hence the headwater elevation.

Starting and stopping of the units is accomplished by contacts attached to the gate-limit device of each governor, and set to correspond to a predetermined headwater elevation.

To illustrate the operation, let us assume that all units are shut down. Rising headwater will start one selected unit, put it on the line, and load it at approximately 50% gate opening. If the pond level continues to rise, the first unit will continue to load up until reaching approximately 70% gate, when a second selected unit will be started, put on the line, and loaded to approximately 70% gate. If the pond elevation remains constant, these two units will continue at 70% gate. If the elevation decreases, the units will unload correspondingly until, at a pre-selected setting, the second unit will be taken off the line.

Starting of the third unit is accomplished in a similar manner to that for the second unit, except that the first two units will have to be loaded to approximately 80% gate before the contacts close to bring on the third unit.

Carrier-current Supervision.—As at Marmet and London, the control of Winfield can be taken over by the remote supervisory control which allows full control of the plant from the central dispatching point without the necessity of having operators on duty in the hydro plant. However, a new type of supervisory control was utilized at Winfield because the control at London and Marmet was made inoperative several times through destruction of the supervisory pilot control cable by thoughtless or irresponsible persons with guns. At Winfield it was decided to use a carrier-current type of supervisory connection between the dispatching and control points.

Automatic Operation.—Classified and operated as an automatic station, this plant has no operating personnel at any time, but enough routine maintenance is carried out to call for the regular services of one maintenance man working a 40-hr week shift. The switchboard stands as a silent but efficient control and watch center, maintaining constant and correct control over the various generators and high-voltage lines.

Automatically Adjustable Vanes

The installation of an automatically-adjustable blade runner of the non-Kaplan type on one of the units at Marmet aroused a great deal of interest since this was the first installation of its kind. On this unit, as load changes occur, the runner vanes simultaneously assume a new angular position so that they are always set for the most efficient position for the particular load and head.

Although to a certain extent this installation was experimental, since only model sizes had previously been tested, the operation of the Marmet runner fulfilled all design expectations, and it was decided to install a similar runner at Winfield. At the same time an identical runner was ordered for London to supplant a Newport New fixed-blade runner. This change was justified on the basis of the increased capacity and output which the adjustable runner made possible. The Winfield automatically-adjustable runner is identical to the Marmet runner except for minor refinements in detail.

3. CLAYTOR – SYSTEM'S LARGEST HYDRO†

ON THE New River in Virginia (recently declared navigable by the U.S. Supreme Court) is the new 104,000-hp Claytor plant of the Appalachian Electric Power Company. While considerably smaller than any of the major steam stations, Claytor is the largest hydro plant on the American Gas and Electric system. It comprises four Francis-type turbines working at 110-ft head to utilize the flow with 12 million kwhr storage capacity of the pond formed behind the 1150-ft dam. Average estimated flow is 3000 cfs, with a July 1916 maximum of 170,000 and a 1940 maximum of about 200,000.

Penstocks are 16-ft diameter with riveted connections throughout, and no expansion joints because position and length give sufficient flexibility.

Headgates are electrically operated. The control is so arranged that the Broome gate must be fully open before the flat gate can be raised and the flat gate must be fully closed before the Broome gate can be lowered. They may be controlled from either the main switchboard in the power house or from the individual hoisting-mechanism housings on top of the intake bulkhead. The push-button control at the main switchboard is for emergency closing only, while that in the hoisting mechanism housing is complete control.

Elevations of the floor and roof of the generator room were determined by the space required for assembling the generating units, servicing the power transformers, and avoiding duplication of crane facilities. Also by depressing the generator stator below the finished generator floor elevation a saving was made in the height of the superstructure.

Hydro-generators

Study of Appalachian Electric's generating requirements demonstrated that it would be best to develop Claytor for neither strictly peak load nor relatively firm base operation.

Each of the four turbines is rated at 26,000 hp when operating at an effective head of 110 ft with a speed of 138.5 rpm. The guide bearings are of the babbitted-sleeve type. To minimize leakage through the wicket gates when the units are not in operation, rubber seals were provided at the top and bottom of each gate. The scroll cases are of riveted, steel-plate construction. A compressible filler was placed between the top of the scroll cases and the concrete substructure to prevent any cracking of the turbine-room floor by expansion of the scroll cases due to water pressure.

Each turbine is provided with a 130,000-ft-lb Woodward cabinet-type governor, in which the actuator, pumping unit, sump tank, piping and restoring mechanism are installed in one compact unit.

Connected to each turbine is an electric generator, rated 20,833 kva, 11,000 v, three phase, 60 cycle, 138.5 rpm, and provided with a direct-connected 250-v

† *Electrical World*, March 8 and 22, 1941, and *Engineering News-Record*, March 13, 1941 (all with H. A. Kammer).

main exciter and a 250-v pilot exciter. To provide for operation as a synchronous condenser, each generator has an overexcited capacity of 13,200 kva at zero per cent power factor and rated voltage, and an underexcited capacity of 18,333 kva at the same zero power factor and voltage.

Ventilation of the generators is accomplished by using two surface-type coolers for each unit, located within the generator foundation and fed with river water by a gravity system. During the summer this ventilation operates as a closed system; in the winter, part of the air heated by the generators is bled from the machines and used to heat the power house.

Electrical Equipment

Transformers.—Two transformer banks, each having a continuous capacity of 43,500 kva, but capable of carrying 50,000 kva when river conditions warrant, connect to the 132-kv transmission system.

Seven single-phase, 132/11-kv, 14,500 kva transformers were installed, six being of the conventional oil-insulated, forced air-cooled type and the seventh embodying a new design using "Pyranol" as an insulating medium, instead of oil, with positive forced circulation and a separate air-cooled radiator system for cooling the "Pyranol". While factory tests indicate that this unit will be at least as efficient as the six conventional units its real economy comes from the savings that it makes possible in space, dimensions, weight and supporting structural requirements.

Fire protection for the transformers has been provided by a "Mulsifyre" system, with a number of sprinkler heads around each transformer. These sprinklers are designed to break a stream of water into a very fine, almost fog-like spray. When this spray, under the proper pressure, is injected into burning oil it breaks the oil into globules and coats them with a thin film of water, forming an emulsion which does not sustain combustion.

Oil-handling.—An oil-handling system is installed for servicing transformer and circuit breaker insulating oil, purifier, filter press, pumps, etc., being housed in a building immediately south of the power house. Adjacent to the oil house are three 5700-gal tanks which store the transformer and circuit breaker oil. Two of the tanks are for reserve clean oil, one for transformers, one for circuit-breakers. The third tank is for storing dirty oil of either type. To facilitate changing oil in substation equipment on the hill, two 3-in. steel pipes with pumping equipment connect the substation to the oil house.

Metal-clad 11-kv Bus.—The 11-kv bus and switching equipment is of the metal-clad type, entirely housed in a steel compartment and provided with safety devices and interlocks to protect the operator from any live parts. All circuit breakers in the factory-assembled 11-kv switching cubicles are of the oil-insulated FH-126 type, and have an interrupting capacity of 1 million kva.

Decision was made to start with oil circuit breakers, but to make structural arrangements to facilitate change to water or air type in the future. Since that time developments have led to the conclusion that the inherent disadvantages of the water breaker discourage any optimism as to its future. Air blast, on the other hand, undoubtedly will move forward. Experience with it elsewhere on the American Gas and Electric system is most promising but it does not warrant any change at this location at this time.

The distance between the power house and switching structure being considerable, it was economical to provide duplicate control batteries, particularly for closing the 132-kv breakers.

Synchronization.—Generators, while not controlled by a float or other mechanical control device, are, nevertheless, automatically started and stopped. The attendant is required to operate only one switch, after which the automatic equipment brings the turbine and generator up to operating speed, synchronizes and loads without further attention. Because considerably less supervision is employed on these generators than would be employed in a manually operated station, certain additional protective devices have been installed. They give relay action in the following contingencies:

> Over-voltage.
> Over-speed.
> Governor oil-pressure failure.
> Stator over-temperature.
> Bearing oil-cooler water failure.
> Field over-temperature.
> Undervoltage.
> Low oil level in turbine bearing.
> Field failure.
> Generator out of step.
> CO_2 fire protection.
> Generator and turbine-bearing over-temperature.

Incidentally, this is the first time that protection afforded by the generator out-of-step relay is being used.

In connection with a co-ordinated system scheme for control of frequency and to provide for possible need for controlling load on the two Glen Lyn–Claytor lines, a complete set-up for frequency and load control has been installed at the Claytor plant. Any one or two of the four generators may be placed under control to accomplish any one of the following results:

> Hold system frequency to a predetermined value regardless of changes in the loading of tie lines or generators; or
> Hold system frequency to a predetermined value provided the change in loading necessary to correct for the demanded change in frequency results in a load change in the proper direction to hold the load on the Glen Lyn–Claytor lines within certain predetermined limits; or
> Hold the load on the Glen Lyn–Claytor lines to a given predetermined value, without regard to the effect of this line loading upon the system frequency; or
> Hold the system frequency to a predetermined value provided the tie-line load does not go outside of a predetermined band; or
> Hold the load on the Glen Lyn–Claytor lines to a predetermined value so long as system frequency stays within limits previously determined.

Range of Energy Output

The water at the Claytor dam drains from an area more than 2400 sq miles. The calculated average flow of the river at the site was about 3280 cfs over the 23-year period from 1907 to 1930. For economic reasons and because the Norfolk & Western R.R. tracks cross the river at the upper end of the reservoir it was concluded that the maximum height of the dam should be such that it would not impound water above about El. 1846. This is about 115 ft above the average river elevation before construction of the dam. On the basis of available streamflow records, a pond level normally at El. 1846 and an installation of 75,000 kw, the expected output is 190 million kwhr for an average year, 400 million kwhr for the best year on record, and about 110 million kwhr for the worst year. The storage capacity of the reservoir is about 12 million kwhr between El. 1846 and that elevation to which the reservoir can be drawn without undue sacrifice of station capacity.

Geology of Site

At the dam site it was found that the native rock underlying the river bed was limestone which changed to dolomite under each of the river banks. The bedding planes in general were almost vertical. Many of the joints and bedding planes were filled with clay to a considerable depth; soluble seams and solution channels were found throughout the area; they were sometimes clay-filled, at other times open. The reservoir rim was found to be adequate, probably requiring no treatment except for a short section about four miles from the development where a small amount of remedial work may be required.

On the right bank of the river and immediately downstream from the dam, there was an important dolomite quarry fully developed and in operation which was purchased.

Grouting.—Because of the geologic formation it was important that a tight grout cutoff be constructed beneath the concrete structures of the dam and continued into each abutment.

Immediately upstream from the upstream face, 3-in. diameter holes were drilled into the foundation, first on 12-ft centers, then on 9, 6, and 3-ft centers. These holes, which extended to 200 ft in depth, were pumped full of grout to refusal at 200 psi.

Naturally the acceptance of grout by the rock foundation was high, and in the case of the north bank the cut-off area averaged as much as 1.88 cu ft of cementitious material per linear foot of 3-in. hole drilled.

Concreting Procedure.—All of the coarse and fine aggregate required for the concrete structures was manufactured from the dolomitic limestone available at the site. The coarse aggregate was graded into a series of particle sizes ranging from ⅜-in. minimum to 3-in. maximum. The fine aggregate was crushed so that all of it passed a No. 8 sieve, and 15% of it passed a No. 100 sieve. Aggregates were graded so that the quantity of each successively smaller size was about an equal percentage of the preceding size. In the crushing cycle the

16a   VEP

material between the No. 8 and ⅜-in. size sieves was screened and recrushed to make the sand.

About 250,000 cu yd of concrete were placed in the dam, powerhouse and appurtenant structures. For the interior mass-concrete of the dam, 0.85 bbl of cement was used per cu yd with a maximum water–cement ratio by weight of 0.75. For the exterior concrete facing the mix contained 1.2 bbl of cement per cu yd and had a maximum water–cement ratio by weight of 0.75. The mass-concrete mix averaged about 3200 lb per sq in. compressive strength; and the higher cement content face mix averaged about 4800 lb per sq in. after 28 days.

The upper part of one of the south abutment blocks was constructed with the mass-concrete mix of 0.85 barrels per cu yd without the higher cement-content facing, so that the relative weathering resistance of the two mixes could be compared as time goes on. The average slump of the mix used for the mass concrete was two inches and for the facing concrete three inches.

All of the concrete was placed with the aid of air-driven internal vibrators, operating at a frequency of 8000 cycles per min.

Design of Dam

All concrete structures making up the dam were designed as gravity-type units assuming the weight of concrete at 154 lb per cu ft; actually it weighed on an average about 157 lb. Uplift was estimated on a somewhat elaborate basis, involving assumption of an irregular gradient between head and tailwater, partial efficacy of pressure-relief drains and varying proportion of area of base subject to uplift. The ultimate result gave about the equivalent of an applied pressure on 50 % of the base and an intensitiy varying as a straight line between maximum headwater pressure at the heel and maximum tailwater pressure at the toe of the dam. This is more conservative than usual normal design assumptions. An ice pressure of 10,000 lb per lineal ft of dam was assumed, applied one foot below the maximum elevation of the reservoir. The structures were so proportioned for stability that under any condition of loading the resultant of the forces was always within the middle third of any horizontal plane through the structures.

4. RESEARCH ON CONCRETE GROWTH†

DISINTEGRATION of the concrete forming the dam and power house of the Buck hydroelectric plant required extensive repairs. This plant of the Appalachian Electric Power Company was built on the New River in 1912. Since that time the concrete surface has failed progressively. By 1928 the condition of the down-stream face of the bulkhead, the power-house walls, and the spillway had become so bad that it was decided to remove and replace the defective concrete surfaces of these structures.

It was originally believed that weathering of the concrete, which had been placed as a very wet mix, was the primary cause of the trouble. Extensive studies demonstrated, however, that there was a chemical reaction between the phyllite aggregate and the high-alkali cement used in the original structures. The products of this reaction caused serious expansion or growth of the concrete. This finding explained to a considerable extent not only the deterioration of the original structures and the concomitant effects on embedded turbine parts, but also the premature failure of the repairs made only ten years previously.

It is the opinion of the authors that the disintegration cycle of concrete at Buck dam is typical of cycles which, once started, move on at a self-accelerated rate. The expansion creates stresses which crack the exterior faces. Water enters these cracks and freezes during the winter, causing further opening and spalling of the face. Further growth of the old concrete widens the cracks, and the next freeze aggravates the situation still more. The cycle goes on, causing the disintegration to extend more and more deeply into the old concrete.

Further expansion of the old structures could be expected. This finding had a major influence on the several schemes considered for repairing the deteriorated concrete. Since nothing could be done to prevent further expansion, and because it would be economically impractical to demolish the old structures entirely, it was decided to make provision for further movement of the old concrete in the plan finally selected for the reconstruction work.

Because of its physical arrangement, the concrete substructure of the power house experienced considerable differential growth, which showed up as a crack 1 in. wide around the periphery of the turbine pit liner, peripheral cracks around the upper section of the draft tube, and breaking of the speed-ring stay vanes. As a consequence the water turbine and generator failed to maintain proper mechanical alignment.

Repair Methods

Concrete which supported the generator and turbine was removed from the generator floor down to a level about 5 ft below the floor of the scroll case. The exposed face of the old concrete was then rebuilt to a smooth surface where it adjoined the new concrete foundation. The facing was built up on the old surfaces to a depth of 6 in. by the "Prepak" method, according to which the forms are first packed with aggregate and then grouted to refusal at about 50 lb per sq in. with a sand–cement grout. All aggregates were dolomitic limestone.

† *Civil Engineering* (with H. A. Kammer), July. 1944.

16a*

Because the old concrete was of inferior quality and quite porous, it was expected that a better bond would be obtained between the old and the new work by the Prepak method than by conventional types of concrete placement. Further, because the new concrete facing was comparatively thin and the height of the lifts that could be used practicably would be too great, it was thought that ordinary placement methods might not give satisfactory results. Cores drilled from the repaired structure show that an excellent bond was obtained with the method employed.

To prevent water leakage through the painted joints a groove, filled with synthetic-rubber expansion-joint filler, was provided around the periphery of the scroll case and the draft tube. As a further precaution, a copper water-stop was also used. After the new facing had been placed on the old concrete and painted, the new foundations for the water turbines and generators were cast and vibrated in place by conventional methods.

The differential movement of the old concrete substructure of the power house had not only caused the rotating parts of the water turbine to wear unevenly, but had also caused fractures in the cast-iron speed ring. Each of the stationary vanes was broken through at either the top or the bottom.

For the repair of the speed ring, a continuous box-type or envelop patch was used, consisting of an envelop of $\frac{5}{15}$-in. steel plate, which was shrunk tight and continuous around the broken joint. It is held in place by $\frac{3}{4}$-in. patch bolts and welded at the upstream and downstream ends of the vane. Before the envelop patch was applied, studs were threaded and brazed into the fractured surface on each side of the break. The fractured ends of all vanes were then built up with bronze to partially close the break. After all the sections of the speed ring had been properly aligned, opposite studs were joined with a full-strength weld. The envelop patch was then applied.

The exterior walls of the power-house substructure, bulkhead sections of the dam, spillway sections, piers, etc., all showed the effects of the unusual expansion of the concrete. The north bulkhead section of the dam was probably in the worst condition, and the method employed to repair its downstream face is typical for all.

The disintegrated concrete was removed to an average depth of 4 in. Wherever it was found that the disintegration extended deeper than that, more disintegrated concrete was removed in one small section to a maximum depth of 18 in. Where leakage was taking place through horizontal construction joints of the old concrete, stage grouting was used effectively to stop the flow. After the leakage had been stopped, the structure was rebuilt to its original lines by the Prepak concreting method.

Because it has been definitely determined that the old concrete within the body of the dam will continue to grow, the new 4-in. facing was divided into panels 3 ft 4 in. high by 8 ft long, separated by 2-in. V-joints. It is expected that as further growth takes place the face will crack at the V-joints. By controlling the future cracking in this manner, it should be comparatively simple to caulk or seal, by conventional masonry-wall repair methods, any cracks that may develop in the future.

CHAPTER 3

NUCLEAR POWER GENERATION

CONTENTS

1. ECONOMIC ASPECTS OF ATOMIC ENERGY
AS A SOURCE OF POWER†

THE possibilities of generating electric power more economically by nuclear piles than by presently used means warrant a thorough investigation to determine the economic aspects of the new means when, and as soon as the necessary data for that purpose become available.

It is a safe guess that the first unit is not going to be a 75,000-kw pile and that as a thermal plant it will leave a great deal to be desired as far as, economy and efficiency are concerned.

With a glaring gap in our technological knowledge and with almost no actual experience on costs, or even engineering cost estimate, it seems to me that an economic study becomes almost impossible at this time. One is therefore forced back on broad generalization and speculation.

Some hydro-electric developments have produced very cheap energy, but so have numerous thermal plants. And probably volumes will be written on the subject of the cost of atomic energy versus the cost of other forms of energy.

An atomic plant cannot be located anywhere. Condensing water, for one thing, must be considered. It is true, moreover, that the gas turbine may come into the picture, although, where coal is as plentiful and generally as cheap as in the United States, the gas turbine has been able to find few enthusiastic backers.

The most efficient straight steam-electric plant in the United States is the Twin Branch station, with which the discussor has been associated both in design and operation. This plant operates with a performance of 10,200 Btu per kwhr, or an efficiency of 33.13%. We have just completed the design of two new stations, one to be located in Indiana and another in West Virginia, each of which will show a thermal performance of 9250 Btu, or a thermal efficiency of 36.9%. This is almost double the thermal efficiency of the average performance in the United States for the year 1945.

Because a nuclear power plant, when and if it becomes technologically feasible, will utilize a freightless fuel at the expense of a much higher capital cost, it is particularly important not to treat this phase of the problem too conclusively. While a thermal power plant using nuclear fuel will be affected to some degree by the efficiency of the thermal cycle, such effect will be minor in comparison to the effect that improved thermal performance will have on a power plant using more conventional fuels.

If we are not going to go astray in our discussions of the economics of nuclear energy, it is necessary to compare new technology with new technology. Therefore, nuclear plants which may represent the latest technology should under no circumstances be compared with anything but the best that can be obtained

† Discussion of paper by Sam H. Schurr, American Enterprise Association, Atlantic City, N.J., January 23, 1947.

with existing technology. But to do that it will be necessary that we first get more reliable information on the investment and probable operating costs of nuclear plants. This appears to me to be impossible until the technology of such plants is further developed.

I hope that I have not given the impression that I undervalue the possibilities of nuclear energy and its development in the future. If nuclear energy will give us more economical power than we can obtain by the use of our hydroelectric resources and particularly our fuel resources, then it certainly should be developed. But obviously it will find application first in locations and in countries that are not so richly endowed as we are with economical normal-fuel resources.

The possibilities that nuclear energy appears to offer are unquestionably pregnant with the greatest economic significance. Those possibilities need to be explored and developed, particularly since it now appears that the most fruitful peace-time application will be in the field of electric power generation. Nevertheless, until we have built the pilot plant at Oak Ridge and perhaps another one after that, and perhaps one plant of a capacity say somewhere between 10,000 and 100,000-kw rating, I am fearful that we will not have the knowledge necessary to carry out an effective economic study of nuclear energy for use in electric power generation. Any such study carried out before then is bound to be more or less a speculation.

2. PROSPECTS FOR INDUSTRIAL APPLICATION
OF ATOMIC ENERGY†

ALTHOUGH it is perhaps possible to conceive of direct conversion of fissionable material into electrical energy, there is no known basis for believing that it can be realized. Therefore, if uranium or plutonium is to be used in industry, it will have to be as fuels and "burned" by bombarding them with neutrons. The products of the combustion will be heat and hot gases. The heat will be absorbed by a coolant which will either generate steam to drive a steam turbine, or heat a gas to drive a gas turbine. The by-product of burning uranium or plutonium will be an ash consisting of fission fragments. The machine in which all this is going to be carried on is known as a nuclear reactor, because it is the means by which the energy in the atomic nucleus is released through the process of fission.

The ash which we end up with in a pulverized-fuel boiler presents a problem, but it is a problem which we know how to deal with effectively and economically. On the other hand the ash developed in a nuclear reactor is highly radioactive and presents problems of infinite complexity and hazard for which no satisfactory solutions are yet known.

The gases discharged from the stack of a coal-fired boiler present difficulties but, again, these difficulties are simple in comparison with the problems of the radioactive gas emitted from a nuclear reactor.

In the last twenty years we have materially improved the efficiency of the steam-turbine cycle. We have raised pressures from about 1200 to present levels of 2100–2400 psi. We have made notable progress in raising temperatures, bringing them up from about 750 F to 1050 F. But it has taken us two decades to do it. One fact will give you an idea of the intricacy of this "conventional" steam turbo-generator: the modern boiler and turbine alone require some 27 different kinds of steel specifications, each particular kind tailored specifically to perform a particular part of the job. The structural materials for a modern boiler producing steam at a temperature not higher than 1050 F now are clearly defined and are reasonably available.

However, no one can tell yet just what materials are needed or available to meet the peculiar requirements of a nuclear reactor. The material must have reasonably low neutron absorption so as not to "quench" the fire; it must be able to stand up under untried conditions of temperature and rates of heat transfer, and it must resist essential changes despite bombardment by neutrons, gamma rays, beta rays and other products of the fission process which the nuclear reactor is designed to propagate.

A most vital factor in the economics of nuclear-power generation is the "breeder". Breeding is the process—theoretically attainable but still to be demonstrated—by which more fissionable material is produced than is consumed in the operation of the reactor. The major obstacles standing in the

† National Coal Association, New York, N.Y., October 7, 1949.

way of developing breeder reactors are of an engineering nature and are concerned with basic conflicting requirements for high neutron economy and high power output for a given material investment. There are also acute chemical engineering problems associated with the treatment of partly depleted fuel. These are problems which confront the development of all reactors—except that for breeders they are even more difficult.

If cheaper electric power can be achieved through the use of atomic energy, several industries might well be stimulated. Examples are: Reduction of magnesium and aluminum, refining of copper, production of cement, chlorine and caustic soda, electric-furnace operations like the production of graphite, carborundum, calcium carbide, phosphoric acid, ferro-alloys and electric steel. All these operations are characterized by a high figure of kilowatt-hours per week per worker and relatively low wages per kilowatt-hour used.

To put it another way, power costs in these industries represent a substantial percentage of total cost, as contrasted with a relatively minor percentage in most industries, so that any appreciable reduction in power costs will materially reduce cost of product.

The displacement of other forms of energy by atomic energy in such industries will be indirect—atomic fuel will displace coal, gas or oil or "white coal" in hydro plants that might otherwise be built.

If and when "breeding" has been successfully developed one pound of ordinary uranium will be able to do the work of 1500 tons of 13,000-Btu coal. Under these conditions, the direct fuel cost will approach the cost of "white" coal—almost zero.

If you start out with a modern steam plant burning pulverized coal, the range in capital costs is perhaps $150 to $200 per kw. The range in fuel costs lies, say, between 10 cents and 40 cents per million Btu. The practical range in intensity of use probably lies between 50% and 68.5% plant factor (500-hr use of capacity per month). Under these conditions using the cost of money as 6%, and allowing for depreciation and local and federal taxes, the range in total costs of energy at the switchboard is between 5.5 and 12.0 mills—a range of 2.2 to 1.

For the nuclear plant, using the rumored range of capital costs—$140 to $1000 per kw—the range in fixed costs of plant, using the same cost of money (but a higher depreciation rate, because of the more rapid obsolescence always present in a new technology) and the same taxes, would be between 3.9 and 38 mills—a range of almost 10 to 1. This wide range of estimates cant be narrowed until the present reactor program develops the data we must have for more informed engineering estimates. The total cost of electric energy at the switchboard will have to include some operating and maintenance costs and some value for cost of fuel. At best, it may be substantially zero. But the actual costs of fuel can not be estimated with any accuracy until the breeding phase of the present reactor program is more fully developed.

Two unknown factors constitute the principal barrier to any sound engineering appraisal of the economics of nuclear-fuel utilization: Lack of engineering knowledge on many phases of reactor design, construction and operation which makes it impossible to estimate the capital costs of a reactor plant; and lack of knowledge and practicability of breeding which makes it impossible to estimate the fuel costs.

What course ought the coal industry follow in this situation? As an outsider, I would not be so bold as to attempt an answer to that question, but perhaps I can offer an analysis of the problem without appearing too brash.

One way of looking at it is this: If nuclear power faces such an array of formidable technical hurdles and if costs are so uncertain, why not forget about the entire atomic business at least for the moment? However, I don't believe that any responsible person in the coal industry can forget about atomic energy, if for no other reason than that the search for peacetime application is being pressed so agressively and will continue to be.

There appears to be every reason for belief that some day nuclear fuel will be burned in industry. But before one can say how long will it take until that actually comes about one needs to distinguish clearly between different possible uses. There may be, on one hand, some limited or special application, such as propulsion of naval vessels, or utilization in areas remote from an economical supply of coal or oil. Or there may be broad applications in industry, with nuclear fuel as a direct competitor of present standard fuels—particularly oil and coal. But even on this qualified basis, the answer to this question of how long, obviously is tied up with the reactor development program under way.

It seems to me that with all the technical problems still to be solved, the first and purely experimental phase of that program will take from three to five years. Within perhaps three to five years after that—six to ten years from now—we may have a reactor producing electric power. Four to five years after that—10 to 15 years hence—we may have some commercial generation of nuclear power. But I anticipate in that time it will be commercial only in the sense that it will be operating more or less regularly day in and day out.

3. REASONS FOR SLOW PROGRESS IN ATOMIC ENERGY†

A FORMIDABLE array of technical obstacles stands in the way of getting ahead with the application of nuclear energy. Until we reach the stage of pilot-plant trials, it will be impossible to assess the exact state of the art. But the technical difficulties in the way of perfecting economical power-producing reactors are only a part of the problem.

Important among the reasons why we are not further ahead is the fact that, relatively speaking, the state of power production and development by conventional means, particularly in this country, is unusually satisfactory. In the period 1949 through 1952 we will have completed installation of new electric facilities to increase our total capacity by close to 29 million kw—more than 50% greater than the installed capacity at the end of 1948. The ability to produce electricity in such relative abundance does not stimulate heroic efforts to develop radical new means of production.

Moreover, the efficiency of conversion in our most efficient steam plants has risen from 13 to 15 percentage points in just a few years to 38% thermal efficiency—9000 Btu per kwhr—in the best plants now being brought in. One might almost say that one of the obstacles to rapid progress with nuclear-power production is the rather bright prospect for conventional-power production.

Nuclear power is not a new kind of power but only a new fuel. As such, it will be used in processes that have been under development ever since the steam engines of Newcomen and Watt. Even after all technical obstacles to perfecting a power reactor have been overcome, the most that we can hope for on costs is the cheapening of one of the elements in power production, namely, fuel. We must expect some—perhaps a large—increase in the capital cost of certain elements of the plant. And there does not seem to be any doubt that operation and maintenance of a nuclear reactor will be more expensive than that of a conventional plant.

Why Atomic Power is Important

Prospect of a radical reduction in the fuel cost for power production, and the hope that this can be accomplished without an increase in capital cost so great as to render nuclear reactors economically unsound, are sufficient reasons to make industry vitally interested.

Development of nuclear power is important to industry not because of expected revolutionary consequences. Rather it is important because it may provide a more universally available, more rapidly portable, and more economical fuel. For instance, if the more optimistic ideas in breeding possibilities are realized, the fuel cost would be reduced to almost zero.

With additions to hydro bound to decrease, as better sites are more and more

† American Association for the Advancement of Science, Cleveland, Ohio, December 28, 1950.

put to use, it is apparent that if another highly economical fuel gives promise, natural pressures for its development will build up, if not before, then surely when the reserves of present highly economical supplies, both oil and gas, begin tapering off.

This will have an important bearing on those industries using fuel directly and on the much larger number using fuel as a source of electric energy. It will not only protect and keep whole the entire base of economical fuel supply, but it might also lead to further expansion of electric use.

If cheaper power can be achieved through the use of atomic energy, several industries might well be stimulated. Examples are: reduction of magnesium and aluminum, refining of copper, production of cement, chlorine and caustic soda, electric-furnace operations like the production of graphite, carborundum, calcium carbide, phosphoric acid, ferro-alloys and electric steel. In all these industries power represents a substantial percentage of total cost.

Some new electric processes might also be developed if atomic power is cheap enough: for example, electric melting of general-purpose glass; direct reduction of iron by electric processes; or certain kiln operations that might be electrified.

Whether or not atomic power will stimulate existing processes or encourage the development of new ones will depend on cost.

Private Industry Participation

Thus far private industry has not had much part in atomic-energy development. This has been only partly because economic advantages of successful nuclear power appear moderate rather than very high. There is already considerable history on the subject of why industry does not have a genuinely important role in atomic work. The report of the AEC Industrial Advisory Group over two years ago gave an excellent analysis and made a significant beginning in suggesting the means by which industry could be brought into atomic energy in an effective way.

That report, while commencing the AEC contract method of operation, demonstrated that that method represented only limited participation as compared with what was required if industry was to make a real contribution. As the report so succinctly states:

> "The difficulty and the danger in the present situation is that industry's part in atomic energy is very limited as compared with the opportunities which exist and always have existed in other fields. The small number of companies which take significant part are selected by the Government and the extent of their role in the work is limited by specific assignments from the Government."

Probing for the cause, the Industrial Advisory Group further states:

> "We think that today the central difficulty in getting a broad industrial attack on the problems of atomic energy is the fact that industry has no way of determining whether important opportunities in fact exist in which to take part."

And again:

> "It has been stated that industrial opportunities in atomic energy are potentially unlimited. But they are at present so shadowy that businessmen neither know where to look

nor what to look for. Today no one can say whether the prospect of profits or other in-
centives exist, because under present conditions the great majority in industry know
little or nothing about the subject.

"The need for Government monopoly in certain important areas, coupled with
secrecy, seems to erect an impenetrable barrier to a wish for knowledge. If industry is to
help in atomic energy and benefit from it, industrialists must first be put in a position to
find out enough about the subject to determine whether and where there are in fact
opportunities to take part. In devising means to this end, only the Government can take
the initiative. The Government, which must exercise a broad monopoly and determine
security regulations, alone is in a position to open the doors. Industry can, and we are
confident that it will, co-operate. But Government must first provide the catalytic forces
that will set more of the normal processes of industry to work."

Even though more than two years have elapsed since the submission of this
report, the basic difficulty so clearly pointed out by the Industrial Advisory
Group has not been much alleviated. Industry still needs the things the Industrial Advisory Group advocated: More information about atomic energy;
more opportunity for direct personal contacts between individual industrialists—technicians and executives—and the atomic-energy program; and greater
utilization by government of the Industrial Advisory Committee concept.

Except for the relatively few people who are now engaged in the various
laboratories, and those among the various manufacturing organizations who
participated and are participating in the atomic-energy program, an almost
complete state of ignorance exists in the atomic field today in the United
States. A few of our college professors and graduate physics students have
some knowledge of the subject; the people of the national laboratories possess
it, and also a limited number of people among the major manufacturers. But,
considering the size of the technical population of the United States, this
group is woefully small.

4. CO-OPERATION BETWEEN AEC AND POWER INDUSTRY†

THE special interest of the electric power industry in the Atomic Energy Commission's activity derives from the prospect of utilizing atomic energy for ordinary power-generation purposes. If this prospect becomes a commercial reality, then the power systems of the country could become the largest potential users of nuclear reactors just as today they are the largest users of conventional fuel-fired steam boilers. Likewise, the power industry could become the largest potential user of nuclear fuel, just as today it is the largest single user of coal.

Whether and when atomic power becomes commercially feasible depends upon the commission's nuclear-reactor program, a program which, except for the piles producing plutonium for bombs, is still in the research and pilot-plant stage. While none of the reactors which are part of the present AEC program have been designed primarily for industrial power production, they will give, as the commission has indicated, "impetus to the ultimate use of nuclear energy" for such purpose.

If, as one member of the commission remarked "the intriguing possibility of nuclear breeding can be made a reality ... it may very well give us a reduction of cost of atomic power to where it can compete with our other [energy] sources." While the prospect of breeding has the most dramatic appeal, there are other possibilities with more limited implications. For example, the ideas now being studied for combining production of plutonium for weapons with utilization of by-product heat energy, although perhaps not economical for power production generally, may be practicable for specific projects under conditions where, as now, the military need of the prime product plutonium gives it a special economic value.

Commissioner Pike accurately described the situation when he said recently, "Because of our natural endowment of large amounts of cheap coal, oil, and natural gas, and our continuous developments and techniques of large-scale power-production units, we have set a very difficult mark for any new source of power to meet." But the same conditions which helped bring the art of conventional-power production to its present stage also foster an interest in a new source of heat energy that potentially is capable of providing a further advance in the art.

Therefore, the power industry should be strongly attracted by any prospect for the production of heat energy which might bring about a material lowering in cost of steam-power production. In saying that, it is not intended to convey the idea that one should expect commercially feasible atomic power would mean a radical reduction in power costs. If nuclear reactors can produce heat energy for power plants, that energy would replace the fuel element in conventional electric generation but not without some increase in capital costs. There is

† Report of Ad Hoc Advisory Committee (Chairman), Electrical Engineering, September 1951.

little, if any, prospect that the overall cost reduction could be revolutionary, although the results could be significant, especially for a number of industrial operations which any appreciable lowering of power costs would stimulate— operations where electric power represents a relatively substantial part of the cost of the finished product.

Perhaps most important from the standpoint of the utilities, through collaboration with the manufacturers and with one another they have become informed of the latest developments as they occur. It is only as part of this process that they are able to evaluate reliably the prospects for still other improvements in the future. All this knowledge in turn becomes a key to the planning which the electric systems do in order to meet the constantly expanding demand for power, to promote new and better uses of power, and to keep costs down.

Areas of Mutual Interest

There is a pervasive community of interest between the Atomic Energy Commission and the electric utility industry, essentially because the electric systems are likely to be the principal medium for distributing to the public whatever benefits are to flow from nuclear fuel.

During our survey of the commission's reactor projects, we noticed striking analogies to a variety of problems which the industry and the manufacturers had encountered as the art of conventional-power production advanced. These common elements included such items as the design of pressure vessels, coolant systems, pumps, piping and heat-transfer systems; the electrical, mechanical and physical aspects of plant construction; fuel preparation and ash handling; control and instrumentation; and the general relationship between reactor operating characteristics and those of conventional electric power plants.

There is, however, a broader and more important area of mutual interest than specific technical problems. As the commission's reactor program proceeds, it will become increasingly necessary to make realistic appraisals of the potentialities of reactor projects as parts of electric power systems. The more closely the commission approaches its goal of a successful power reactor, the more compelling will be the need for these overall engineering-economic judgments.

These analyses will require not only the talents of the physicists, chemists, and chemical and mechanical engineers who are engaged in the commission's reactor program but also the talents which now reside in the system-planning engineer, power-supply engineer, transmission engineer, and in the utility management experts, who have responsibility for utility economics.

A Basis for Continuing Co-operation

We recommend a permanent industry committee that would be broadly representative of the electric power industry. Membership of 10 to 15 individuals would accomplish such representation, while keeping the size within manageable limits. Members should be drawn from the top executive ranks of the electric systems and at least a substantial portion of them should have some engineering background. Staggered terms of membership are desirable so that over a period of years a large number of industry executives could have the experience of membership. All members of the committee should be cleared. Presumably they would meet according to some more or less regular schedule and would have the assistance of minimum staff facilities. If the commission wishes, we are prepared to discuss the details concerned with the setting up of this group.

The committee members should be given an opportunity to familiarize themselves with the reactor program and other relevant phases of the commission's enterprise. A visit to commission installations similar to that afforded the Ad Hoc Committee would be informative and educational to them as it was to us. We believe that this could be arranged in such a way that it would not cause serious inconvenience to the commission's staff.

5. COMPETITIVE ATOMIC POWER†

WHAT may we look forward to in the way of competitive atomic power over the next five to sixteen years—say by the end of 1970?

Today, more than nine years after Hiroshima, we have two full-scale prototype atomic plants either in actual service or about to go into service. The power plant of the first—the U.S.S. *Nautilus*—consists of a water-cooled, water-moderated, highly enriched atomic-fuel reactor. The reactor for the U.S.S. *Sea Wolf* uses liquid sodium as the coolant and as the medium for generating steam to drive a turbine and provide electric power. Essentially, these two reactors are similar since both use the heat from the reactor to heat water or sodium, which in turn is used to heat water to make steam to generate electric power.

But to make the ships practical and reliable the designers had to confine themselves to very low steam temperatures and pressures which were discarded 30 years ago in conventional-fuel power plants. From the standpoint of power-production technology, the low temperatures and pressures which had to be adopted in the submarine reactors represent a 30-year retrogression.

In addition, there is under way a series of projects embracing five different reactors designed as experimental developments along the road to competitive or economical atomic power. The Westinghouse-Duquesne Light project, known as a pressurized-water reactor, now under construction at Shippingport, Pa., near Pittsburgh, will have a capacity of something over 60,000 kw and should be completed some time in 1957. While this will be the first American large turboelectric unit to be operated by heat of fission, it also will operate on a cycle and an efficiency that were discarded as too inefficient something like 25 or 30 years ago.

The other four reactors embody still other designs in various stages of development, none of which is likely to lead quickly to economical or competitive electric power. Each, however, is expected to increase our store of knowledge about the subject and teach us things that will help get costs down—but costs of nuclear power have a long way to go before they can be competitive.

Capital Cost Excessive

The cost of power at the point of generation consists of two factors: 1. The fixed or capital charges which include the interest and dividend requirements of capital, and the taxes, insurance and depreciation charges on the plant: 2. operating expenses—labor, fuel, and maintenance material and supplies.

A reactor of the pressurized water type, as at Shippingport, has an initial cost per kilowatt of capacity of the order of two or three times that of a conventional power plant.

This is a burden on the ultimate cost of power that can be compensated for only by savings in fuel, materials and supplies, and labor. However, in a reactor of this type (non-regenerative) where fissionable material is consumed and

† Public Utilities Advertising Association, New York, N.Y., December 2, 1954.

only heat energy is produced you have a relatively high fuel cost to start with; this is aggravated by the high costs of fabrication and reprocessing.

The exact cost at which fissionable material is available is not known today, in part because of security, in part because the material being a government monopoly has in it elements of subsidy.

Because the Shippingport operation is a technically involved one, the labor and maintenance costs are not likely to be any lower than in a conventional plant. It is evident, therefore, that no easy road to economical power is to be found along the route of the pressurized-water reactor although that is not to say that quite a bit will not be learned from it.

There are promising possibilities of reducing both the capital and operating costs of nuclear reactors, which seem to lie along the road of the homogeneous reactor, where the fuel is incorporated in an aqueous solution, and the fast-breeder reactor. In the homogeneous reactor, fabrication and reprocessing should be materially simplified. In a breeder reactor fuel cost can be reduced, theoretically at least, to a point where it can become almost a negligible figure; through that route, a breeder reactor could be made to carry the burden of the high capital costs involved.

But these are long-term developments. I do not know whether the time within which they can be brought to fruition in the form of large-scale economical operations is of the order of 5 years, or 10 years, or, what is probable, an even longer time.

In the meanwhile, what is going on in conventional-power development? Just within the past year technical developments have raised the efficiency in conversions of heat energy into electric energy in the best thermal plants to 37.5%. Looking ahead only as far as 1970, efficiencies of over 45% are clearly in view. In 1925 our most efficient plant used 14,400 Btu per kwhr; in 1970 it should only take 7400. Although the fuel supply needed for the gigantic amounts of electric energy we are producing is not as limitless as was once thought, there still is every reason for believing that the conventional fuels available will be adequate to supply the requirements of the United States for at least several hundred years to come.

Long Range View Point

Why consider atomic power? Considered in terms of the immediate future there is perhaps very little if any necessity for contemplating a change in our present methods of energy production. But that is a short-range view. While fuel supplies are now adequate they are also definitely finite and subject to rather constantly rising costs.

Economic and technical difficulties show up now in limited areas. If we look ahead two or three or more decades, it is possible to visualize such difficulties occurring somewhat widely. As a means of anticipating these conditions and finding a form of energy that can take its place in competition with coal, oil, gas and water, the prospect of perfecting the technology of nuclear power is a challenging one.

Even apart from the areas which now or in later years may encounter difficulties in securing an adequate, cheap source of conventional energy, nuclear reactors as they become perfected might be able to edge out conventional thermal stations in those industries which use electric power as a raw material—the electro-chemical and electro-metallurgical industries—or in the industries with a very high load factor.

The complex of economic factors that determine which of a number of competing sources of energy will be utilized to fulfill the requirements of the electro-metallurgical and electro-chemical industries could tend to favor nuclear power at an earlier date than general industrial, residential and commercial needs for power.

Many Problems to Solve

These prospects influence the determination to persevere in research and development designed to solve the many problems that must be cleared up before competitive atomic power can become a reality.

Are there many such problems? The answer is yes—altogether too many.

One of the basic problems in reactor technology is that relating to materials which is a far tougher challenge to the metallurgists even than those presented by some of our most advanced heat-transfer pressure-temperature operations in conventional-power technology. In a reactor core we need materials that will stand up while subjected to intense radiation, that will offer high resistance to corrosion and yet will possess the property of low neutron absorption. This is not an impossible combination of characteristics, but it is an extremely tough one to achieve.

A second problem is waste disposal. Difficult as ash disposal is in a large modern, coal-burning power plant, the problem of atomic ash is much harder. If an atomic power industry is to expand on an economical basis, other and less-expensive methods of disposing of atomic ash than are now known will have to be developed.

A third problem is that of reprocessing partly spent fuel material. In a good many of the economic studies, particularly those related to solid-fuel reactors, the problem presented by reprocessing is solved by a proposal to discard the irradiated fuel elements.

Finally, there is the problem of inherent hazards. Atomic reactors potentially are hazardous operations—more so than any industrial operation with which we currently have to contend.

While all of these developments are going on, what kind of a program should we in the United States adopt and follow in the expansion of this country's electric power-producing facilities in the next decade or decade and a half? Is the additional required capacity to be built with the type of facilities that we now know about, or with those that are still in an early experimental stage?

As additional new, more efficient generating capacity is brought into a system, it naturally commands the position of highest load-factor use, thus permitting the system to take maximum advantage of its greater efficiency. By the same token, as it gets older, it is relegated to lesser load-factor use.

To justify investment in new and more efficient power facilities, it is necessary to foresee a sufficient use for them—a sufficiently high load factor—to
justify the capital expenditure. This is a test that we take for granted in all our
projections of new capacity. It is a test which will be more rigorous in the case
of nuclear plants than in conventional installations so far as we can foresee
now, because we have not yet got to the point of visualizing any nuclear project
that will not involve a significantly higher investment per kilowatt than conventional stations.

The hope is that lower or even negligible cost of fuel in a perfected nuclear
reactor, as compared with a conventional plant, will compensate for this
higher capital investment. But even with the lower fuel costs, the higher capital
investment will not be compensated unless we can visualize high load-factor
use. This is but one of a number of complications which will act as a brake
upon any replacement of existing conventional power plants by nuclear
plants.

In judging whether and to what extent nuclear plants will be built in the
future it needs to be kept in mind that the nuclear development will always be
competing with a constantly improved, more efficient conventional alternative.

Atomic Power Prospects

The enormous requirements for electric power that will have to be satisfied
by plants which are built in the next 5, 10 or 15 years will, in the great
majority of cases, be supplied by conventional fuels or water power.

In the next 5 years it would be surprising if installation of nuclear generating
capacity exceeded 500,000 kw. In the next 10 years or so, by which time the
country will need in the neighborhood of 200 million kw of installed electric
capacity, I cannot visualize development of reactor technology that would
enable us to justify economically a total of as much as 3 million kw of atomic
power. By 1970, when our overall national requirements for electric power
will be around 275 million kw, I would not expect the proportion of nuclear
power to be as much as 10 million kw.

In considering the prospects of competitive atomic power, we must not
forget that not only do we have many general technical problems still to be
worked out, but we have ahead of us the task of exploring a great number
of possible concepts of atomic power plants and finding by such a process the
good and the bad features of each. Out of this work will come a selection of
the most promising candidates for further development. All this will take
much time, money, highly skilled effort, and a continuing determination to
move ahead with the program.

We have set ourselves to this task and I believe we are going about the
business of carrying out just such a program. The reactor-development program now sponsored by the Atomic Energy Commission is well conceived
and a broad one. As time goes on it doubtless will be extended. Some of the
electric utilities are actively co-operating in it, and the number will grow.
Because of the promise and the prospects of this new fuel, because of the hopes

it holds out for reducing cost of generation of electric energy, and because of their responsibilities for seeking and finding new and improved methods of serving their various areas, the utilities are bound to invest more heavily in study, research, development and eventually the installation of nuclear reactors.

It is for this reason that we in American Gas and Electric Company have associated ourselves as a member of the five-company group calling itself the Nuclear Power Group, or NPG, which is currently spending in research close to $400,000 per year and is hopeful that it will find a sound basis in the near future for more extensive activity.

While I believe we should persevere vigorously in the development of this new and promising source of power, it would be most tragic if we did anything in such efforts that would lead us to neglect our more conventional power technology. Clear opportunities exist to carry it forward to new levels of technological and economic improvement.

6. UTILITY THINKING WITH RESPECT TO ATOMIC ENERGY FOR POWER†

AMERICAN Gas and Electric Service Corporation of New York City, Commonwealth Edison Company of Chicago, Pacific Gas and Electric Company and Bechtel Corporation, both of San Francisco, and Union Electric Company of Missouri at St. Louis, have joined to form the "Nuclear Power Group" commonly referred to as NPG.‡

There are some twenty-one reactor-study teams in existence. One of these is NPG, the basic philosophy of which can be described as follows:

Since no reactor can be built at present that will produce power in the United States which is commercially competitive in the electric utility field, NPG believes that one of the most fruitful areas for development of practicable reactors is where fossil fuels are high in cost or difficult to get.

In developing much larger reactors there should be two main considerations: First the larger developmental reactors should be so diversified as demonstration projects as to yield maximum new knowledge. Second, they should be for applications in locations either where the demand for power cannot be met or as economically by conventional means, or on systems that are large enough to integrate them readily with fossil-fuel units to warrant their installation for developmental experience.

Implementation of this large-reactor program, will require government participation in some of the research and development expense. Any development in power reactors has a broad public interest. Also there is a considerable relationship between reactors for peacetime power and reactors for defense purposes. The recent announcement by the AEC of a reactor-demonstration program is a recognition of these facts and is a significant step forward in bringing about the conditions necessary to proceeding with larger reactors. NPG believes that both public and private power groups ought to participate in these development programs.

Today we do not know which reactor designs will prove to be best. We need to try out a number of the most promising types. We recognize that we must keep abreast of all important developments in reactor design. We must contribute our fair share in manpower, development cost, and capital investment in helping to bring about competitive nuclear power, but we must limit the risk to avoid undue hardship to our customers and our investors, and we must not jeopardize our position in the financial market.

Further, we must keep in mind that reactor development will not come to an end with the construction of the first series of demonstration reactors. In a few years reactor development may become an even more rapidly advancing

† American Power Conference, Chicago, Ill., March 30, 1955.

‡ Formed with the approval of the Atomic Energy Commission in October 1953 to permit one group of outside interests to study the commission's reactor-development activities with a view toward assisting in the research and development necessary for application of nuclear energy to production of electric power for public use.

process than at present. Hence the investment in our first reactor plant must not be so great that we will be unable to afford investment in additional developmental reactors springing from the new technology yet to come.

In our search for competitive nuclear power, we must not neglect our more conventional power technology that is serving us so admirably and is destined to reach goals far surpassing present attainments. In the final analysis the plants—whether powered by conventional fuels, nuclear energy, or even solar energy—which can give us the combination of abundant energy supply and maximum economy are the ones we must help to develop.

Therefore, NPG is proceeding with the preparation of a proposal covering the construction of a boiling-water reactor. In its first design such a reactor will not yield competitive power. It will represent a design significantly more promising than any yet constructed or under construction. It will be a significant step toward finding the right design for competitive power.

7. GOVERNMENT AND INDUSTRY IN NUCLEAR POWER DEVELOPMENT†

I WANT to record my appreciation for the invitation to testify on the measures now under consideration—the Gore Bill, S. 2725, and the companion measure, H.R. 10805.

The Gore Bill starts with the premise that our national prestige and welfare require an all-out program to develop large-scale atomic power at the earliest possible date. From this premise the bill proceeds to the assumption that progress will be more rapid if, instead of relying on the development of the necessary demonstration projects by private industry and other non-federal interests, the government embarks on an ambitious construction program of its own. Involved in the Gore Bill is perhaps the belief that the costs and risks of an accelerated reactor program are too great for private industry to undertake enthusiastically and vigorously.

Underlying the bill, I take it, is the assumption that if specific construction goals are established and sufficient money provided by the government for projects, the technical problems will somehow be solved more rapidly than by purely private projects, or by work in government laboratories carried out only to the prototype stage, as distinguished from the later stage of major, large, demonstration plants.

I would like to discuss the issue from the standpoint of five reactor types.

The first two are the natural-uranium, graphite, gas-cooled reactor and the pressurized-water reactor. The others are the boiling-water reactor, the aqueous, homogeneous reactor, and the liquid-metal fuel reactor.

Natural-uranium Reactor

The natural uranium, graphite, gas-cooled reactor is exemplified by the famous Calder Hall plant now nearing completion in Great Britain which is planned as one of a series of the same type for British construction.

This reactor has probably fewer technical problems than any reactor that is being proposed today for commercial-scale generation of electric power. But it has many disadvantages. It has the disadvantage of size. It takes up much more cubage, and therefore involves higher costs than the reactors that operate on enriched fuels. It has the further disadvantage that, in the transfer of heat, it utilizes a gas-to-gas transfer surface which is extremely inefficient from the viewpoint of engineering material and therefore from capital cost. We found this out in our study of analogous circuits with the gas turbine.

For Great Britain, however, the choice of this particular reactor for its first series seems natural and wise because the British are finding it difficult to obtain the necessary coal and oil to take care of their growing energy requirements. Foremost in the British thinking has been the necessity of building re-

† Testimony on Gore Bill before Congressional Joint Committee on Atomic Energy, Washington, D.C., May 25, 1956.

actors that can start to work successfully almost from the first day they go into operation.

In the United States we do not have a fuel crisis. We do not need to confine ourselves to natural uranium because the great physical-chemical operations at Oak Ridge and Portsmouth give us ample supplies of enriched uranium which the British do not have. Hence, it is not surprising that there is no proposal in the United States for development of this type of reactor. There is no need for us to spend our resources of manpower and materials to build reactors of this kind which, in my opinion, will not be efficient power producers in the long term.

Pressurized-water Reactor

The pressurized-water reactor is a more advanced design, successfully used in the power plant that drives the U.S.S. *Nautilus*. This reactor in large sizes already shows great momentum of development in this country. The Shipping-port (Pa.) installation is expected to be completed and producing power sometime in 1957. Consolidated Edison's pressurized-water reactor is to use thorium as a principal fuel element. It is hoped to have that plant in operation before the end of 1960. Yankee Atomic Electric Company is undertaking a third variation of the pressurized-water reactor, using a different type core. Other projects of this species are bound to come.

Boiling-water Reactor

Commonwealth Edison is installing a 180,000-kw boiling-water reactor plant at Dresden, Ill., with the General Electric Company as the prime contracting agency. This reactor was the next logical step after Shippingport for large-scale demonstration projects, because our current technological knowledge gives us confidence that it can be brought into practical operation, and because is affords a reasonable expectation of advancing the atomic industrial art further along the road toward commercially competitive atomic power. Compared with Shippingport, the Dresden plant will make possible higher pressures, higher temperatures and use of materials for heat-transfer purposes which can bring about economics in the generation of steam to propel a turbine. These savings will show up in the ultimate cost per kilowatt of capacity and per kwhr at the switchboard.

By the end of this year (1956), Commonwealth Edison will have invested $6 million in the project and the Nuclear Power Group† will have paid in an additional $3 million to bring the project into being. In addition, Commonwealth Edison will have spent in the neighborhood of a half million dollars for site and acquisition costs so that there will have been spent very close to $10 million by the end of this year. Obviously, the money expended is not productive until such time as the reactor goes into operation.

The 22,000-kw boiling-water reactor proposed by the Rural Co-operative

---

† The Nuclear Power Group (NPG) is composed of four utility systems and one manufacturer with AEC approval to engage in atomic energy research for general power purposes.

Power Association of Elk River, Minn., will use another variation of this concept. Compared with the 1000-psi pressure and 545 F temperature of the initial steam in the Dresden reactor, Elk River proposes to drop the pressure to 600 psi but raise the temperature to 825 F with a separately fired superheater.

Additional reactors of this type are bound to come along.

The two most advanced types of reactors that I have been associated with are the most exciting because of their promise in leading to commercially competitive atomic power. They are the aqueous, homogeneous reactor and the liquid-metal fuel reactor.

Homogeneous Reactors

All reactors described up to now are the so-called heterogeneous variety, utilizing fuel elements in a lattice or frame of the moderator. The fuel elements frequently have to be built with very fine dimensions and tolerances. When subjected to radiation they change form. And, of course, fuel elements have to be replaced as the number of fissionable atoms are used up and as poisonous by-products form and slow down the reactor. Reprocessing of fuel is an expensive operation.

Aqueous Type.—All of these problems become less serious in a homogeneous reactor where the fuel is in a solution. The aqueous, homogeneous reactor is one that the Nuclear Power Group has been studying for a long time.

The first homogeneous reactor to produce power, known as HRE-I, was built at Oak Ridge and operated from 1953 to 1955. In 1955, it was dismantled to make way for HRE-II; this reactor, designed to produce between 5000 and 10,000 kw thermal power, is now nearing completion.

Although Commonwealth Edison and NPG decided to go ahead with the boiling-water project as our first plant, we did not drop the idea of the more advanced homogeneous reactor. In November 1955, a special team of NPG engineers undertook a one-year study of the aqueous, homogeneous breeder reactor in co-operation with Babcock & Wilcox. B & W has eight engineers assigned full-time to this job and all of the specialists of that company are available when needed.

The Oak Ridge National Laboratory development of this reactor concept has the reactor-fuel solution housed in a thin-core vessel surrounded by a second heavier vessel. The space between the two is occupied by the fertile breeding blanket material in a slurry form—a finely divided solid suspended in a liquid. The problems of core-vessel corrosion and slurry stability are proving very difficult to solve.

The Atomic Energy Commission's Reactor Development Division has asked industry to assist in the evaluation of other concepts of this homogeneous reactor type and that is precisely what the NPG–B & W study team is doing. We have adopted as our objectives the following:

   1. Devise a reactor concept that, if possible, circumvents the major problems of the Oak Ridge design.

   2. Set the requirements for plant availability on the basis of electric power system operating principles.

3. Design the plant components and their containment to simplify maintenance.

4. Visualize maintenance practices and their effects on plant availability, and estimate their costs, including the required tools for handling inherently radioactive components.

5. Estimate kilowatt-hour costs on a realistic basis.

At this stage of our study, we have developed a reactor of unique design in which the core tank and the slurry have been eliminated. Plant arrangements are being prepared to minimize the disadvantages in maintaining a reactor system in which the fuel is in solution.

Out of the work that is being done at Oak Ridge on HRE-II or by the NPG–B & W group, it is possible that sufficiently encouraging results will be obtained to warrant a large-scale demonstration project. Development work on the homogeneous reactor is under way by Pennsylvania Power & Light Company in collaboration with Westinghouse Electric Corporation.

Liquid-metal Type.—In the liquid-metal fuel reactor the fissionable fuel material is carried in solution in liquid metal. Bismuth has been found best for that purpose; it can act in part as a moderator. For the full moderator effect required, additional graphite becomes necessary. As with the aqueous, homogeneous reactor, thorium is proposed as the blanket to surround the core and end up with fissionable isotype of uranium U-233. The Brookhaven National Laboratory began work on such a liquid-metal fuel reactor as early as 1947.

This concept of the homogeneous reactor appears to be the most daring and the most promising. But it is the least advanced and least ready for exploration in a large-scale demonstration reactor.

The liquid-metal fuel reactor (LMFR) in my judgment is not ready for a demonstration project. While the AEC's second-round reactor-demonstration program contemplates development and construction of a reactor of this type to produce between 25,000 and 40,000 kw of power, I believe the technical obstacles are such that some time will pass before this project can be realized.

General Observations

It is one of the basic characteristics of our economic system that, in the advancement of any technology, those concepts and designs which show promise for advances in efficiency or economy tend to be selected quickly as they become ready for development. There is every reason to believe that this process will continue to work in the atomic power field.

It seems clear that the nation's best interests will be served by giving every encouragement to private industry and other non-federal interests to use their initiative and their resources to the fullest in the development of atomic power. The Atomic Energy Act of 1954 provided such encouragement and opportunity. Legislation like the Gore Bill would tend to discourage such non-federal initiative and progress.

In this country we are undergoing a critical shortage of engineering and scientific manpower. Any premature, large-scale construction of projects would mean a waste of a great deal of money, and even more important a wasteful drain on engineering and other technical personnel which we cannot afford. Devoting the resources of the United States to a premature development program sets up competition for technical personnel which the rest of the economy will find it very hard to meet. Such competition cannot help slowing down the great momentum being built up in reactor development by private industry and other non-federal interests which the Atomic Energy Act of 1954 did so much to get started.

8. ATOMIC POWER SITUATION†

RECURRENT crises in the Near East oil situation not only aggravate our concern in relation to oil, but may also increase our reliance on coal. Coal, which offers much more in the way of reserves than gas and oil, will be subjected to heavy strains by our expanded use and by needs of other parts of the world. This strain will be accelerated as the better reserves are exhausted. The best hydro sites, of course, we know to have been pretty well put to use.

As we look ahead, therefore, nuclear energy may very well be the fuel which, in the long run, will make it possible for our industry to thrive when competitive-energy suppliers are encountering severe difficulties. In the field of atomic power, therefore, industry must be prepared to spend money, and to spend the talent of its best people on gathering know-how. We must, of course, be realistic. While there is a limit to what the utility industry can do, I believe that limit is much greater than most people in management today seems to have assumed. Government may still have to play an important and direct role in a program that fulfills our national interests.

The competitive but co-operative spirit within the electric industry, its reaction over many decades to what Toynbee has called a challenge and response, the availability to each member of anything new in the technology or practice of utility engineering operation, the bringing together of individual ideas of many groups—all integrated directly and through manufacturers—has been of immeasurable value in pushing the industry into new frontiers of service and to new levels of efficiency. However, I have some doubts that we have succeeded in carrying over effectively into the field of atomic energy that same combination of co-operation and competitiveness.

Our industry needs to get into atomic power to a much greater extent than it has up to the present time. The industry needs to do this on the basis that it is something not only for today but for a long period into the future.

It is true that many people in our industry are now tied into some kind of an atomic program. But this is not true for the great majority of companies and I question whether the extent of involvement is as deep as it can be for many. Provided prudence and realism are applied, I believe every utility in the country that grosses as much as $1 million a month should start training some of its people in atomic matters and taking an active part in research and development work.

We are not spending enough effort or enough money on research and development of atomic reactors. While it is true that we have had a number of groups organized into working entities it still seems that there simply are not enough of them in the active business of researching, developing, spending money, and learning about atomic-power projects, their design, construction, operation, and performance.

What we should aim for is a larger number of smaller groups—each with its own project—rather than a small number of very large groups. Otherwise

† Edison Electric Institute (Presidents' Conference), New York, N.Y., September 19, 1956.

large groups may exert a sort of gravitational pull for outsiders to join up, and we shall lose the advantage of more numerous smaller groups. We need to encourage an opposite tendency to this gravitational pull in order to get advantage of the creative thinking that will be generated by more subdivision and more individual exploration of prospects and possibilities.

We must re-examine our position in relation to the government's role in atomic power. Even though the industry undertakes more extensive programs of research and development and embarks on additional demonstration reactor projects, it seems clear to me that the United States Government is going to stay in atomic power.

It is going to do so because Russia will continue to stay in the atomic business, the British government will continue to stay in that business, the French government will continue to stay in it, and because the American government is already in the atomic business right up to its neck.

We have been too timid in our approach to the expenditures for research and development.

Assuming ordinary good judgment and prudence, the leeway that a utility has in making expenditures for such purposes is much greater than the industry has thus far used. So far as availability of money is concerned, I think our industry can become perhaps ten times more effective in this field than it is today. From a regulatory standpoint, the history of utility organizations which did not heed technological challenges is full of wreck and ruin. It seems to me it ought to be no problem for utilities to authorize expenditures for research and development one-half to one percent of their annual gross a year in a manner that will be supported and even encouraged by the regulatory bodies. Such action would give a source of funds for training, education, and laboratory or demonstration projects that would be justifiable research and development expense.

Success in conducting our own atomic-power program and accommodation to government programs without being swamped by them are not easy goals. But the prospects open to our industry over the next 20 to 25 years and beyond are so bright that difficulties should not discourage us. It is worth working hard to make sure that we will be ready to make a transition from conventional to atomic power as opportunities occur.

9. OUTLOOK FOR NUCLEAR-POWER REACTORS†

LOOKING ahead, say two decades and beyond, one sees pressing economic forces increasing the cost of conventional fuels as demand grows and supply becomes harder to obtain. During the same interval the cost of nuclear power, both capital and operating, should decline, possibly even sharply, as many of the difficult technological problems in reactor design, construction and operation are solved. In the long term, therefore, we would expect nuclear power to take its place side by side with more conventional methods of generation and, in many cases, supersede some. We must be mindful that in any long period many radically suprising things are likely to happen, contrary to our present expectations.

Whatever the long-term prospects, we already know too much about the present and the immediate future to deal with the next decade in broad brush strokes. A decade represents a short period when one observes the existing gaps in the crudely estimated nuclear-power costs and the large gaps in experience and technical knowledge of large-scale power reactors. A decade is a brief time to fill gaps of these kinds. To such aspects there must be added another factor critically relevant in utility planning—the long lead time necessary for any large power plant.

Today we cannot reliably estimate nuclear costs prior to a commitment that might involve expenditures of $65 million to $75 million for a single 200,000-kw plant.

Small reactors now operating do not give indices of capital costs for a large reactor suitable for a commercial enterprise. Shippingport (Pa.), Dresden (Ill.), Indian Point (N.Y.), Yankee Atomic (Mass.), and Monroe (Mich.)—the five largest reactors under way in the United States—are not yet ready to furnish reliable information on capital costs. Except for Shippingport, none will be able to do so for several years. We do know that costs of nuclear projects have been rising steadily compared with earlier estimates, and that the chances of bringing down costs to the utilities by contributions from manufacturers have decreased to the vanishing point. Except for government aid, cost reduction can now be achieved only through solid technical advances.

Fuel-cost estimates in reactor operations involving frequently recurring hot-chemical processing, and significant but uncertain plutonium credits against original fuel cost are still too unreliable to provide a sound basis for a capital expenditure where estimated life may be 20 to 30 years.

Operating costs are perhaps least difficult to appraise although they are still not succeptible of the precise determination customary in conventional plants. Operating-cost factors weigh against atomic plants because of the more complicated codes of regulations, the complexity of carrying out certain work, particularly maintenance, and the need for more highly trained people.

† *Nucleonics*, September 1957.

For a long time to come operating costs of atomic power plants are going to be higher than those of conventional plants.

Conventional-power Costs

On the basis of present costs, disregarding inflation, it seems reasonable to consider that typical conventional power plants can be constructed for $150 per kw. With the entirely reasonable assumptions of a thermal efficiency of 8500 Btu per kwhr in relatively large-size units of 200,000 kw or more, gross fixed charges of 14.5 and 80% annual use factor (7000 hr use of full capacity), you get the costs shown in the following table:

Item	Conventional production costs (mills/kwhr)				
	Fuel per million Btu				
	15 cents	20 cents	30 cents	40 cents	50 cents
Fixed charges	3.11	3.11	3.11	3.11	3.11
Fuel	1.28	1.70	2.55	3.40	4.25
Operation and maintenance	0.39	0.39	0.39	0.39	0.39
Total	4.78	5.20	6.05	6.90	7.75

The table shows that the economic target for a nuclear plant to be completed in this country in the next ten years is a total cost roughly between 4.75 and 7.75 mills per kwhr.

Atomic-power Costs

How close can an atomic plant visualized today meet this target? It is impossible to develop comparable capital, fuel or operating costs. The degrees of accuracy of nuclear figures forming the counterpart of conventional-power costs are not of the same order. We simply have no solid basis for accurate nuclear-cost projections. Therefore, the rough nuclear-cost figures tabulated below have an element of tentativeness that makes them quite different from the highly accurate conventional-power figures; and because of that I have purposely favored nuclear plants in the costs given.

It will be assumed that the capital costs of a nuclear plant of about the same size as the conventional plant cited would be $325 per kw and the fixed charges, 14%, taking into account the higher depreciation of some components but the lower carrying charges of certain items. The fuel cost (based on the large sets of figures that have been developed in the past year or two for heterogeneous reactors) will be taken at 2.00 mills per kwhr. This figure could be taken as derived from a fuel cost of 18 cents per MBtu and 11,000 Btu per kwhr or 20 cents per MBtu and 10,000 Btu per kwhr. The components of nuclear-energy cost, again at 7000 hr use of full capacity, then appear as follows:

ATOMIC PRODUCTION COSTS

Item	Mills per kwh
Fixed charges	6.50
Fuel costs	2.00
Operation and maintenance costs	1.25
Total	9.75

The differences between the atomic and conventional costs are 4.97, 4.55, 3.70, 2.85 and 2.00 mills per kwhr for 15, 20, 30, 40 and 50-cent conventional fuel, respectively.

If it is assumed that this represents as good a reactor as can be put together for continuous operation commencing not later than 1966, which seems reasonable, then the burden or extra cost per year of operating a nuclear instead of a conventional plant of 200,000-kw rating and generating a net of 1.4 billion kwhr with 7000 hr use becomes $6,960,000, $6,370,000, $5,180,000, $3,990,000 and $2,800,000 for 15, 20, 30, 40 and 50 cents per million Btu conventional-fuel costs.

I have been discussing optimum costs that may be expected of a 200,000-kw nuclear power plant built for initial operation toward the end of the coming decade. That means plants on which initial construction would begin some years from now and which would take advantage of knowledge gained from research and development currently in process.

In the light of today's limited knowledge, probably such a nuclear plant could not be built for operation in 1960–61 to produce nuclear energy at a cost much less than 11.5 to 12.5 mills per kwhr. If the mean of these two values is taken, 12.0 mills per kwhr, then the annual cost burden of a reactor built on today's technology, to operate late in 1960 or early in 1961, becomes $10,100,000, $9,500,000, $8,350,000, $7,150,000 and $5,950,000 for 15, 20, 30, 40 and 50-cent fuel respectively.

Prospects for Atomic Power

For the remainder of the decade, that is the years 1961 through 1966, we need to anticipate 70 to 75 million kw of new generating capacity.

What part of this generation should be nuclear? Considering economics alone, we should scarcely look to nuclear plants at all for power supply in this span of years. The improved nuclear plants that we expect toward the end of the decade should be compared not only with a conventional-fuel cost of 50 cents per million Btu in some important areas but, in all likelihood, with a continuously rising conventional-fuel cost during the life of the atomic project.

Whether technical considerations, especially reliability of service, would permit much of the total installation from 1961 to 1966 to be nuclear depends upon the outcome of research and development projects now in progress or yet to be started.

We shall never get the knowledge we need to project large nuclear plants with confidence as to cost and reliability until we build and operate large demonstration plants. This is the course that my own company naturally follows and which has recently given us gratifying results in a new approach to conventional steam stations. I refer to our large demonstration project at Philo, the success of which has led us to project our new coal-fired supercritical steam pressure 450,000-kw units at the Sporn and Breed plants.

From the foregoing evaluation of annual economic burdens imposed by the operation of a 200-Mw nuclear reactor it is obvious that we need much improved reactors. There is no dearth of ideas. The trouble is that none of the advanced ideas can be translated into practice until we first fill the big holes in present knowledge. To fill these gaps calls for basic research, or research along a specific embodiment of a principle, or for engineering development, or for all of them.

Problems

The end sought is an advanced reactor, say of 200,000-kw capacity. Since a deficiency of basic or engineering knowledge exists, the obvious shortest path to the objective is to remedy each of the difficulties in the shortest time, using the smallest scale equipment consistent with the end sought and calling for a minimum draft upon the limited national pool of specialized manpower and manufacturing capacity. The practical implementation of this approach would be a program of research and development leading to the smallest useful prototype reactor. It should be a preliminary projection of the full-scale 200,000-kw demonstration plant.

Let me illustrate this point. A typical advanced reactor design to which we have given serious consideration embodies a number of well-known components put together in a new form. The essential elements are a ceramic fuel, a graphite moderator, and sodium cooling to produce steam at a temperature around 1000 F. Before these apparently simple elements can be put together successfully in a large-scale demonstration reactor, a research and development project of vast proportions will be required.

Ceramic fuel has been known and worked upon for at least ten years. Since detailed knowledge of its thermal conductivity and structural stability is lacking, such fuel cannot be used without extensive investigation. Furthermore, each reactor design involves a new fuel-element structure of unproven feasibility and requiring unknown fabrication procedures. Hence the actual production costs are impossible to estimate.

Graphite as a moderator has been used extensively but many problems must be solved before this material can be put together in specific shapes involving untried geometrics. When combined with sodium there are corrosion problems. If graphite should come in contact with the sodium, it may act as a carburizing agent on some of the hardware such as the stainless steel piping. Canning the graphite would lead to difficult problems in detection of leaks.

Sodium as a coolant has been experimented with since the early days of power-reactor development, but design of leak-free heat-exchangers has yet to

17 a*

be proved out. When the sodium coolant is used to generate steam, leaks present a major hazard since sodium and water are not compatible. Although large sums of money have been spent and some progress has been made on this problem, it appears if future heat-exchangers are not to be unduly complicated and therefore unduly expensive, that a great deal more research and development need to be carried out.

Many other problems are raised by using sodium as a coolant. For example, there is the induced radio-activity in sodium and the problem of handling fuel when it is brought out of the sodium coolant. Sodium technology with all its attractions still presents a host of difficult problems which must be solved before subjecting a large-scale reactor to its many pitfalls.

Both engineering and construction changes, usually called for as a design moves along, take longer on a large-scale project than on a small prototype and require additional material and manpower. Dismantling and modifications after a partial or full-operation period take longer.

The lack of a prototype makes the problem of designing and building a large unit much more difficult than going from prototype to demonstration size. Even when this is done there are still plenty of pitfalls and uncertainties in extrapolation from a small to a larger one. This is especially so when the extrapolation is beyond a modest ratio, say $1\frac{1}{2}$ to 1 or 2 to 1, since the likelihood of encountering unknown limitations or unknown factors due to change in size beyond a modest limit is much greater. This has been common experience in successfully developed new technologies, and there is no reason to expect different experience with the much more difficult technology of atomic power.

Financing

Although many utilities of this country have been carrying our research and development in power technology for many decades, the fact remains that a great deal of the basic research on equipment was carried out for them by the manufacturers who added the developmental expense to the price of the equipment. That this procedure is not fully suitable in the case of atomic power has been recognized by the utility industry, which has organized regional and other groupings of electric companies to finance and carry out research and development directly. This is one of the most promising developments in the assurance of eventual competitive atomic power and continuing American leadership in the field of power generation in the nuclear age to come.

Government aid can be used more effectively and to better purpose than it now is. With so vast a program of research and development ahead, government aid is justified as part of the "Atoms for Peace" program, if for no other reason than to reduce the normal time to get results.

10. IMPACT OF ATOMIC ENERGY ON POWER INDUSTRY†

THERE is no reactor or reactor concept presently known which gives promise of producing energy in the continental United States at a cost equal to or less than that obtained with conventional fuels. Our general situation in this respect is not comparable with that in certain important parts of other countries where present reactors or reactor concepts may be expected to yield power fairly soon at no substantial cost disadvantage.

No reactor now contemplated can give assurance of competitive atomic power in this country unless major technical improvements of a kind which we do not yet know how to make are brought about.

With fuel costs in this country now ranging from 15 to 50 cents per million Btu we are able to build a conventional power plant of 200,000-kw size that will generate energy at about 4.75 mills per kwhr in favorable fuel areas or at about 7.75 mills per kwhr in the higher fuel-cost areas. The energy costs are based upon 7000-hr use, 14.5% fixed charges and a heat rate of 8500 Btu per kwhr.

An optimistic projection at the present time would show that by 1966 we might have a nuclear plant generating at perhaps 9.75 mills per kwhr; a more conservative expectation would be somewhat in excess of that. Even the optimistic figure represents an annual cost disadvantage for the nuclear plant, running from $2.8 million in a 50-cent fuel area to close to $7 million in a 15-cent fuel area.

Extent of Industry Participation

Many utility organizations, probably well over 100, are already engaged in research and development or in the construction of prototype or demonstration-reactor projects. The most important plants to be completed by mid-1962 will total nine, the largest of which is a reactor of 180 Mw capacity and the smallest 10 Mw. Together they aggregate some 800 Mw. While this total capacity is relatively small, it does involve a large number of different types including pressurized-water, boiling-water, sodium-graphite, fast-breeder, organic-moderated, sodium-cooled heavy-water-moderated, and controlled-circulation boiling-water designs.

We must seek improvements in the art along the general lines that have given us the great advances achieved in conventional power—larger units, higher steam pressures, higher steam temperatures, reheat and perhaps multiple reheat. These are the conditions necessary for higher thermal efficiencies and consequent reduction in fuel cost per kwhr. Attainment of the conditions which characterize our most modern conventional-steam stations will reduce capital costs. Because capital costs for atomic plants are higher than for conventional plants, these improvements are just that much more important in this new technology.

† Massachusetts Institute of Technology Club, New York, N.Y., February 6, 1958.

We could and should concentrate on low-enrichment fuel and the possibility of using natural uranium. Research should include the breeder in the heterogeneous and homogeneous forms. These lines of effort look primarily to reduction in fuel costs apart from capital costs. There is another line of effort which promises benefits in both fuel and capital costs, namely simplified cooling systems, especially the exploitation of gas and steam, if the technical problems do not prove too serious.

Looking Further Ahead

Now let me turn to the decade 1965 to 1975. In that interval we expect to add to the systems of the United States about 200 million kw of generating capacity. By 1975 around 5% of the then total capacity, or roughly 20 million kw, may very well be atomic. Since nuclear capacity should operate at a higher load factor than conventional capacity, we can expect that 7.5% of the 2000 billion kwhr to be produced that year, or approximately 150 billion kwhr will come from the new source. The impact of such amounts of nuclear power on the utilities and other industries will thus be sizable.

Looking still further ahead, in the five years following 1975, perhaps as much as 120 million kilowatts of additional capacity of all kinds will be installed, of which as much as 25% may be nuclear. This would represent an average annual growth of atomic power for those future years of 6 million kilowatts—an industry of quite significant proportions even on the basis of domestic requirements alone.

11. ATOMIC POWER IN FRANCE†

APART from the United States and Great Britain, France perhaps has had the most active nuclear program in the Western world. Last summer the French National Assembly approved a five-year nuclear development program which was expected to cost about $1.1 billion in indirect expenditures by 1962. A little over 60% of this sum will be for the central program of the French Atomic Energy Commission, including research, development, construction and operation of experimental prototypes, and the exploration and production of uranium. The remainder will be spent on outside projects such as an atomic submarine, mining equipment, and chemical and metallurgical development. An allocation of $366 million will be for the atomic power program of Eléctricité de France, and some amount, not yet determined, will be contributed to the Euratom program. It is now expected that the total expenditure for the five-year program to 1962, including Euratom, will total some $1.43 billion.

Included in this program are a 1000-thermal-kw "swimming-pool" reactor, Melusine, to be in operation this year at the nuclear-study center in Grenoble and a similar one, Triton, to go into operation this year at Chatillon. Also expected in operation this year at Chatillon will be a reactor, known as Minerva, for the measurement of neutron absorption. A zero-energy pile for the study of fast, plutonium-breeder reactors is planned for 1959. A prototype reactor, cooled by liquid sodium, moderated with either graphite or organic liquid, and using slightly enriched fuel, is contemplated for 1961. Also under consideration are a high-flux highly enriched uranium reactor in the medium-power range to explore higher temperatures and the possibility of a reactor linked to a gas turbine. An organic-liquid-moderated, gas-cooled, natural-uranium prototype and a fast-breeder prototype are also planned.

In addition to the G-1, a combined power and plutonium reactor of about 500 kw gross output which went critical in 1956, there is now under construction the substantially larger G-2 and G-3 reactors at Marcoule. They are scheduled to be completed this year and have a combined capacity of 50,000 kw and 50 kg of plutonium. The high-flux 15,000-thermal kw reactor for testing materials at Saclay, using heavy-water supplied by the United States and enriched uranium supplied by Britain, went critical last summer. All this is in addition to the extensive facilities that had been installed over the previous ten years.

Eléctricité de France also is building a natural-uranium, gas-cooled, graphite-moderated, 60-Mw power reactor at Avoine in the Loire Valley which is scheduled for operation in 1959 and expected to be followed some 18 months later by a similar reactor with 180-Mw capacity.

The proposed target for 1965 is 850 Mw of nuclear capacity. All this adds up to a well-thought-out, well-organized, research and developmental program which is so essential to a rational development of competitive atomic power.

† Société des Ingénieurs Civils de France (American Section), New York, N.Y., March 25, 1958.

France has been the leading uranium producer in Western Europe; its production of natural uranium has increased steadily. Most of it has been within metropolitan France. Known reserves of uranium are estimated at 100,000 tons. This year, production is expected to be 550 metric tons, and that figure is expected to rise to over 1000 tons by 1960 and 3000 tons by 1975. In addition, thorium ores from Madagascar are being processed at Le Bouchet, near Paris.

France has concentrated on natural-uranium reactors because of the desire to be free of dependence on foreign sources of enriched fuel.

However, an agreement reached with the United States last year made available to France 2500 kg of enriched uranium. An isotope-separation plant is contemplated but because of its cost is likely to be a joint Euratom undertaking rather than an individual effort.

Competitive nuclear energy in France, as in the United States, will not arrive by way of any quick solutions. Substantial research and development remain to be done.

12. ISRAEL AND THE ATOM†

HOPES for a better world through the atom captured the imagination of the world before the dust cleared following the first atomic explosion. For good reasons, such hopes were nowhere raised to a higher level than in Israel. Although our outlook has been tempered by second thought and a more realistic appraisal of the technical difficulties still to be overcome, the initial surge of optimism had a solid basis that persists and encourages us to undertake the hard work ahead.

I believe that it is most important to concentrate on a program for the development of nuclear-fueled electric power in Israel. But to do so requires a process in which there are many steps. First it is necessary to observe what has happened in Israel so far in the field of conventional power, and to project requirements rather far into the future. Next, it is necessary to analyze future power costs, including fuel and its availability, and appraise likely advances in conventional power, in atomic-power technology and in the economics of each. Only in this way is it possible to formulate a nuclear-power program suited to Israeli needs, a program which from the standpoint of national policy and economics will best serve the country.

Growth of Conventional Power

Israel started with virtually no power supply in 1922. In the ensuing 25 years some 70,000 kw of fuel-generated capacity was installed; but the 20,000 kw Jordan hydro capacity was lost during 1948. That year, despite the now-historic dislocation of ordinary activity and limited endemic resources, close to a quarter billion kwhr of electric energy were produced and distributed. This was a great achievement, possible only because of the enterprise of the people and their mastery of the relevant technology.

Generating capacity to June 1957 had increased almost fourfold to 270,000 kw, and energy sales almost four and a half times, to more than 1.1 billion kwhr in 1956. At this rate of increase capacity and consumption doubled every five years, approximately twice the rate of long-term increase in the United States.

This expansion in the use of electric power undoubtedly played a key role in the rapid growth of population and industry in the country before and after independence. But it must be understood that it is only one of many factors and not the dominant one.

Atomic power is not an end in itself. To understand its role there must first be a clear view of the role of electric power in modern society. Therefore, the future of atomic power in Israel is dependent largely upon the future of electric power in Israel. What is that future likely to be in terms of amounts of power required and its cost?

To project electric-energy use very far into the future, even a decade or two,

† Israel Institute of Technology (Technion), Haifa, Israel, May 30, 1958.

is a hazardous undertaking for any economy, no matter how long established; it is especially hazardous in a society as fluid as that of Israel today. But granting the infirmities of forecasts, projections ahead as much as 20 years are meaningful, considering our past experience.

Figures of growth of electric power use in Israel over the last 35 years have been compared with the records of other countries. There is reason to believe that the Israeli rate of increase in the last decade can and probably will continue for the next five years. By 1963, the 1957 peak demand of approximately 270,000 kw may be expected to increase to 600,000 kw and that year's total generation to some 2.8 billion kwhr. By 1968, the probable peak will be 1,000,000 kw, and total generation over 4.5 billion kwhr. Those figures should be doubled by 1978 when there may be a peak demand of 2 million kw and energy generation of the order of 9 billion kwhr, or about eight times last year's figures.

The performance of the Israeli power system to date with respect to operating costs, including maintenance and supplies, has been disappointing in comparison with high standards of performance elsewhere. But I regard this as a transient condition to be corrected by the installation of larger and more modern units. I am therefore assuming reasonable improvements in generating costs.

This leaves only the question of fuel. There seems to be no doubt that over the next 20 years there will be an adequate supply of fuel oil to meet the anti-

TABLE 1. ESTIMATED ANNUAL COST OF NUCLEAR POWER
IN ISRAEL BY 1965†

(50-Mw gas-cooled D_2O-moderated pressure-tube reactor)

	Annual cost Optimistic (Dollars)	Annual cost Realistic (Dollars)
Capital‡	1,776,950	2,132,340
Depreciation ††	654,933	785,920
Working capital ‡‡	66,000	79,200
Liability insurance	100,000	100,000
Fuel †††	750,015	684,255
Heavy water	179,200	179,200
Operation	758,000	758,000
Total	$4,285,098	$4,718,915
Annual generation	350,000,000 kwhr	305,000,000 kwhr
Cost per kwhr (mills)	12.24	15.47

† Figures rounded.

‡ Interest 6.5%, insurance 0.5%. Total plant cost optimistic, $25,385,000; realistic, $30,462,000.

†† Depreciation rate 2.58% (6.5% 20-year sinking fund).

‡‡ Working capital taken as 4% total plant cost at 6.5% interest.

††† Annual consumption 5010 kg U-235 for 350×10^6 kwhr; 4380 kg U-235 for 305×10^6 kwhr. See Table 2 for details.

TABLE 2. ESTIMATED FUEL COST FOR 50 Mw NUCLEAR REACTOR†

	Cost for 350×10^6 kwhr Annual Generation (5010 kg fuel consumed)	Cost for 305×10^6 kwhr Annual Generation (4380 kg fuel consumed)
U-235 burnup ($98.90 per kg)	$495,000	$434,000
Chemical processing ($25 per kg)	125,000	109,500
Conversion Pu to metal ($12.20 per kg)	61,200	53,400
Chemical loss ($0.73 per kg)	3,660	3,200
Fabrication cost ($42.01 per kg)	211,000	184,000
Use charge on the core and inventory‡	114,306	114,306
Annual charge for fabrication of core inventory††	16,849	16,849
Shipping cost	90,000	90,000
Plutonium credit‡‡	− 367,000	− 321,000
	$750,015	$684,255

† Figures rounded.

‡ Fuel cost $95/kg, interest rate 6.5%, fuel in core and inventory 1.5 times core loading of 14,000 kg UO_2. Ratio of uranium to UO_2 of 238/270. Charge = $14,000 \times 238/270 \times 1.5 \times \$95 \times 0.065 = \$114,306$.

†† Annual charge: $\$42.01 \times 14,000 \times 238/270 \times 0.5 \times 0.065 = \$16,849$.

‡‡ $73.20 per kg of fuel consumed.

cipated rapid growth in world demand. Barring political upheavals of catastrophic proportions, Caribbean resources coupled with Near East reserves, when combined with more efficient coal production in the United States and in parts of Europe, ought to result in a plentiful supply of fuel oil available in Israel.

Comparative Costs of Atomic Power

What is the present and prospective situation on nuclear-fuel-generated energy? In the past dozen years great strides have been made in the technology necessary to generate electric energy successfully by atomic means. The same cannot be said about costs. We must be candid in acknowledging that in the United States, for example, there is no reactor or reactor concept presently known which gives firm promise of producing energy within the continental limits of that country at a cost equal to or less than that produced by conventional fuels. The situation is somewhat different in degree in certain important locations in other countries. But there is nothing yet to warrant a conclusion that a reactor could be erected at the present time in Israel, for example, without a substantial cost disadvantage compared with conventional fuel.

I realize I express an opinion that runs counter to highly optimistic views of others about imminent attainment of competitive atomic power. Many factors enter into the development and peacetime use of atomic energy, particularly generation of electric power. Among these factors none requires more rigorous analysis, yet has been dealt with more superficially than the matter of economics and time scale. Until one gets down to specific details, lays out engineering plans covering the full job from site through maintenance and repair facilities, and studies the cycle from fuel purchase or rental through retreatment and handling of atomic ash or waste, one is in no position to talk of atomic-power costs with the precision and confidence that begins to approach the reliability of cost estimates of conventional power plants.

Prospects for Lower Costs

Beginning with about 1965, atomic power costs should start on a course of significant reductions: Capital costs should come down; fuel costs should reduce as the natural-uranium technology develops with its lower fuel-fabrication cost; and operating and maintenance cost should improve over the years for a variety of reasons. However, I do not expect that the overall cost, even by 1978, will reach a much better figure than that obtainable by conventional fuels at that time. But I do believe that by then atomic and conventional power could be closely competitive in this country.

TABLE 3. ESTIMATED GENERATING COSTS IN ISRAEL

Mills per kwhr in best available conventional steam plants
1963, 1968, 1973, 1978

Year	Btu per net kwhr	Fuel‡	Maintenance labor	Supplies	General expenses	Capital††	Total
1957†	13,200	6.6	0.8	0.8	0.9	–	–
1963	9300	4.65	0.5	0.5	0.6	2.25	8.5
1968	8900	4.45	0.5	0.5	0.6	2.25	8.5
1973	8700	4.35	0.5	0.45	0.55	2.25	8.1
1978	8500	4.25	0.5	0.45	0.55	2.25	8.0

† Average.
‡ Fuel costs 50 cents per million Btu.
†† Capital cost $160 per kw, 6.5% interest cost, 1.16% depreciation (30-year 6.5% sinking fund), 0.74% general administration. Total carrying charge rate 8.4%.

National safety might be obtained by introducing a moderate amount of nuclear power at the earliest time that this can be done without too heavy a drain on the Israeli economy. The assurance of nuclear fuel supply may be counterbalanced by the greater likelihood of loss of capacity from major technical difficulties which must be anticipated in the first atomic plants. But I believe this risk is worth taking and that there is a valid reason for making an atomic start before economic considerations alone fully justify it.

The technology and the economics of atomic power present novel problems. Learning how to cope with these problems can best be accomplished by starting early and by acquiring the necessary new techniques gradually—provided this can be done without paying too high a price for the process of education.

Suggested Program

What appears to be a rational program for Israel in atomic power? If developments in atomic and conventional-fuel power proceed as suggested then an initial step toward achieving competitive atomic power in 1978 ought to be made now. By *now* I mean a program that will get the first atomic plant built sometime between 1963 and 1965, and preferably the later date.

The size of the initial plant? By 1965 the peak demand of the Israeli power system ought to be about 750,000 kw. For that situation an atomic unit representing 7% of the total capacity might very well be considered as not too great a risk from an operating standpoint. A 50,000-kw unit would be logical to aim for as an initial atomic installation. With 1965 as the goal, seven years are available for the necessary steps: 1. To study different designs available in the United States, England, France, and possibly other countries; 2. to study locations for a site (no easy task in Israel with the close proximity of condensing water to large centers of population); 3. to obtain proposals; 4. set up and train design and construction organizations as needed; 5. train supervisory, operating and maintenance personnel; and 6. actually build the first plant.

With such a program under way there need be no hurry in proceeding with a second atomic unit. Rather, maximum advantage ought to be taken of the first installation to obtain every benefit and experience that possibly can be obtained from it. That will yield the optimum in the education of design, construction, and supervisory personnel, and in the training of operating and maintenance personnel. Particularly important will be the opportunities which arise for improvements in core construction and other features which should lead to redesigns with substantial improvements in power, burn-up and reduction in the annual burden of operation.

13. DISPOSAL OF RADIOACTIVE WASTE†

SOME prediction about the growth and scale of atomic power installations is necessary to get an idea of the size of the waste-disposal problem and the timing of measures to cope with it.

Between now and 1965 our industry is going to be confronted with the need of installing 100 million kw of capacity. Since no reactor or reactor concept presently known gives firm promise of producing energy in the United States at a cost less than that produced by conventional fuels, it has seemed unlikely that very much of this generating capacity will be atomic. It might be of the order of 1 to 2 million kw.

Between 1965 and 1975, there is a much more optimistic atomic picture. In that interval it can be expected that the power systems of the United States will add around 200 million kw of generating capacity and the total capacity will rise to the order of 400 million kw. Of that total I believe 5%, or 20 million kw, may be atomic.

More important, however, is the period beyond 1975. The chances are that in the 5 years following 1975 there will be as much as 100 to 120 million kw of new electrical capacity of all kinds installed in the United States. Of this about 25% may be nuclear.

In other words within 20 to 25 years from now, we are in for an industrial development in atomic energy of as much as 30 million kw in a period of 5 years, or at the average rate of 6 million kw per year. The rate in 1980 may well be 10 million kw.

If you assume that roughly 1.5 kg of fission products will be produced for each electric megawatt of nuclear generation then 75,000 kg of fission products will be produced a year by 1980.

On these assumptions, the volume of atomic waste represents a very small fraction of 1% of the wastes from fossil-fuel plants which we have learned to handle. The quantities involved would offer no barrier to storage in relatively temporary leak-proof vaults at plant sites, if that appeared to be the best solution.

Relatively temporary storage of this kind could go on for many years. But considering the character of the wastes such a solution is hardly likely to prove good enough for the long term. We have a difficult problem to solve and we have not very much to guide us. Yet there are some fortunate aspects:

1. In dealing with reactors designed to generate power we are working with something that is under rigid and tight control, and the system of safeguards at the present time is certainly on the conservative side.

2. The problem is not going to become acute overnight. It seems to me that it will take something like two decades for atomic power to get rolling in the

† Congressional Joint Committee on Atomic Energy, Subcommittee on Radiation, Washington, D.C., February 2, 1959.

power industry. Therefore, there is time—not to squander—but I believe adequate to work on these problems and to bring them under effective control.

With this background, I should like to discuss briefly three questions which I believe are especially pertinent:

Where should responsibility lie for disposal of waste products created by the power industry?

It seems to me that if atomic power is to develop to as significant a point as indicated, it is most important that industry have under its responsibility the complete operation starting with the planning of the reactor to the end of its life. The responsibility should include working out all arrangements for the reactor installation, licensing and contracts for the fuel including processing and fabrication of the fuel elements and the design, construction and operation. Unless it becomes possible to work out readily such arrangements there is grave danger that industry would be hesitant to go into atomic power on a major scale. If these conditions are not met, industry may make anti-atomic decisions in many critical cases just because the problems, the responsibilities, and the economics of the atomic operation might not be possible of clear visualization. The atomic industry would not develop with the necessary virility if somewhere in the total cycle a major governmental operation (other than regulatory), even the most benevolent kind, is introduced.

We are fortunate to have time on our side to work out an effective program under which we can undertake the complete responsibility advocated. But we should recognize the principle now so that our studies and our pertinent research and development programs will be conducted with this goal in view.

Would the development of re-processing operations by industry enhance effective responsibility and the program of development of the peaceful uses of atomic energy?

One of the things that has been too long delayed in the development of the peaceful use of the atom is the introduction of the economic factor; that is an extremely difficult one to introduce into governmental operation. If the economic welfare of the country as a whole is to be enhanced, it is important that atomic power be economical; that means that it must be competitive. The omission of private industry responsibility from so important a phase as disposal of its waste products would delay the introduction of the necessary factor of economy in that phase. That could not help having a retarding effect. So from that standpoint also, I believe it is essential to turn the job over to private industry.

One of the things to which we are going to have to devote a lot of study during coming years is precisely who should do the reprocessing. In most cases it had best be done by a separate fuel-processing industry. In some cases, however, it may best be done on site by the power producer.

For the time being, the government must continue and expand its efforts in the field of reprocessing and waste disposal. But these efforts of the government should be carried out in such a way that industry is brought into the picture to as great an extent as possible and should be terminated when the technology and the economics have developed to the point where industry is able to take on the full responsibility on a commercial basis.

In the field of regulation where should ultimate responsibility lie—the local, state or federal government?

For most purposes the responsibility will be discharged best if it is placed in state or local hands. I say this with some conviction because our experience in more conventional technologies (even in their most advanced form, for example very high-pressure boilers) has shown that the present situation in the United States, where the codes, inspections or licensing approvals are handled on a state basis, gives on the whole a most satisfactory arrangement.

Our experience leads me to the judgment that the existing state agencies are the ones to be given the responsibility for passing upon the safety and adequacy of programs of disposal worked out under state laws with the help of such federal guides as may be needed and which will surely be forthcoming. Situations not clearly visualizable at the present time may develop which will have a strong interstate character; that will disqualify jurisdiction by any one of the two or more states that might be involved in the matter. But as has been observed in many other cases—irrigation, pollution, water withdrawal—the state compact is an ideal instrument for that purpose and can be used here.

I cannot emphasize too strongly that I have simply attempted to give expression to the principles which I believe should govern our approach to the problem of waste disposal from the industrial standpoint.

14. RESEARCH MUST PRECEDE COMMITMENTS†

I AM appearing today as chairman of the research and development committee of the East Central Nuclear Group (ECNG) to present a status report on the nuclear power project of that group and the Florida West Coast Nuclear Group (FWCNG).

Since early 1958 ECNG and FWCNG have been conducting with the Atomic Energy Commission a research program aimed at the development of an advanced high-temperature, gas-cooled, heavy-water moderated reactor of the pressure-tube type capable of satisfactory operation on natural uranium fuel leading to the construction by FWCNG of a 50,000-kw prototype.

This concept is just as attractive today to these two utility groups as it was three years ago. Some of the reasons for our continuing interest follow:

1. Our concept would perform attractively on natural uranium even though better economics may dictate slight enrichment. The fuel cost would be relatively unaffected if the price of plutonium should be reduced, or the cost of chemical reprocessing increased, since our present estimates show that they just offset each other. In this event, spent-fuel would be stored instead of being reprocessed. This concept would suffer less than most others from the sale instead of lease of U-235.

2. Steam can be produced at conditions conforming to modern utility practice while the capability of refueling under load should provide greater unit availability.

3. The reactor is not capacity limited as is a concept employing a pressure vessel. Pressure tubes replace the pressure vessel. The size of power plants is constantly increasing, and any limitation on the maximum unit size presents an obstacle to the reduction of capital costs.

In summary, ECNG and FWCNG are convinced that the reactor concept we are pursuing if technically feasible, offers many advantages over a number of other concepts. We know of no inherent reason now why this concept in a large size should be more expensive to build or operate than other types. But, of course, the main function of the research, development, and engineering program we have under way, particularly with regard to fuel, is to enable us to affirm or modify this judgment.

The project was not conceived as a typical "third round" project, but as a partnership effort in which AEC and ECNG would share the research costs for this advanced but promising reactor concept. All parties hoped that the program would lead to the construction of a prototype plant. But all recognized that the technical and economic feasibility of the concept should first be determined through research and development work, and that results of such work are not always predictable nor necessarily favorable. Under our arrangement with the commission, construction of the prototype was expressly con-

† Joint Congressional Committee on Atomic Energy "202 Hearings", Washington, D.C., February 24, 1961.

ditioned on a determination of technical feasibility, and an economic test designed to limit the burden which FWCNG would be expected to bear in the construction and operation of the plant.

Through December 31, 1960, a total of over \$3 million has been spent by AEC and ECNG in connection with this program. FWCNG provides 2.5% of these total research and development expenditures and additionally has spent some \$300,000 on other aspects of the project. The conclusions reached as a result of the work to date can be summarized as follows:

1. While much research and development work remains to be done, the work has uncovered no indication of technical infeasibility.

2. The estimated costs of the prototype, both capital and operating, have increased substantially over first estimates, and there is no prospect of meeting the economic test prescribed in our AFC contract.

3. The ultimate objective of the project—a large-scale plant capable of operating on either natural or slightly enriched uranium—continues to look attractive, particularly from the standpoint of lowering fuel cycle costs.

Desired Beryllium Cladding Altered Program

Up until the March 1960 report to AEC the research and development program had concentrated on the use of stainless steel cladding for the first core of the prototype, so as to permit early construction of the plant. However, it has been recognized from the beginning that the attractiveness of a large-scale plant depends ultimately on the successful development of beryllium or other low-cross-section cladding material. Following the March report, the AEC staff recommended, and we concurred, that the program should be reoriented toward development of beryllium cladding for the first core. Accordingly on May 2 we presented to AEC a revised program in which further work on steel cladding was eliminated and the effort on beryllium substantially enlarged. Both we and the AEC staff recognized that the reoriented program would necessarily delay the determination of technical feasibility at least eighteen months and therefore the date of prototype construction and completion.

The reoriented research and development program necessitated a modification in our contract with AEC and was submitted by the AEC staff to the commission for approval early in May. It came as a complete surprise to us to learn on May 11 that the commission had not accepted the staff recommendation and instead had instructed the director of reactor development to get together with ECNG and FWCNG to look into the problem of immediate mutual termination of the AEC–ECNG–FWCNG contract. We expressed both our dismay at the commission action and our desire to proceed with the project in a telegram to the chairman of the commission on May 12, and requested a meeting with the commission.

In response to this request we met on May 27 with the then chairman and other representatives of the commission. As a result of this meeting, and in line with our understanding of the basis desired by the commission for continuation of the project, ECNG and FWCNG agreed to use their best efforts to formulate a new proposal to the commission involving a commitment by FWCNG to construct the prototype. This commitment would be contingent on later proof of technical feasibility and certain other conditions, but not subject to any economic test. We understood that the commission would consider paying all of the research and development costs of the project subsequent to January 1, 1961. Funds previously committed by ECNG for its share of the research and development program could then be applied by ECNG as a contribution toward the construction of the prototype.

Since our May 27 meeting, research and development work has proceeded on the basis of using beryllium cladding in the first core. Meanwhile ECNG and FWCNG have devoted a great deal of time and effort to developing a proposal along the lines desired by the commission. Our first step was to request from Combustion Engineering, Inc. a fixed-price proposal for construction of the prototype. That manufacture was a natural choice because of the important role which General Nuclear Engineering Corporation, its wholly-owned subsidiary, has played in the research and development program and the interest which its president, Dr. Walther Zinn, has had from the outset in developing this reactor concept.

Method of Financing

ECNG and FWCNG also held a number of discussions with respect to the financing of the project on the basis that FWCNG would proceed with the project, as it has always been prepared to do, if the estimated cost of energy from the prototype, after taking into account ECNG's contributions, did not exceed by more than 50% the estimated cost of power from a conventional plant in Florida. Furthermore FWCNG was prepared to take reasonable risks that actual costs might somewhat exceed this figure. Discussions proceeded on the basis that ECNG would contribute up to $11 million to the capital cost of the plant, over and above amounts already expended by it prior to January 1, 1961, to the extent such contribution might be necessary to bring the estimated cost of energy to FWCNG within the target figure, and on the assumption that such contribution would receive the same regulatory treatment as the expenditures of ECNG members under our present AEC contract.

Despite strenuous efforts on the part of ECNG, FWCNG and Combustion Engineering, we have been forced to the conclusion that it is not feasible to put together at this time a proposal containing a firm commitment to construct a prototype plant along the lines desired by the commission. Technological uncertainties remaining to be resolved in the research and development program create greater economic risks than we feel FWCNG can reasonably be expected to accept until more of the research and development work has been completed. We also encountered serious difficulties in our efforts to firm up the cost of energy from the prototype because quotations from suppliers had to be obtained and operating costs had to be estimated so far in advance of final plant design and actual commencement of construction. The main areas where we experienced difficulties are as follows:

1. Combustion Engineering's proposal, received on January 10, 1961, did not include a price on the first reactor core nor any guarantees on fuel costs or core

performance, because it was not feasible to develop a proposal covering the cost and performance of the nuclear fuel until the design and behavior of beryllium-clad fuel elements have been established through the program of research and development now underway. The inability to place any ceiling on the cost of fuel to FWCNG is the greatest single obstacle to a present commitment by FWCNG to construct and operate the prototype reactor. Conversely, we believe that when the design and technical feasibility of beryllium-clad fuel elements have been established through future research and development, the uncertainties as to fuel costs will have been largely resolved.

2. The Combustion Engineering proposal leaves open the possibility of substantial additional costs to FWCNG for changes in the present plant design which may be necessary to comply with unforseen AEC safety requirements. The difficulty of anticipating possible safety requirements can be a troublesome item in arrangements for the construction of any nuclear power plant, and particularly for a new reactor concept. Moreover, in our case, questions of safety are intimately related to questions of technical feasibility—for example, the integrity of the pressure tubes and of the fuel cladding. The problem of safety requirements will, we believe, be reduced to manageable proportions once questions of technical feasibility have been settled. Undoubtedly, we will come to know more in the next year or two about AEC safety requirements applicable to our reactor concept from the experience of the Oak Ridge gas-cooled reactor and the project of the High-Temperature Research Development Associates.

3. The price proposed by Combustion Engineering for the plant is substantially in excess of the estimate previously made by ECNG. We have been advised by Combustion that the greatest single difficulty encountered in developing their proposal was the inability to obtain firm prices for major plant components for which there could be no firm ordering date. This factor necessitated the inclusion of substantial contingencies in its proposal against future pricing by major suppliers. Combustion Engineering has expressed the belief that, were it possible to place firm orders today for the major items of equipment, its price for the plant would have compared favorably with ECNG's own estimates.

4. Due in large part to the difficulties of estimating fuel, maintenance and other operating costs prior to final plant design, complete agreement was not reached between ECNG and FWCNG on estimated operatings costs. The differences in their respective estimates of annual costs were not large—about 8 % of the total annual costs. However, capitalizing this small annual amount results in a difference of several million dollars in the amount of capital contribution which ECNG would have to make to the project. We believe these differences could have been substantially eliminated by further discussions between ECNG and FWCNG had not other more important factors already mentioned brought us to the conclusion that a proposal along the lines desired by the commission was not feasible at this time.

While we are disappointed not to be able to make a firm proposal for construction of the prototype, we are nevertheless anxious to continue the research and development work in the belief that a firm proposal can be developed after further work has been completed. To this end we submitted to AEC on February 14, 1961, a proposal for continuation of the work on a new basis which, in our opinion, will be advantageous to the government.

New Basis for Continuation

Briefly, we propose that the research and development program now under way continue until the end of 1962 with costs to be shared initially between AEC and ECNG in accordance with our existing contract. Our estimates indicate that approximately $5,492,000 will be required during 1961–62 for the joint research and development program. By October 31, 1962 we shall report to the AEC on the results of the work and our then judgment as to the technical feasibility of the proposed reactor. If AEC determines that the plant is technically feasible or that there is reasonable assurance of technical feasibility, either FWCNG will agree to proceed with construction of the prototype (with financial assistance from ECNG as indicated earlier), or ECNG and FWCNG will refund to AEC its matching contributions to the research and development program since January 1, 1961. If FWCNG agrees to proceed with construction of the prototype, AEC will reimburse ECNG for ECNG's matching contributions to the research and development program since January 1, 1961.

If the FWCNG construction proposal is accepted, AEC will be responsible for the future costs of the pre-operational and post-construction research and development work, and all amounts so reimbursed to ECNG will be paid over ECNG to FWCNG as a research and development contribution to be applied by FWCNG toward its construction of the prototype plant.

In the event AEC determines that the project is technically unfeasible or is unable to conclude that there is reasonable assurance of technical feasibility, the project will terminate, unless otherwise mutually agreed, and each party will bear its share of the costs incurred prior to termination in accordance with our existing contract.

Since AEC has not had time to consider and act upon our February 14 proposal, I do not believe it is appropriate for me to go further into details of the proposal at this time. Of course, if our proposal is accepted by the AEC, it will be presented to this Committee in accordance with existing legislation before revisions in our present contract are put into effect. But I do want to take this opportunity to explain our philosophy.

The original concept of our partnership arrangement with the commission seems to us as valid today as it did at the beginning. We believe it is important to conduct conceptual design work concurrently with other research and development. We believe the research program will be most effective when it is oriented toward a specific reactor project. We believe it is important to bring to the research and development work the practical engineering judgment of men and organizations experienced in power plant construction and operation.

There is need for flexibility in arrangements with the government to promote the participation of industrial organizations and resources in the development of new reactor concepts. In the case of the more-advanced reactor concepts, we do not think that these arrangements should be confined to the typical "third

round" contracts where government help is predicated on a commitment to
build a plant. We recognize, of course, that in establishing government policy,
the AEC and this committee must make balanced and sometimes difficult
judgments. While government policy should not encourage delay in the construction of desirable prototype and demonstration plants—whether they be
public or private projects—once necessary development and design work has
been accomplished, it does not make sense to us for government to insist on
firm construction commitments where research and development work must
first be done to establish the technical and economic feasibility of the reactor
concept. In our own case, we are convinced of the advantages of keeping intact
a going program of research and development and the participation of a group
of utilities which have worked well together on this project and which would be
in a position to proceed expeditiously with plant construction.

We believe that it is important in our total national nuclear-power program
to pursue this reactor concept. Our willingness to continue the expenditure
of large sums on the research and development program is evidence of our
sincerity in undertaking the work and of our determination to see the project
through to completion if it is at all feasible to do so.

15. A POST-OYSTER CREEK EVALUATION OF NUCLEAR ECONOMICS†

IN THE two years that have elapsed since I last submitted a statement to the Joint Congressional Committee on Atomic Energy, several exciting and highly gratifying developments in atomic power have occurred. With somewhat less public attention, there has also been some remarkable progress in conventional fuel technology, as well as a reduction in the delivered cost of coal in those areas of the United States where transportation accounts for a large portion of the total cost.

Because of the important effect of these many changes on energy conversion economics, I appreciate the opportunity afforded by Senator Pastore's invitation to follow up my 1962 appraisal with an analysis of the current competitive positions of coal-fired and nuclear electric energy generation, especially in the light of the nuclear cost reductions reflected in the announcement of the Oyster Creek project. This analysis has led me to certain conclusions on several important related problems on which I shall comment also in the course of this statement, including the question of the appropriate degree of urgency for the development of advanced converters and breeders, the advisability of mandatory private ownership of nuclear fuel, toll enrichment, and government buy-back of by-product fuel. In our approach to these problems, and indeed to all aspects of the transition to this new primary energy source, it is essential that all those actively involved be guided by a strong and undeviating sense of responsibility to preserve a sound energy economy.

In order to evaluate the present competitive position of nuclear power, it is necessary first to summarize the present status of conventional technology and, more specifically, the technology of coal-fired electric energy generation.

Coal-fired Cardinal Project

In my 1962 "202" statement, I estimated that the total cost of a highly efficient 580–600 Mw coal-fired generating unit completed in 1965 might be reduced to $120 per kw, and for a similar-sized unit completed in the period 1973–78, a figure of $100 per kw might be attainable. However, at the time I submitted these estimates to the Joint Committee an exhaustive study of potential improvements in conventional plant design and thermodynamic cycles had already been initiated by the American Electric Power Service Corporation in response to my challenge that the availability of the $100 per kw conventional plant, projected for the seventies, might be accelerated to as early a date as the mid-sixties if the vital ingredients of long-term experience and engineering imagination could be blended to create a greatly improved, low cost, but highly efficient and reliable plant design. As a direct result of that program, announcement was made last October of the joint plans of the American Electric Power system's Ohio Power Company and Buckeye Power, Inc., a corporation con-

† Joint Congressional Committee on Atomic Energy, Washington, D.C., July 17, 1964.

sisting of Ohio's thirty rural electric co-operatives, to construct the Cardinal plant, consisting of two 615-Mw coal-fired, supercritical generating units to be completed by 1967 near Brilliant, Ohio. This plant will have a capital cost of approximately $97 per kw, and an anticipated heat rate of 8650 Btu per kwhr.

Among the many technical innovations that have contributed to the advance in conventional technology represented by the Cardinal project, are (a) the pioneering use of 615-Mw, single-shaft, 3600-rpm tandem turbogenerators; (b) a new ac excitation system; (c) 825-ft stacks—the tallest in this country—to ensure the proper diffusion of gaseous effluent in a region of difficult prevalent environmental conditions whereby use can be made of locally available high sulphur-content coals; and (d) highly centralized control and automation to make possible plant operation and maintenance with an efficient manpower to capacity ratio of approximately 0.08 man per Mw.

With a looked for delivered coal cost of 17 cents per million Btu, Cardinal energy costs are expected to be about 3.59 mills per kwhr† computed on the basis of 13.5%‡ annual carrying charges, a very gratifying end result of almost two years of intense development effort.

Recent Developments in Atomics

At the same time that our engineering effort at American Electric Power was responding to the challenge in conventional technology, other members of our technical staff, with the close co-operation of three major reactor manufacturers, devoted more than 18 months to a study of those reactor systems most likely to be competitive in 1967 at an actual site in the northwest portion of the AEP system where coal costs are relatively high—about 23 cents per million Btu. In the course of this internal competition between coal and atom, we examined the turnkey nuclear plant proposals of two manufacturers and completed an overall plant design around the nuclear steam supply package of another manufacturer.

The results of our comparative studies, which were completed in mid-1963, indicated, as shown in Table 1, that a coal-fired plant would be substantially more economical than the best atomic plant at that site. Nevertheless, we were much encouraged by the new low energy cost (5.92 mills per kwhr) quoted for the best of the light-water reactors and by the impressive programs of the manufacturers which, with the support of the Atomic Energy Commission, were aimed at further significant reductions in cost through plant compaction, simplifications and improved fuel cycle performance.

Because of its important relation to the Oyster Creek situation, I would like to comment specifically on the BWR development program, as outlined to me and some of my colleagues in May, 1962, by representatives of the General

---

† Based on $97 per kw plant cost, 80% capacity factor, and O & M costs of 0.25 mill per kwhr—operating as a 2-unit plant.

‡ While earlier studies were based on the long-and-well-established ground rule of 14% annual carrying charges, all figures presented herein have been adjusted to 13.5% carrying charges to reflect the effect of the 1964 reduction in corporate income taxes and further reduction scheduled for 1965, as provided in the 1964 tax legislation.

TABLE 1. COMPARATIVE COAL-FIRED AND NUCLEAR ENERGY COSTS†
(1962 AEP study)

Item	Cardinal-type coal-fired plant			Boiling water reactor (BWR)
Plant capacity (Mw)		615 ‡		500 ‡
Capital cost ($ per kw)		102 ††		172
Coal cost (cents/million Btu)	23	30	40	
Energy cost (mills/kwhr)				
Capital (13.5% ‡‡ fixed charges, 80% capacity factor)	1.97	1.97	1.97	3.32
Fuel	1.99	2.60	3.46	2.05
O & M	0.30	0.30	0.30	0.55
Total	4.26	4.87	5.73	5.92

† All plant costs modified to reflect constructions conditions at the site in northwest part of AEP system. Evaluation ground rules were indentical for both plants.

‡ No adjustment was made for the capacity difference between the two plants because the disparity between the nuclear and coal plant costs already appeared so wide.

†† Capital cost increased by $5 per kw and O & M by 20% to account for higher cost of single unit.

‡‡ Reduced from old standard of 14% to reflect 1964 federal corporate income tax reduction.

Electric Co. I want to emphasize, however, that in our 1962 study, we found the pressurized-water reactor development programs every bit as impressive.

In the area of fuel cycle performance and economics the overall program sponsored by the commission in fuel technology, coupled with the continually increasing BWR manufacturing and operating experience, is leading to closer definition and understanding of the problems of optimizing manufacturing quality control and design of fuel elements for in-pile performance. Thus, not only has the quality and reliability of the fuel been improved markedly, but it has been accomplished at the same time that a substantial reduction in the cost of fabricating uranium-oxide fuel for these reactors has been achieved. Since 1962, work has continued with emphasis on improving the in-pile performance of fuel with respect to higher specific power and burnup, with feedback from operating reactors providing information vital to the closer determination of permissible limits of these design parameters.

The development programs, sponsored jointly by the commission, GE and reactor owners such as Commonwealth Edison, Consumers Power and Pacific Gas & Electric have led to BWR plant compaction and simplification with lower capital cost per kw as a result of the following technological achievements:

1. Complete elimination in large units of secondary BWR steam circuits, making the entire operation a direct-cycle system in which all the steam utilized in the turbine is generated without the intervention of subsequent secondary heat exchanger surface.

2. Elimination of external steam separators and dryers in even the largest

BWR units, by the development of compact separation equipment for in-vessel use, thereby removing the need for drums and the concomitant large containment volumes.

3. Reduction thereby of the volume of the steam-supply package to where pressure-suppression, instead of dome-type containment, can be employed with significant economic gain.

4. Material reduction in the number of control rods required, which has an important impact on many aspects of performance and size for a given power output.

Concurrently with the above progress, technological and production advances by the container manufacturer have made it possible to project realistically and to offer 1000-Mw water reactors in a single vessel.

While the above multi-faceted program was being vigorously pursued, the most optimistic report from GE, even as late as May 1962, was that the payoff from most of these developments would not be evident in the price of reactors sold before 1966. As a typical example, a $142 per kw price was projected for a 500-Mw reactor sold in 1966, which figure includes GE's estimate of 10% for "interest during construction and site costs".

One might therefore ask what, in view of the BWR development program summarized above, and the 1962 GE projection of $142/kw for a 500-Mw reactor sold in 1966, permitted the sale of Oyster Creek in 1963 at the announced cost of $129/kw on the 515-Mw guaranteed rating, or $104/kw on the fully-stretched output of the finally selected single cycle plant—640 Mw?

Competition Pressures

The answer, it seems to me, lies in the integrated effect of four distinct but nevertheless closely interrelated factors. These are: (a) the unusually intense competitive pressures prevailing in nuclear reactor manufacturing prior to and during the Oyster Creek bidding, (b) the earlier than expected payoff of the development programs described above, (c) the lower conventional equipment prices prevailing at that time (which had a favorable effect on Cardinal prices as well), and (d) the perhaps over-optimistic manner in which such items as contingencies, stretch-out, etc. were treated.

Since 1962, there has been intense competition among two—and sometimes three—of our technologically most advanced and sophisticated manufacturers of reactors on a number of atomic jobs, including Connecticut Yankee, Los Angeles, Niagara-Mohawk, Jersey Central, and the recently withdrawn Consolidated Edison Ravenswood project. Being committed to a nuclear future, they are understandably determined to maintain strong and close competitive market positions. Even a short-term pulling-ahead in the race by one of these organizations in today's embryonic nuclear market-place brings an imaginative and challenging competitive response from the other. Such competitive pressures were very strong at the time of the initiation of bidding on the Jersey Central Oyster Creek job.

Given this intense competitive background, the opportunity to bid for a nuclear plant at Oyster Creek called forth the effort to incorporate simultaneously in one project every available technological development, a number of which I have described, and every optimistic market projection based upon the new pricings made possible by these developments, none of which, however, had passed the test of experience. This is not to say that not one of these developments has behind it creative engineering and sound technical foundations, but rather that none is so striking or vital a departure from previous lines of development as to merit the term "breakthrough". Furthermore, it is generally known that GE priced the Oyster Creek plant on the altogether reasonable assumption that three or more very similar units could be sold to absorb some of the preparatory costs. It is the combined impact of these technical and market factors that represents a good portion of the progress implicit in the Oyster Creek prices. The above, I believe, are quite clear.

However, there are a number of factors that are not so clear. With all good will toward all the organizations involved in the Oyster Creek project, it is still my personal judgment that one of the effects of the competitive pressures of the marketplace, was to induce the manufacturer to risk somewhat greater uncertainty in the costs behind the turnkey price than might be tolerable repeatedly, when it decided to combine the effects of these innovations with that of a substantial size extrapolation. In addition, I am extremely doubtful, even in the case of the most fortuitous minimization of the effects of the manufacturer's financial riks, whether the Oyster Creek sale will yield the same financial return as the manufacturer would normally require as a matter of sound business; the kind of return required of conventional turbine business, for example.

In saying the above, I do not mean to detract in any way from the effective job done by the power company, General Public Utilities, in negotiating and putting together the Oyster Creek project. However, the easing of commercial pressures coupled with recent price increases in both nuclear and related equipment make it unlikely that another Oyster Creek job could be ordered today without a substantial price increase. In other words, while Oyster Creek indicates that nuclear power now stands ready to compete seriously for the electric energy market, the competitive level may not be quite as low as the initial announcement had seemed to indicate.

Economic Effect of Stretch-out

Before presenting what I believe to be current costs of a nuclear plant such as Oyster Creek, it is important that I comment on one item having a direct effect on costs which perhaps has not been given adequate consideration. I refer to stretch-out, which is being projected at two different levels for the Jersey Central job, i.e. "intermediate stretch-out" from 515 Mw to 565 Mw and then to the originally projected full or "expected" stretch-out of 620 Mw in the dual-cycle plant for which a full economic analysis has been published.

Stretch-out play an important role in the economics of this project as shown in the cost data provided in the Jersey Central Power and Light Company's *Report on Economic Analysis for Oyster Creek Nuclear Electric Generating Station*, February 17, 1964. That early report, which is based on a dual-cycle

18*

plant rather than the later-selected single-cycle plant, shows that the 30-year averaged annual present-worth advantage of that nuclear unit versus a mine-mouth coal-fired unit is negative ($800,000 disadvantage) for the rated output of 515 Mw and relatively small ($100,000) for the 565-Mw "intermediate" stretch-out. At 620 Mw there is a substantial advantage for the nuclear unit ($1,100,000). Similar figures for the direct-cycle plant would show a greater advantage for the nuclear unit, but still would affirm the economic importance of stretch-out.

If the additional capacity (above the 515 Mw guaranteed output) derived from stretch-out is not achieved at startup, the advantage of the nuclear unit in a present-worth comparison between the Oyster Creek nuclear and the mine-mouth coal-fired units is materially reduced. This can be illustrated by an analysis of the following table which appears on page 16 of the above mentioned Jersey Central economic analysis.

ANNUAL DOLLAR ADVANTAGES AND DISADVANTAGES

Years	620,000 kw expected capability (millions)	515,000 kw minimum capability (millions)	565,000 kw intermediate capability (millions)
1–5	$0.8	$(1.1)†	$(0.2)
6–10	1.7	(0.3)	0.6
11–20	1.2	(0.6)	0.3
21–30	0.6	(1.0)	(0.2)
30-year present-worth average	$1.1	($0.8)	$0.1

† Bracketed figures represent disadvantage.

If we were to assume that the Oyster Creek reactor achieves intermediate stretch-out (565 Mw) at startup, but does not reach ultimate capacity (620 Mw) until 5 or 10 years after startup, the table would then appear as follows:

ANNUAL DOLLAR ADVANTAGES AND DISADVANTAGES
565 Mw at Startup

Years	620 Mw at startup (millions)	620 Mw after 5 years (millions)	620 Mw after 10 years (millions)
1–5	$0.8	$(0.2)†	$(0.2)
6–10	1.7	(1.7)	0.6
11–20	1.2	1.2	1.2
21–30	0.6	0.6	0.6
30-year present-worth average	$1.1	$0.8	$0.5

† Bracketed figures represent disadvantage.

A rough calculation shows that the achievement of full stretch-out after five years operation, rather than initially, reduces the annual cost advantage of the dual-cycle nuclear plant by $300,000, from $1,100,000 to $800,000, or about 27%, and if full stretch-out is delayed 10 years, the cost advantage is then reduced by 55% to only $500,000. This, of course, ignores any additional costs that may have to be incurred to achieve the full stretch-out, and those may be substantial.

With Oyster Creek scheduled for startup in 1967–68 (and based on my own experience with stretch-out in conventional plants), it is my judgment that there is a reasonably good chance that the nuclear operating experience now being accumulated can be employed fruitfully toward the earlier attainment of the intermediate stretch-out of 565 Mw—perhaps even at the beginning of Oyster Creek operating life. The schedule and details of the program for achieving either the intermediate or fully stretched output, have not to my knowledge been outlined publicly. However, on the basis of recent technical discussion held with General Electric representatives, I have concluded that to exceed a stretch-out of 10% in a current BWR system would involve alteration of some of the reactor equipment. Inasmuch as it is difficult to assess the need, if any, for extra shut-down time and costs required to make such modifications, and in view of the reduced discounted-value of any stretch-out that is not achieved soon after startup, I have limited the stretch-out allowance on nuclear units for the purpose of economical analyses to 10% above guaranteed capability: In other words, for the Oyster Creek unit, which carries a guaranteed rating of 515 Mw, I have assumed an output of 565 Mw throughout life and for the post-Oyster Creek units which have a guaranteed output of 550 Mw, I have used 605 Mw.

Current Nuclear Power Costs

Table 2 presents nuclear energy costs for three periods which I have designated as pre-Oyster Creek, Oyster Creek and post-Oyster Creek. For the pre-Oyster Creek figures, I have used the BWR numbers obtained in the study made by American Electric Power in 1962, based on prices and design information supplied by GE. The Oyster Creek figures are taken from the publications and releases of Jersey Central Power and Light and supplemental information presented to me and some of my associates by General Electric this past May. General Electric's presentation is the source of my post-Oyster Creek figures. I do not expect these current numbers to differ greatly from those to be published shortly by General Electric in a nuclear price handbook. And while near-future actual negotiated prices may be somewhat lower than the figures shown in Table 2, they are even more likely in my judgment, to be somewhat higher. But, on the whole, I believe these figures are perhaps as reliable evaluations of atomic power costs as can be obtained today.

In each column of Table 2, I have added to the manufacturer's direct cost the AEP-estimated overheads, contingencies and land allowance which we have determined would be about 16% on a turnkey job. These overheads have been analyzed item by item and represent the minimum estimates based on our extensive experience with power plant design and construction.

It should be noted that the capital costs are based on fixed charges of 13.5%. This departure from the previously established usual nuclear ground rules of 14% carrying charges, as noted earlier reflect the corporate income tax reduction provided in the 1964 tax legislation.

TABLE 2. RECENT NUCLEAR POWER ECONOMIC ANALYSES †

Item	1962 AEP study pre-Oyster Creek BWR	Oyster Creek	Post-Oyster Creek BWR
Capability basis	Guaranteed	10% stretch	10% stretch
Capability (Mw)	5000	565	605
Cycle	dual	direct	direct
	Dollars per kw		
Unit capital cost			
Direct (GE price)‡	148	104††	117
Estimate overheads, contingencies, land allowance	24	17	19
Total	172	121	136
Energy cost ‡‡ (mills/kwhr)			
Capital	3.32	2.34	2.62
Fuel (present worth levelized)	2.05	1.45	1.45
O & M (including nuclear liability insurance)	0.55	0.55	0.55
Total	5.92	4.34	4.62

† Based on data supplied by General Electric Company, AEP estimates and Jersey Central report.

‡ These figures do not include the cost of a switchyard (approximately $3 per kw) which has been included in all conventional capital cost figures given in this statement.

†† Based on reduced Oyster Creek price—$58.5 million.

‡‡ Adjusted to normalized ground rules for comparative listing, as follows: 80% capacity factor, 13.5% annual charges.

Fuel costs have been levelized by present-worth techniques. In addition, the Jersey Central capital cost has been translated from the 88% capacity factor it used, to an 80% capacity factor which, in the absence of long-term, high-availability (including refueling time) experience with large power reactor-plants, represents in my opinion, a more soundly based figure for current analyses. Moreover, this question has been thoroughly argued and debated by and with AEC over a period of many years, and the currently established ground rules of 7000 hr per year and the use of equal figures for nuclear and conventional plants, are still valid.

Table 2 indicates an impressive 1.58 mills per kwhr reduction in total energy costs from the time of completion of the 1962 AEP study, to the Oyster Creek numbers as normalized, and based on the intermediate stretch-out capability. The reasons for this choice of capacity have been discussed earlier.

Since Oyster Creek, the price structure for nuclear power reactor plants has changed again. Assuming the achievability of a 10% stretch-out early in reactor life, the estimated cost of energy from a post-Oyster Creek BWR to be sold in 1964 for criticality in 1967–68, is shown in Table 2 to be an exciting 4.62 mills per kwhr based on the ground rules outlined earlier. This figure is 1.3 mills lower than that computed in the 1962 AEP study, and represents an achievement in which the joint committee, the commission, the reactor equipment manufacturers, the electric power industry—indeed, the entire nation—can take great pride. It represents, in my judgment, a remarkable tribute to American technology and our competitive economic system.

Comparison with Nine Mile Point Costs

Although the Oyster Creek story justifies high optimism with respect to the future of nuclear electric generation, the apparently sharply contrasting projected economics of Niagara-Mohawk's Nine Mile Point project have raised doubts. Whereas the GE turnkey price plus Jersey Central Power and Light overheads yield a $66.4 million total for Oyster Creek, the Nine Mile Point plant, which is being put together by Niagara-Mohawk's own staff with GE supplying only certain equipment, including the reactor package, is estimated at a total cost of $90.2 or $23.8 million in excess of the Jersey Central estimated costs.

However, Table 3 indicates, after normalization of the Oyster Creek price to a basis suitable for comparison with Nine Mile Point, that the apparent price difference is $13.2 million. Almost all this remaining difference can be broken down, according to the April 2, 1964 issue of *Nucleonics Week*, into two major elements—deliberate conservatism in the preparation of the direct cost estimates amounting, possibly, to $9 million and further conservatism in the form of an allowance for contingencies that is $3.5 million higher than that for Oyster Creek. In my opinion, there appears to be little doubt that when the two plants are completed, the final difference in total cost will be substantially less than the current estimates would seem to indicate.

Turnkey Versus User Design

In any case, I am confident (and in this I concur fully with Niagara-Mohawk) that whatever cost differential may remain in favor of Oyster Creek does not represent an inherent cost penalty associated with Niagara-Mohawk's participation in plant design and construction as its own architect-engineer as contrasted to the purchase of a turnkey job. The record shows that those utilities having the manpower and resources to support their own architect-engineering organizations, by incorporating their unique operating experience as a guide to new plant design, have been able to do an unexcelled job in putting together the most efficient and economical advanced conventional power plants. I am sure that in the long run, this will be the case in nuclear plants as well. That an

TABLE 3. COST DIFFERENCES BETWEEN OYSTER CREEK
AND NINE MILE POINT †

Oyster Creek total cost	$66,400,000
Difficult construction conditions at Nine Mile Point	5,000,000
Switchyard (not included in JCP & L cost)	2,000,000
Escalation (not included in JCP & L cost)	2,600,000
Higher land cost at NMP	700,000
Higher interest during construction at NMP on above items	700,000
Lower training cost at NMP	(−400,000)
Total	$77,000,000
Total cost for Oyster Creek corrected for NMP conditions	$77,000,000
Total Nine Mile Point cost	90,200,000
Apparent difference	$13,200,000

† Table derived from February 20 and April 2, 1964 issues of *Nucleonics Week.*

estimate for an initial utility-engineered atomic unit is being made with considerably more conservatism than a manufacturer's turnkey estimate may reflect the limited utility experience along these lines, or it may be due to nothing more than a particular management cost estimating philosophy.

The concept of a turnkey electric generating plant deserves more comment. In my judgment it is a concept which, if generally adopted, can lead only to an eventual decline in the technological and economic well-being of the electric power industry in both the utility and manufacturing segments. For the utilities, it will lead to a decline in the quality of its manpower; it will discourage qualified people with imagination from joining its ranks. Furthermore, it will eliminate the invaluable contribution of utility operating experience as an indispensable element in providing the direct feed-back to plant design and to manufacturing organizations that for a half-century and more has stimulated so much progress in electric power technology. The manufacturing segment of the industry, deprived of the technical contributions of knowledgeable and expert user organizations, will gradually lose its vitality which, to a significant degree, has been maintained by the challenges and ideas presented by user technologists. I fear the net result would be the decline of US electric power technology from its present position of world leadership and, eventually, would lead to higher cost power to the consumer.

I mention this matter because a superficial look at the recently announced nuclear-plant price might leave the erroneous impression that the turnkey job offers a means of reducing nuclear costs. Not only is this untrue, but the turnkey plant short-circuits the development of badly needed utility technological organizations and this in the long run can only be detrimental to the development of nuclear technology.

Competitive Status of Coal and Atom

In Table 4, I have shown comparative energy costs for approximately equivalent atomic and coal-fired units based on our Cardinal experience and the reactor-plant cost information supplied by General Electric. As in Table 2, the overheads have been estimated by AEP, and energy costs are based on 80% capacity factor and 13.5% annual charges for each plant. The reactor fuel costs have been levelized by present-worth techniques.

TABLE 4. PRESENT COMPETITIVE STATUS OF COAL-FIRED
AND NUCLEAR ENERGY
(Based on Cardinal experience and GE
reactor cost presentation)

	Cardinal-type coal-fired unit			Post-Oyster Creek atomic unit
Capability basis	Guaranteed			10% stretch-out
Capability (Mw)	615			605
Unit capital cost ($ per kw)	107			139†
Heat rate (Btu per kwhr)	8650			—
Delivered coal cost/MBtu	20 cents	25 cents	30 cents	
Energy cost (mills per kwhr)‡				
Capital	2.07	2.07	2.07	2.68
Fuel	1.73	2.16	2.60	1.45††
O & M	0.30	0.30	0.30	0.35
Nuclear liability ins.	—	—	—	0.20
Total	4.10	4.53	4.97	4.68

† Includes $3 per kw allowance for a switchyard which is included in the $107 per kw coal-fired plant price.
‡ Based on 80% capacity factor, 13.5% annual charges.
†† Levelized by present-worth techniques.

Three representative coal costs—20 cents, 25 cents and 30 cents per MBtu—are presented. Please note that the unit capital cost of the conventional plant has been raised from the $102 per kw figure shown in Table 1 to $107 per kw for comparison with a post-Oyster Creek atomic unit, to take care of equipment price increases since Cardinal equipment was purchased. These figures indicate a significant advance in nuclear economics to a position of competitiveness with coal priced at about 27 cents per MBtu.† It is particularly gratifying to me to find that progress both in atomic and in conventional power has proceeded farther and more rapidly than anyone had any basis for projecting in 1962.

The tough competition between coal and the atom that has helped spur this recent progress must now be allowed to proceed on a more equitable basis so

† A question naturally comes to mind: How accurate does one judge this 27 cents per MBtu determination to be? The answer: Of necessity, this cannot be too precise. Certainly no better than ± 1 cent MBtu; it may very well be no better than ±2 cents MBtu.

that the natural forces of our free market can function in this fiercely competitive race. Ultimately this will provide not only generally competitive atomic power, but at heretofore-unreachable low cost levels—levels that will stimulate the further electrification which must take place if the atom is to supply a major portion of our overall energy requirements.

Need for National Energy Policy

With the coming of age of the atom, we have entered another historically significant period of transition among our primary sources of energy—the third such transition in this nation's history. The first of these transitions occurred between 1850 and 1895 when coal replaced wood as the dominant energy source. Between 1910 and 1955, a second transition occurred as oil and gas found their place in our energy picture, both in unique applications, such as transportation, and as substitutes for coal. And, roughly in 1609, the increasing importance of nuclear power began to usher in a third great transition, one that will take us at least to the end of this century.

In my judgment, there is a grave question whether this current transition can be made without serious disruption of our energy economy in the absence of a sound, responsible, carefully developed and effectively implemented national policy on energy. Moreover, I regard the formulation and implementation of such a national policy as one of the salient challenges that confronts us today in the energy field. For such a policy to reach formulation and, later, implementation, it will be necessary to have the co-operation of the coal, oil, gas and uranium industries, the conventional and nuclear energy conversion equipment manufacturers, the utilities, the Atomic Energy Commission and the joint congressional committee.

Although my great optimism with regard to the coming role of the atom in our over-all energy economy continues unabated, I am deeply concerned over the welfare of our domestic fossil-fuel industries on which we shall have to depend heavily for the major portion of our over-all energy supply for the foreseeable future—certainly to the end of this century. Our national policies with regard to the further advancement of nuclear technology must be shaped in response to the nation's need not only for atomic power, or even electric energy, but for the entire spectrum of energy applications, including transportation, industrial processing, and manufacturing.

Against the background of the outlook for energy resources and requirements which I have presented previously, especially in my 1959 "202" testimony, and in light of the analysis of future energy requirements presented by AEC in its 1962 report to the President, both of which clearly indicated the very large future requirements for fossil fuels, at least for the balance of this century, it seems to me that some very fundamental questions now have to be answered with regard to the future direction of our national atomic development program.

Of these, perhaps the most important question is whether the coming pace and the final outcome of the competition between nuclear and fossil fuels are to be influenced and predetermined primarily by government nuclear policy. Is the commission going to continue to support the further development of light-water reactors following each progressive improvement in conventional technology in order to assure that the atom remains competitive in the fossil-fuel cost zones it has now reached and extends its competition into lower cost zones?

Because such a policy could result only in discouragement and, ultimately, the premature demise of conventional technology to the detriment of fossil fuels and, probably, the atom also, it seems clear to me that light-water reactor technology has matured to the point where it can safely be cut loose from governmental apron strings and be permitted from now on to make its own way, and carve out its own destiny.

With the current level of competitiveness for the atom at 27 cents per MBtu, as indicated by the numbers in Table 4, it would appear that the atom can now compete in important regions of this country. Thus we have entered the period of vigorous competition between fossil fuels—primarily coal—and uranium, for the dominant share of the future market for electric generation. There is no doubt in my mind that the coal industry, the transportation industry, and conventional fuel technology are far from the end of their record of progress toward lower costs. I know that, given the opportunity, they will continue to do competitive battle for the electric energy market and thereby will motivate further improvements in nuclear energy as well.

In my judgment, the first phase of nuclear power—the converter phase, particularly as exemplified in the boiling and pressurized-water reactors—has now reached the point where it is capable of joining this battle armed only with its own remarkable record of achievement and the promise of advancing further the established record of cost and performance without justification for, or need of, federal assistance to help counter each advance made by its prime competitor. The objective of government assistance now needs to be shifted to more basic research and development aimed at bringing to fruition the more advanced reactor types, notably breeder reactors, which undoubtedly we shall need in the future, but for which a high priority national program fortunately is not necessary at this time.

Steps to Implement Policy

The implementation of a policy to encourage free competition between fossil and nuclear fuels would require the following significant steps:

1. *Enactment of legislation to permit private ownership of special nuclear material.*

    With nuclear power at its present state of development, it is time, in my judgment, to remove an influence that has retarded and could continue to inhibit the growth of nuclear power in the United States. Prior to the construction of a coal-fired plant, for example, a utility can negotiate long-term agreements for fuel and, subject to reasonable escalation, be assured in advance, of fuel costs over plant lifetime. In the case of a reactor, however, the present requirement for exclusive government ownership of special nuclear material, and guaranteed buy-back of fuels, with credits and prices set by legislation or AEC policy rather than by free market conditions, leads to a degree of uncertainty in long-term nuclear fuel costs which, despite optimistic projections by reactor manufacturers, can deter a utility from the purchase of a nuclear plant. I would like to see this retarding influence on atomic power removed by the early enactment of legislation embodying the private ownership program recently announced by the commission. Specifically, I advocate legislation which would:

(a) Establish on January 1, 1969, a toll enrichment service for domestic uranium;

(b) Discontinue after January 1, 1971, the distribution by leasing of special nuclear material for facilities licensed under either Section 103 or 104b of the Atomic Energy Act of 1954, as amended;

(c) Establish June 30, 1973 as the date of the termination of all leasing arrangements and the institution of mandatory private ownership;

(d) Limit government purchase guarantees on plutonium and U-233 to that which is produced through the use of special nuclear material that has not been toll-enriched but which has been leased or purchased from the commission; and

(e) Terminate by June 30, 1973 all AEC purchase guarantees on plutonium, U-233 and uranium enriched in U-235.

2. *Finding by the Atomic Energy Commission that boiling and pressurized-water reactors are of practical value and therefore subject to licensing under Section 103 rather than 104 of the Atomic Energy Act of 1954, as amended.*

This would have the effect of precluding the waiver of lease charges on nuclear fuel; it may cause the commission to reassess immediately its standard for determining reasonable charges with respect to lease of uranium for reactors licensed under Section 103, and, although this is not altogether clear, may preclude the commission from entering into further R and D contracts on PWR and BWR systems. Since these steps are vital to the establishment of free competition between fossil and nuclear fuels, I am somewhat troubled by the position taken by the commission as outlined in Dr. Seaborg's letter to Chairman Pastore dated May 15, 1964. Specifically, I doubt that we serve the national interest by supporting the further subsidization of PWR and BWR systems until they can compete "in areas of the U.S. consuming a significant fraction of the nation's electrical energy". Is not the 27 cents per MBtu level of competitiveness of nuclear power sufficient demonstration of the practical value of light-water reactors and a solid base from which to construct a viable, totally unsubsidized nuclear power industry? I am concerned over the likely effect of continued subsidization of the atom on the welfare of the nation's vital fossil fuel industries, particularly coal. There would be little or no incentive for the conventional equipment manufacturers, the fossil fuel producers and the transporters of fossil fuels to make an effort to improve their technology and reduce costs, if they were warned in advance that whatever their efforts the federal government will guarantee the advantage to their nuclear competition.

3. *Establishment of a solid long-term program of AEC research and development on a deliberate, rather than an urgent or crash schedule, for the long-term development of advanced converter and breeder-reactor technology.*

This program should be aimed at the development of advanced reactors which make more efficient use of nuclear fuels than do our present light-water reactors. Selection of specific concepts and the setting up of development schedules in this program must be consistent with the overall and developing energy needs and policies of this country. In particular, it is my judgment that we need to focus both on reactors which can utilize thorium efficiently and on fast breeders. Moreover, I believe that the potential economic performance of each of these reactors must be considered every bit as seriously as their ability to extend our nuclear fuel resources. We can afford neither the waste of our fuel resources nor of our manpower and other resources all of which must be integrated in the final figure of economic performance.

In this connection you may be interested in the fact that the East Central Nuclear Group, of which my company is a member and in which I serve as chairman of the Research and Development Committee, has recognized both the need for a more diverse national breeder-reactor program and the potential advantages of fast-reactor systems cooled with supercritical steam, or gas. As a result, we are supporting financially and participating technically in a joint study with the Babcock & Wilcox Company to investigate the problems and potential of a supercritical steam-cooled breeder reactor.

Orderly Transition

The three-point program I have outlined above would help to assure the orderly transition toward an eventual major role of the atom in our energy economy. It also would permit the continuation of competitive opportunities for our existing primary-energy resources, and the full interplay of competitive forces to spur the further development of all energy technologies so that this country may continue to enjoy the benefits that have been made possible by its abundant energy resources and highly developed energy conversion technology.

In my earlier discussion of the third great transition I have placed much emphasis on the need for responsible planning of our overall energy economy at the national level, but I also wish to emphasize the vital need for responsible leadership by the utility sector of our energy industry in particular during this coming period of transition.

Atomic energy is beginning to thrive on the impressive record of development that has been achieved through American power technology, enormously aided by the contributions of the Atomic Energy Commission and of the Joint

Congressional Committee on Atomic Energy. However, we cannot afford to forget that for the balance of this century, fossil fuels will still be a vital factor in our economy. We must make certain, therefore, as we proceed toward the nuclear economy of the future, that the decisions which will affect the rate at which nuclear energy enters our economy are based on sound, reproducible economic achievement and solid fact, not transitory price changes or special methods of economic analyses.

To avoid potentially disruptive influences on every segment of our energy economy, we need the wholehearted co-operation of the primary-energy producers, the transportation industry, the manufacturing industry, the utility industry and government. But, a particularly heavy responsibility rests on the utility industry.

CHAPTER 4

PROMISING NEW WRINKLES IN POWER GENERATION

CONTENTS

1. MAGNETOHYDRODYNAMIC CONCEPT FOR LARGE-SCALE POWER GENERATION†

DOORS have opened on three avenues that may lead to advances in energy conversion and may affect electric generation of the future (the next 100 years or so) more significantly than atomic fission. These are:

1. Thermionic generation;
2. Thermoelectric generation; and
3. Exploitation of the magnetohydrodynamic (MHD) principle of generation.

While these highly intriguing methods depend on different principles, they have this important fact in common—all three offer prospects of direct conversion of heat energy into electric energy.

Thermionic Generation

Thermionic conversion makes use of the phenomenon of electron emission from the surface of an electron-conducting material owing to thermal energy of the electrons within the material. This phenomenon was first observed by Edison (*Edison effect*) and later, in 1901, fully described by Richardson.

New developments along two lines have been reported in the past year for the direct conversion of heat to electricity by this means. Work at Los Alamos Laboratory culminated in the design of a cesium-filled thermionic converter which, when operated in a reactor, produced about 40 watts. Similar work has been carried out with high vacuum and close-spaced diodes producing outputs of several watts using various heat sources, one being the concentrated rays of the sun.

Thermoelectric Generation

The discovery of thermoelectric generation was made by Seebeck in 1820 who, by accident, achieved thermal efficiencies with his thermocouple that were comparable with the then-contemporary steam engine. But the long-range significance of his accomplishments could not be foreseen. As steam engines improved in efficiency the thermocouple was relegated more and more to the background until it became nothing more significant than a temperature-measuring device.

A number of theoretical unfoldments in solid-state physics and quantum mechanics and the development of devices like the semiconductor have now made it possible to think seriously of the thermocouple again as a means for converting heat to electricity and vice versa.

Magnetohydrodynamics

In magnetohydrodynamics (MHD) no such solid achievement has been realized as in thermionic generation, but the possibilities for new developments in generation of electric energy are more exciting. Faraday's discovery of elec-

† *Power* (with Dr. Arthur Kantrowitz), November 1959.

tromagnetic induction brought forth the law that when a conductor and a magnetic field move with respect to each other an electric voltage is induced in the conductor.

Faraday's law does not tie down the conductor to any special form such as a loop of copper wire, for example. The conductor could be a fluid—gas or liquid. The idea of using such a fluid with a magnetic field constitutes the phenomenon of magnetohydrodynamics.

Concept Not New.—This is not a new idea. Patents dealing with the idea of producing electric energy without moving parts by the motion of a conducting fluid (in one case, salt water) through a magnetic field date back at least 50 years. But the various forms of the idea previously visualized never got very far because of inadequate understanding of the phenomenon involved. There was a lack of knowledge of the performance of gases at high temperature.

Recent work, however, in the study of missile re-entry problems has developed a good theoretical understanding of the electrical conductivity characteristics of gases at high temperatures in ranges above those normally encountered in thermal-power plants. This new understanding has raised the question of whether or not a technically feasible and economically attractive plant for the large-scale generation of electric energy can be built using MHD principles.

For more than three years, operating under an air force contract, the Avco-Everett Research Laboratory has been investigating the behavior of high-temperature gases in shock tubes and correlating the results with modern theoretical physics and aerodynamics.

This research has led to re-examining various aspects of generating electric energy by utilizing MHD concepts. Avco-Everett has proposed that electric energy could be generated by passing a high-temperature high-velocity gas stream through a strong magnetic field. To obtain adequate gas conductivity the machine must operate at high temperatures. These present handling problems but promise high thermal efficiencies in a device that acts like a gas turbine but is completely static and yields electric power directly. This break-through in the search for higher thermal efficiency in converting heat to electricity offers opportunities to the electric-power industry for continuing profitable exploitation of the MHD concept.

To evaluate the potential gains and the problems inherent in a thermodynamic cycle using an MHD generator, a 3-month study has recently been carried on jointly by the American Electric Power Service Corporation and the Avco-Everett Research Laboratory. The theory of the MHD generator performance is fairly well developed, but its practical design and accommodation in a power-cycle for large-scale power generation remain to be studied.

The joint study has indicated the potential performance of a coal-burning and nuclear-fueled MHD plant, as well as areas of development which would be required before such plants finally could be designed.

Coal-fired MHD Plant

To obtain an estimate of the fuel and capital cost which might be needed by a coal-burning MHD plant, a specific preliminary design has been made. An open cycle was chosen in preference to a closed cycle because of its greater simplicity. This selection reduces problems of multiple heat transfer even though leading to more difficult problems in the MHD generator and associated equipment because the products of combustion pass directly through the generator.

In making this preliminary design the knowledge and experience gained in studying the missile re-entry problem provided the prime guide in placing quantitative values on such unknowns as disassociation of gases at high temperatures (particularly of coal-combustion products), heat-transfer rates at these extremely high temperatures, and the effect of additive seeding agents. All these factors must be checked by test before an actual plant lay out or even an effective cycle arrangement can be made.

The accompanying schematic diagram (Fig. 1) shows the preliminary coal-burning cycle, in which the MHD generator is combined with a conventional reheat steam turbine. The cycle performs like a combination steam-turbine gas-turbine cycle in which the MHD generator takes the place of the high-temperature gas turbine.

By virtue of the high operating temperatures, a low heat rate of about 6200 Btu per kwhr is indicated—a 25% improvement on the heat rate of the most efficient plants now projected. Further improvement may be expected as experience is gained.

MHD Plant Cycle Features

The preliminary study is by no means adequate to define fully the best operating conditions for peak efficiency, but the diagram shows reasonable values. Flow rates have been chosen to give a net plant output of about 450 Mw. In practically all features the MHD generator is more attractive in larger sizes.

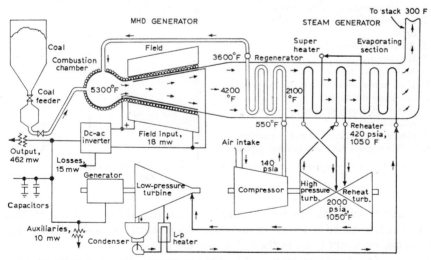

FIG. 1. Preliminary coal burning MHD cycle combined with conventional steam turbine at net heat rate of 6200 Btu per kwhr.

The cycle compresses atmospheric air to about 140 psia and heats it to 3600 F in a regenerative air heater. Coal combustion raises it to 5300 F. This high-temperature gas with its conductivity increased where necessary by "seeding" (adding a small amount of vaporized metal such as potassium), then passes through the MHD generator to develop about 360 Mw of dc energy.

Hot exhaust gas from the MHD generator at 4200 F passes through the regenerative heaters which cool it to 2100 F and then flows to the boiler. Steam generated in this boiler by cooling gases from 2100 to 300 F drives a conventional steam-turbine that drives the compressor and also generates an additional 107 Mw of ac energy.

In this cycle the steam turbines, boiler, compressor ac generator, feedwater heaters and their auxiliaries are essentially conventional equipment and represent no problem in development. The MHD equipment, however, involves special problems which require further research and development effort, particularly on the behavior of high-temperature gases in magnetic fields and the adequacy of high-temperature materials.

Economics

It is much too early to prepare a reliable study of the economics of MHD electrical-energy generation. At the present writing an MHD generator designed and tested by Dr. Richard Rosa of the Avco-Everett Research Laboratory has

COMPARISON OF INVESTMENT COSTS,
CONVENTIONAL VERSUS MHD GENERATING UNITS
(preliminary)

Equipment	Conventional unit (dollars per kw)	MHD unit (dollars per kw)
1. Steam generator and accessories	31	10
2. Steam-turbine-generator and compressor drive	38	12
3. Air compressors	–	4
4. Regenerator	–	7
5. MHD generator	–	7
6. High-temperature valves, ductwork and combustion chamber	–	4
7. Electrical converters and accessories	–	33
8. Reactive supply	–	5
9. Piping, pumps, heaters and condensers	23	10
10. Building space for conversion equipment	–	2
Totals	$92	$94

produced 10 kw of electrical energy. Its performance, which approximates that calculated, is the basis for the design estimates given above, and for the rough economic estimates given in the accompanying table.

Many of the fundamental principles need to be more thoroughly investigated and established. Yet a start has been made on sketching in the economic parameters. More particularly, since successful development of an MHD generator would displace certain heretofore indispensable items in conventional plants, we could put together a table of costs that might be eliminated and new costs might be incurred. The MHD generator displaces heat-absorbing (as distinguished from heat-releasing) parts of the boiler, such as steam generating, superheating and reheating, and also steam piping, steam turbine, turboalternator, steam condenser, boiler-feed pumps and a host of additional associated equipment.

The table shows a preliminary comparison of major equipment that would be different for the two types of units. These costs are based mainly on experience, with the exception of items 4, 5, 6 and 7 in the case of the MHD unit. In many areas the MHD unit has great economic advantages over the conventional setup, but it has a major handicap at present—generation of direct current instead of alternating current. For all practical applications this requires the conversion of dc to ac. Generating at 2500 v and conversion by mercury-arc rectifiers would be perhaps the most practical means of converting the 365 Mw of dc power, used as the basis of the table. It appears that this conversion can be done at a capital cost of $40 per kw and at an efficiency of approximately 96%.

At present we are nowhere near certain that a MHD plant like this can be made technically feasible. While the economic prospects look promising, there is no way of telling whether the economic-feasibility test could be met if the technical feasibility were established.

Formidable problems must be faced before a plant like this can be designed and operated. They consist mainly in understanding the stability of current paths in the gas, electrode effects, and other MHD phenomena. In addition, materials must be developed to handle gases at the very high temperatures— 3000 to 5000 F range—which would be encountered in a MHD plant.

The preliminary investigations carried out in the above-mentioned 3-month joint study indicated that there is fair expectation that the theoretical and practical problems can be solved. Before accepting this, however, it is felt that a program is needed to explore more fully this new concept in power generation consisting essentially of the following:

1. A study of the basic principles of electrical energy generation by a hot conducting-gas stream interacting with a magnetic field;
   (a) Construction and operation of experimental generators.
   (b) Theoretical study of generator types.
2. A search for and development of high-temperature material suitable for use in an MHD generator and in the heat sources and heat-exchangers associated with it;
   (a) Tests of heat-exchanger, liner and electrode materials.
   (b) Design of heat-exchanger and electrode configurations.
3. A study of electrical conductivity in gases;
   (a) A search for superior seed materials.
   (b) Refinement of present values for collision cross-sections in gases of interest.
4. Engineering and economic studies of over-all power-plant design.

Completion of this program should determine the practical validity of the basic theory on which the MHD generator is based and also develop generalized solutions for the critical-materials problem. If this first stage were concluded satisfactorily, it might open the way to the projection of a smaller prototype plant which, in turn, would have to carry with it a research and development phase, the successful completion of which would lead to the building of a demonstration prototype plant.

It is obvious that all of this represents some highly intriguing and exciting prospects. Not only does it appear conceivable that an MHD power plant may be produced at a competitive cost, capital-wise, which will produce much higher thermal efficiencies (of the order of 60% as against the present best of 40%) than any existing power plant, but it would open up a wide avenue of prospects for future improvement. The fact that it makes use of a new and heretofore unexplored phase of a well-founded principle leaves open the hope that further developments or new ideas will be forthcoming, even though they cannot now be foreseen with certainty.

Because of this, a small group of ten electric-utility companies has been discussing sponsoring this research in a co-operative arrangement with Avco-Everett Research Laboratory.

On November 17 the following ten electric utilities and Avco announced an agreement to proceed with a program of research on a magnetohydrodynamic power generator. American Electric Power Service Corporation will serve as agent for and will represent the utilities, which are:

Appalachian Power Company
Central Illinois Light Company
The Dayton Power and Light Company
Illinois Power Company
Indiana & Michigan Electric Company
Indianapolis Power & Light Company
Kansas City Power & Light Company
Louisville Gas and Electric Company
Ohio Power Company
Union Electric Company of Missouri.

2. EXCITING PROSPECTS OF MHD†

THERE is a distinct possibility that magnetohydrodynamics may come into being in sizes above mere watts before the other four potential direct conversion processes, namely photovoltaic cells, thermoelectric, thermionic, and fuel-cells.

Ten utility companies and the Avco-Everett Research Laboratory have joined in the development of an MHD generator. I speak as the chief executive of three of them and as representative of the other seven. The reason why these companies are so enthusiastic is that MHD is the first development in mass generation of electric energy that promises a change both in the basic cycle and in the basic equipment for producing centrally generated electric energy at efficiencies higher than any previously contemplated.

Despite all the progress that has been made in developing the modern boiler, including those operating at supercritical pressures and temperatures, and despite everything that has contributed to the modern turbine, including units up to 500-Mw size with 800-Mw units undoubtedly in the relatively near future, the fact remains that the present heat cycle is essentially no different from that utilized in Edison's first central station at Pearl Street in 1882. Of course, we have made many improvements in the 78 years that have elapsed since then, and have reached a point where today we have installations that operate at thermal efficiencies of 40%, or possibly slightly higher.

However, to bring that achievement very high pressures and steam temperatures are necessary and large-size units must be exploited. While I am confident that large modern coal-fired units will be built at lower costs it is a fact that today in most parts of this country most plants do well if they end up at a cost of about $140 a kilowatt when built for coal as the primary fuel. The ability to reduce the capital and operating costs from existing levels is, however, greatly reduced by the difficulties of developing more efficient use of materials to cut capital costs, and higher temperatures to bring about lower fuel costs.

With MHD, on the other hand, there open up the possibility that *eventually*, and please note the emphasis, *eventually* the following improvements may give promise of being realized:

1. *Lower Capital Costs.*—This will come about through reducing the amount of equipment and simplification of much of it. For example, the heat-absorbing surfaces will be materially simplified and the use of an intermediate fluid, such as water or water plus some other fluid in a binary cycle, is entirely eliminated.

2. *Simpler Heat Cycle.*—Elimination of most of the boiler, the intermediate fluid, the turbine, the regenerative system of heaters, and the condenser plus its auxiliary equipment should make a major contribution to capital-cost reduction and make a simpler plant.

† Introduction to three MHD Papers at American Society of Mechanical Engineers, New York, N.Y., November 29, 1960.

3. *Fuel Costs.*—There are clear indications that thermodynamic efficiencies in the range of 50–55% eventually should be realizable. This admittedly is not going to come over night, but when one considers how long it took, and with what effort, to reach the present top level of 40% thermal efficiency one can see what an exciting prospect the promised breakthrough to a plateau above 50% represents for the entire art and technology of mass-energy generation.

4. *Lower Operating Costs.*—This will come about primarily by the elimination of a great deal of complicated rotative and heat-absorbing equipment.

5. *Lower Maintenance Costs.*—The lower maintenance costs will come about by virtue of the simpler nature of the equipment and the elimination of a great deal of that now required.

Of course, all the above are merely promises, but very exciting promises. It seems to me that they are exciting enough to warrant all the effort being put into exploration and research in this particular project. In fact, I believe the promises are exciting enough to warrant an enlarged program of research and development on a step-by-step basis, picking up each new step just as soon as each preceding step has cleared the way for the next step to follow.

CHAPTER 5

TRANSMISSION PROGRESS

CONTENTS

1. A DECADE OF TRANSMISSION LINE PROGRESS †

ALTHOUGH no revolutionary changes in the art of electric-transmission have taken place in the past decade considerable evolution has come forth. Some has been fundamental and has resulted in major improvements in the technique and economy of transmission and since these changes are evolutionary, they are still going on.

The early part of the decade saw considerable optimism in the future of dc transmission, but developments failed to substantiate even the moderately optimistic predictions. On the contrary, the end of the decade definitely saw a more intelligent appreciation of the intrinsic merits of the ac transmission system and a more realistic appraisal of the difficulties involved in dc transmission.

Marked progress was made in the development of the established system of ac transmission. No mean contributor was the development and more widespread use of more accurate methods and facilities for faster analysis of transmission problems, especially the ac network analyzer. Developed prior to the beginning of the decade, its full potentialities were not realized until the middle part of the decade and were not used to maximum advantage until the last part of this period.

By a thorough study of stability it has been possible to take complete advantage of the inherent possibilities of a system and to design with a fairly certain knowledge of the limitations. Thus, while taking advantage of maximum capability of a system, no situation would be created that would result in disturbances to the service rendered.

Development of the line as an effective and reliable element in the scheme of ac transmission in the last decade has been due primarily to the marked progress made in the study and understanding of the lightning problem. This, in turn, led to preventive and remedial measures that have been very effective in almost entirely eliminating lightning as a cause of outage of high-voltage transmission lines.

Magnitude of Currents

No less important was the establishment of the magnitude of the lightning currents encountered in transmission-line structures as a result of direct strokes.

Magnitude of current and attenuation having been determined, knowledge of distribution of the current was necessary for an intelligent study of the lightning problem. The surge-crest ammeters installed at hundreds of locations on lines yielded definite results so that with a given stroke-current the part that actually flowed within the tower structure, and therefore the voltage rise of the tower structure, could readily be determined.

† *Electric Light and Power*, June 1941.

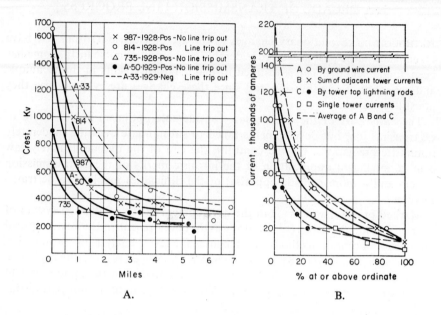

A.

B.

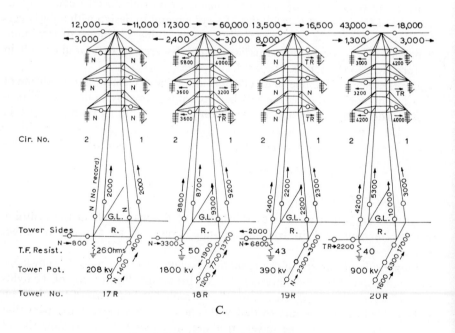

C.

Fig. 1. Lightning stroke characteristics.
A. Voltage extenuations.
B. Magnitude of currents.
C. Distribution of current in tower structure.

Tower Designs

With a better understanding of the nature of lightning strokes, their effect on tower structures, and the mechanism of flashover, better tower designs have been developed. Much more effective placement of the ground wire and lower tower-ground resistance have been made possible. Where low tower-footing resistance was not available, resort has been made to counterpoises.

Footing Resistances

Various schemes have been used to obtain lower tower-footing resistance. Multiple grounds, treated grounds, and particularly grounds driven to much greater depths than usually considered technically and economically feasible have been used successfully. In many cases grounds have been driven to a depth of 100 ft or more; although this would appear physically difficult, the development of special methods of driving brought that practice within sound economic range.

The counterpoise was studied extensively in this period theoretically, investigationally, and experimentally.

Therefore the use of counterpoises grew to a considerable extent although their application in lowering tower-footing resistance has not been as great as the driven ground rod and soil treatment. The counterpoise has found particular application where the use of additional or deeper ground rods was most difficult.

Protective Tubes

Among remedial measures the most notable devices have been the expulsion or deion protective tube, high-speed breakers and relays, and ultrahigh-speed reclosing. The expulsion protective tube was a direct outgrowth of lightning

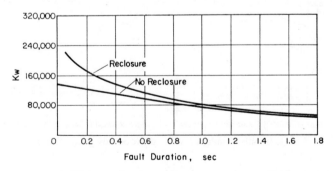

FIG. 2. High-speed reclosure improves service continuity.

studies and knowledge obtained in the past decade. Although the expulsion tube tended to become obsolete by more recent developments, particularly ultrahigh-speed reclosing, it has performed, and is still performing today an unusual service in improving service continuity of lines subject to severe lightning storms.

High-speed Breakers and Relays

High-speed clearance of faults was consummated in this period by the development of the high-speed breaker and of the carrier-current differential relays. With the first breakers the tripping time was eight to six cycles, but that has been reduced within the last few years to five and then to three cycles. By using carrier-current differential relays operating on one cycle, a total fault-clearance time of four cycles has been obtained with 3-cycle breakers, and six cycles with 5-cycle breakers.

Even more important was the by-product of the high-speed breaker and relaying system—ultra-rapid reclosing. When it was seen that the duration of the shock caused by a line short circuit could be reduced to as short as four cycles, it was evident that fault elimination and circuit reclosure could be performed so quickly that the net effect on the system and equipment served would be substantially no different than if no fault had occurred.

Wood Insulation

Although wood had been known to possess excellent mechanical and insulating properties it was not until recently that the lightning or impulse value of wood was determined and utilized knowledgeably in the design of transmission structures.

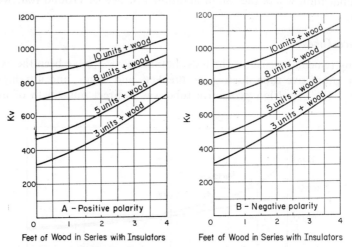

FIG. 3. Impulse insulation value of wood now determinable.

Cable Developments

Considerable progress has been made in the development of oil-filled cables, which originally appeared in the 1920's. A number of installations of this design have been made with single-conductor cable for 66 to 230 kv. Improvements in oil treatment and insulating paper have made it possible to reduce the thickness of insulation, resulting in smaller and lighter cables. Oil-filled construction has been extended to 3-conductor cables operating at 38 to 69 kv.

The design of both normal and stop joints has been progressively simplified, resulting in smaller and more easily made joints.

Shortly after the beginning of this period a new cable design for high-tension circuits, based on a new principle, appeared which makes use of the fact that oil and other fluid dielectrics increase in dielectric strength with increase of hydro-static pressure up to about 200 psi. The so-called "compression" type cables which applied this fact were first used for a 132-kv, single-phase circuit installed some years ago in Baltimore. In this case, the conduit consisted of a 7-in steel pipe with all joints carefully welded. Two paper-insulated cables were drawn into this duct so that the complete circuit was in the one duct. These cables were shipped from the factory with a light lead sheath which was removed as the cable entered the duct. Upon completion of the cable installation, the duct was filled with a high-viscosity oil (600 Saybolt) and maintained at a pressure of 200 psi by an accumulator station. The same principle has also been applied on three-phase circuits; all three conductors of the same circuit are in the same steel conduit.

STANDARD BASIC IMPULSE
INSULATION LEVELS

Reference class (kv)	Basic impulse level (kv)
1.2	30
2.5	45
5.0	60
8.7	75
—	95
15	110
23	150
34.5	200
46	250
69	350
92	450
115	550
138	650
161	750
196	900
230	1050
287	1300
345	1550

A recent development of this same principle has appeared in Europe and is now being tested in this country. In this application the fluid is not oil but compressed nitrogen gas. The cables are insulated with oil-impregnated paper, for which no satisfactory substitute has been found so far, but the thin lead-sheath which is applied over the insulation at the factory is allowed to remain when the cables are installed. Two variations of this development have been used. In each case, the three single-conductor cables are twisted together at the factory after a protective coating of canvas or other suitable material has been applied over the sheath.

Transformer

During the past decade major design and construction improvements were made in the impulse or lightning strength, lightning protection and impulse testing of transformers. All have resulted in a significant improvement in the reliability of transformers and, particularly, have made possible closer designing of their insulation strength to meet application requirements.

Outstanding in technical improvements was the development of transformers having superior and definite lightning characteristics, that is, definite abilities to withstand impulse voltages of certain kinds. These designs embody principles and ideas tending to give more equal distribution of the impulse voltage impressed upon a transformer. This is accomplished by remedial devices such as shields, or arrangement of coil sections and their insulation to produce as even distribution of stress as possible between line and ground and to minimize, insofar as possible, the development within the transformer of oscillatory voltages of a magnitude greater than the impressed voltage.

A considerable stimulant behind these developments was the extensive work carried out on insulation co-ordination. First propounded in 1928, the idea was developed on both the theoretical and practical fronts in the decade 1930–40, and was particularly applied most successfully to transformer design.

One phase of insulation co-ordination culminated in 1940 with the adoption of standard basic-impulse insulations levels. For the first time there has been provided an agreed-upon series of steps in impulse insulation strength so arranged that within the present operating voltages almost any grade of insulation strength required by any local conditions can be obtained in technically feasible and economical steps. Application of these principles to other links in the transmission-equipment chain will have further beneficial effects in the future.

Lightning Arresters. — Another contribution toward transformer operation, from reliability and economic standpoints, has been the development of the lightning arrester, particularly the so-called block-type valve arrester. Not only does it cost much less but it has reasonably consistent characteristics.

One of the outgrowths of this work was an understanding of the importance of reducing circuit length between the arrester and the transformer. In its final development this took the form of having the arrester mounted directly on the transformer.

Impulse Tests. — The means of testing lightning arresters, transformers, and the two in combination resulted from the work done on the impulse generator originally undertaken in the laboratory for the study of lightning and impulse phenomena. Within the past decade it has been developed to a consistency of performance and ease of operation that makes entirely practical the carrying out of extensive tests on every piece of high-voltage equipment where assurance is desired as to its actual impulse strength. The availability of such assurance has made possible the closer design of insulation strength for any particular installation, which in many cases has produced significant economies in transformer cost.

Recent developments of new high-permeability, continuous cold-rolled steel strip, having the ability to carry flux densities as much as 35% in excess of standard steel, is removing previous limits to progress in transformer design. This new steel is bound to have a major effect on cost where it can be used with the flux in line with the roll. When this development is coupled with new

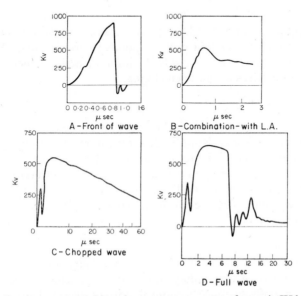

FIG. 4. Routine commercial impulse tests on power transformer in IIJ kv class.
A. Test to crest of 900-kv wave rising at rate of 1000 kv/μsec.
B. Lightning arrester holds 1500-kv impulse down to 530 kv.
C. Wave of 650 kv chopped at approximately 7 μsec.
D. Ability to stand 1½ × 40 wave of 550-kv crest.

methods of cooling which have not been sufficiently exploited, and perhaps with new dielectrics, it is bound to yield less costly and more reliable transformers.

Communication

No extensive transmission system can be operated without proper communication and metering facilities. Progress in communication has been largely one of details.

During the past decade carrier coupling to a high-voltage system was developed by insulating the ground wire at one end of a span and carrying it down along the tower to a disconnecting clamp which normally maintains the grounded condition of the ground wire, but upon being opened leaves an insulated coupling conductor. This idea made possible the more extensive use of coupling

19 VEP

points and added materially to the ease of maintenance and operation on long lines or those difficult of access.

Of significance also is the development of short-wave space-radio communication, operating in the 30- to 40-megacycle range, between fixed stations and on truck-mounted mobile units. With fixed stations at strategic points on a system this combination provides a communication system which plays an important part in coping with emergency situations and hazardous conditions arising from lightning, high winds, sleet, fire, flood or other emergencies.

Telemetering

As a system grows in size and complexity, it becomes more and more important to have available at certain strategic points records of power flows at other points. This problem was entirely solved within the period under review by the development of carrier-current telemetering.

Frequency and Load Control

Frequency and load-regulation techniques have become recognized, as fundamental to successful transmission system operation. At the beginning of the decade, a comparatively insignificant number of generating stations or lines were equipped for frequency and load control; today several score are now equipped. The value of these developments has been the elimination of erratic or uncontrolled power flow, resulting in more complete and economical use of transmission facilities.

Hot-line Maintenance

With the growth of standards of service reliability, the problem of maintenance and repair of high-voltage transmission systems, including lines and substations, naturally assumed a more important aspect. During this period considerable work was done on development of live-line maintenance methods for high-voltage lines. At the present time live-line maintenance is carried out not only satisfactorily on high-voltage lines but in many cases more safely than if the work were done on de-energized circuits.

2. A NEW TRANSFORMER CORE MATERIAL†

MAJOR design and construction improvements in the impulse strength, light-ning protection, and impulse testing of transformers have affected cost only indirectly. Even more notable is the fact that the old design trend of continually reducing losses has not been re-examined.

The development of a new steel, Hipersil, marks a definite step in altering that condition. Its outstanding characteristic is its ability to carry approxi-mately one-third more magnetic flux in the transformer core. With the same exciting ampere-turns the flux measured in a Hipersil core was 38% greater than in a silicon-steel core. This pushes aside some of the design limitations which heretofore have held transformer designers in check. Although for the present, the advantages of the new material must be limited to where it can be used with the magnetic flux in line with the direction of rolling it is bound to have a material effect on transformer design economics.

This paper presents specifically the results of an analysis to see what the possibilities of this new material would be in the design of large transformers for a base-load station. Three units at Philo station of the Ohio Power Com-pany rated at 40,000 kva, 138 kv with non-solidly grounded neutral were used for the study. The relatively lower cost of energy at the station bus was fully considered in evaluating the transformer losses. Full use was made of the in-herent properties of Hipersil in reducing dimensions and the amount of active material required.

It was found that the smaller weights and dimensions resulting from the greater flux-carrying capacity of Hipersil could be reduced still further through the use of forced-oil cooling, also such cooling was favored because little over-load capacity is needed in a base-load unit operating on the unit principle—generator and transformer switched as a unit.

It was found possible as a result, to reduce weights and dimensions to a point where these large transformers could be assembled completely in the factory to be shipped in oil with bushings in place. Moreover, due to the smaller dimensions and the relatively small amount of oil involved, it was possible to weld the cover on and seal the unit permanently against deterioration of the oil and insulation. There is no provision for breathing.

When operating at full load, these transformers have oil-circulating pumps and fans in continuous operation, but much thought has been given to the question of reliability and protective devices. It is believed that with the arrange-ments as made a degree of reliability comparable with that of other major apparatus in the generating station has been obtained.

Compared with a standard self-cooled air-blast unit of the same rating these transformers with Hipersil cores, forced oil circulation and fan coolers have 25% lighter-weight cores, 10% lighter copper windings and 5% higher losses.

---

† AIEE Summer Convention (with H. V. Putman), Toronto, Can., June 16, 1941.

19*

Cooling Systems.—The transformer is of shell-type construction with horizontal mounting in the tank. Fin-type coolers employing round tubes of substantial cross-section are mounted on each side of the tank wall and provided with circulating pumps and cooling fans. The two radiator assemblies are operated as independent units.

Operation by Copper Temperature.—Since the thermal capacity of these transformers is small and comparatively little oil is required, excessive copper temperatures could be reached in a relatively short time if difficulty should develop in the cooling system. For this reason relays are provided in each unit which operate on copper temperature. At 105 C copper temperature (measured by resistance) a warning signal is sounded to the station operator indicating that steps should be taken to reduce the load.

Fans and Pumps.—Several different blower arrangements were attempted during the course of the design but simple air-blast fans equipped with motors having "sealed-for-life" bearings were finally adopted as the simplest and most economical arrangement.

Conclusions

The development of Hipersil has opened up many new avenues for the design of transformers. Coupled with other design ideas this material has made production of large units more economical through reduction in size and amount of material required. Weights and dimensions become particularly important in very large units. Using standard magnetic steel transformers have had to be shipped in nitrogen gas and special containers to clear bridges and tunnels. Upon arrival they required a considerable amount of erection work and expense prior to being placed in service. Use of Hipersil automatically extends the rating size before these steps become necessary. The same development will extend the practical shipping size of transformers, whether on car or trailers. This development is another proof of the vitality of the electric power art and of the ability of modern technology to move forward rapidly even in a well-developed field.

3. REASONS FOR EXTRA-HIGH-VOLTAGE TEST PROJECT

WITH increased uses and new loads and with increased generation facilities systems operating over an extensive area will require vastly increased transmission facilities to provide the proper capacity in their integrating and coordinating networks. Also the ties between adjoining systems will have to be on a much bigger scale. To do that will require voltages higher than have been utilized heretofore, if the economics of transmission are to be maintained in balance.

But before the necessary transmission lines can be designed with a degree of knowledge and precision to bring about economical transmission, more technical information is needed about the characteristics and performance of the materials involved in building lines above the highest voltage at which lines are now operated, namely 287,000 v, as will other phenomena which will be encountered. In particular we need to know the performance and behavior of the air surrounding a high-voltage conductor when the conductor is raised to such high voltages. We need to know the performance of the insulators which keep the electric pressure from breaking through and going to ground, instead of staying on the conductors. We need to know what kinds of conductors and what arrangements can best accomplish the transmission purpose at these higher voltages. When we know all these things, we can design our supporting structures with the degree of accuracy that economy dictates. In addition we need to have much more information than we have on the performance of a good deal of other equipment that enters into the operation of a high-voltage system—the circuit breakers, and the transformers.

The project we are starting here today was conceived and initiated by the American Gas and Electric Service Corporation for the benefit of the associated companies. We were able to enlist in this project the enthusiastic support and co-operation of a group of outstanding manufacturers—Aluminum Company of America, American Bridge Company, Anaconda Wire & Cable Corporation, General Cable Corporation, General Electric Company, Locke Insulator Corporation, Ohio Brass Company and Westinghouse Electric Corporation.

The field laboratory includes a full-size model of towers, lines and equipment. Nothing is scaled down because we are operating in a field here where full-scale measurements are essential. We are equipped to operate the experimental lines with voltages from 264,000 up to 500,000 v. This perhaps is higher than any voltage considered practicable at the present time, but to determine the factors of practicability, it is necessary to go beyond the practicable range. It is for this reason that we are planning to go up to 500,000 v. Earlier today you saw the line energized at 350,000 v; before we finish these proceedings, we expect to bring it up to 500,000 v. When we do this you may observe some phenomena which would certainly not be part of a practical 500,000-v line, but they will

† Remarks at initiation of 500-kv test project at Tidd plant, Brilliant, Ohio, October 1, 1947.

be of extreme interest and profit in determining the practicable limits. What these limits are is not known at the present time, but they will be influenced to a considerable extent by the results of these tests.

This entire project is an outstanding example of co-operation between segments of private enterprise. There does not seem to be any question that both technologically and economically the development of electric power, and the art of electric-power generation, transmission and distribution has gone farther in the United States than anywhere else. It has done so, largely under a system of private enterprise—the power systems and the manufacturers.

4. EXTRA-HIGH-VOLTAGE TRANSMISSION STUDIES†

MUCH interest is being shown in the high-voltage investigation now being conducted on the American Gas and Electric Company system near the Tidd station of the Ohio Power Company at Brilliant, Ohio.

Some logical questions are: Why any tests? What tests are contemplated? How are they to be conducted? What are the design characteristics of the equipment to be tested? What equipment will be used for carrying out the testing program?

Past experience in the development of power systems and analyses of the economic considerations involved in transmitting increasingly larger blocks of power indicate the desirability, in some cases, of going to higher voltages than have been used heretofore. However, increased voltages require effectively larger conductors, larger towers, higher insulation levels and more expensive terminal equipment. With these costs rising very rapidly with higher voltage, it becomes increasingly important to design the system so that maximum use is made of the capabilities of the equipment. This is another way of saying that as voltage increases it becomes more than ever necessary to design for the very minimum margins of safety permissable. But before extra-high-voltage systems can be designed with such close margins without sacrificing reliability, engineers need more information than is now available.

The projection of higher-voltage transmission is becoming a pressing problem that is beginning to be more or less universal. In some localities where fuel is scarce and remote undeveloped hydro energy is plentiful, long-distance transmission will be used. Also, it is quite apparent that there are systems in the United States where transfer of much larger blocks of power than any heretofore involved will be required for base load, for co-ordination of large integrated systems and for interchange between them and contiguous systems.

From a consideration of load, distance, reliability, and other influencing factors, the voltage level, number of circuits, and circuit arrangement can be determined. All of these factors will affect the economics of the project, but perhaps the one factor that will affect costs the most is the insulation level adopted for terminal equipment and line and the spacing for the line, and these levels naturally will be influenced by the type of system grounding.

It has been the practice on some lower-voltage systems to operate lines with ground-fault neutralizers. However, it is the opinion of the authors that in the case of extra-high-voltage transmission, above 230 kv, the lines ought to be grounded permanently and solidly at all transformation points to permit the use of reduced-voltage arresters. Coupled with the use of interrupting devices which will prevent more than a single restrike, solid grounding will limit the magnitude of switching transients impressed on the line and equipment, and hence will permit insulation of a lower level than previously thought possible.

Reliable information is available on lightning and switching surges, and line

† AIEE Midwest General Meeting (with A. C. Monteith), Chicago, Ill., November 3, 1947.

reactance and capacitance, but insulation levels need to be considered thoroughly. Much information is needed on corona and radio influence, and how they are affected by size and type of conductor, spacing, height of ground wire, and atmospheric conditions. It seems appropriate, in this introductory paper to the symposium, to review briefly the known factors in line design and by the process of elimination develop in a little greater detail the significance of the elaborate set of tests now in progress.

TABLE 1. COMPARATIVE INSULATION CHARACTERISTICS OF STEEL-TOWER TRANSMISSION LINES

Kv class	Number of insulators †	Minimum impulse level, kv, $1\frac{1}{2} \times 40$ wave †	60-cycle dry flashover, kv RMS †	Times normal line-to-ground operating voltage †
69	(4–8) 5	(475–780) 550	(280–500) 320	(7–12.5) 8
138	(8–12) 10	(780–1100) 940	(500–710) 600	(6.3–8.9) 7.5
230	(14–20) 16	(1270–1770) 1450	(810–1140) 910	(6.1–8.3) 6.8
287	24	2110	1350	8.1

† Range of values is in parentheses. Most common construction is figure following parentheses.

TABLE 2. COMPARATIVE LINE INSULATION CHARACTERISTICS FOR TRANSMISSION AT EXTRA-HIGH VOLTAGES

Number of standard insulators	conductor spacing (ft)		Minimum impulse level (kv)	60-cycle dry flashover in kv RMS ‡			
				Actual	times normal line-to-ground operating voltage		
	Actual	Equivalent			345 kv	400 kv	460 kv
20	27	34	1770	1140	5.7	4.9	4.3
24	33 †	41.5	2100	1350	6.8	5.8	5.1
27	38	48	2380				
30	43	54	2600	1700	8.5	7.3	6.4
35	51.5	65	3100				

All single-circuit flat-steel constructions with 1000-ft spans.
† For certain tower construction and types of conductor these spacings might be as high as 33 and 37 ft, respectively.
‡ Wet values will be approximately 70% of these ratios.

To make some specific comparisons for extra-high voltages, three voltage levels are used, namely, 345 kv, 402 kv, and 460 kv. Table 1 gives typical insulation characteristics of higher-voltage steel-tower lines used in the United States. Table 2 gives comparative insulator characteristics for various lines over the range of extra-high-voltage transmission. In Fig. 1 the straight line representing the most common insulation levels has been extended to indicate the levels that would result if the practice of proportionate increase in insulation levels were continued. This is too conservative in the light of present-day knowledge.

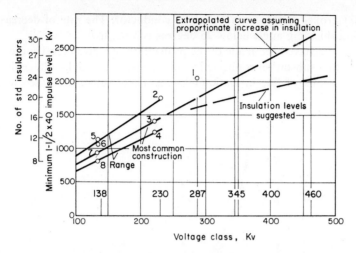

FIG. 1. Insulation levels for typical transmission lines. 1—City of Los Angeles Department of Water and Power, Boulder Dam–Los Angeles. 2—Pennsylvania Water and Power Company, Safe Harbor–Washington. 3—Pennsylvania Power and Light Company, Wallenpaupack–Siegfried; 14 to 16 insulators averaged at 15.4—Pacific Gas and Electric Company, Tiger Creek–Newark. 5—Pennsylvania Water and Power Company, Safe Harbor–Perryville; Ohio Power Company, Lima–Fostoria. 6—Union Electric Company of Missouri, Osage–Cahokia; Ohio Power Company, Philo–Canton. 7—Ohio Power Company, Lima–Fort Wayne. 8—Northern Indiana Public Service Company, Michigan City–State Line.

Lightning Performance

For the voltage classes so far used in the United States, lightning protection has been the primary consideration in the choice of transmission-line insulation levels. However, sufficient knowledge and experience has now been gained to show that there is a definite upper limit of insulation required for lightning protection, which already has been exceeded by numerous higher-voltage lines in use today. Normal steel construction with spans of the order of 1000 ft or less and effective tower-footing resistances of 20 ohms or less should experience substantially no flashovers from direct strokes when the phase wires are shielded by overhead ground wires and the equivalent of 16 or more standard suspension insulators are used. Some benefit in reducing the phase-conductor voltages is obtained by increasing the transmission-line dimensions which decrease the coupling factor.

However, the principal source of the higher surge voltages that can appear at the terminals of a highly insulated line is thought to be strokes that actually contact the phase-wires through lack of complete shielding. Both theoretical considerations and model studies show that, although the frequency of direct strokes to phase-conductors can be made quite small with shielding angles of 25 to 30 degrees, such contacts cannot be eliminated entirely with only one or two overhead ground wires. The model studies indicate that with a shielding angle of 25 degrees on a conventional 2-ground-wire 230-kv steel line about one out of every 900 strokes might contact a phase conductor. The rate of decrease with decreasing shielding angle below 25 degrees is quite low.

Calculations have been made of the probability of a given crest surge-voltage appearing at a substation connected to well-constructed high-voltage lines with various amounts of line insulation—Fig. 2. These values are based upon our

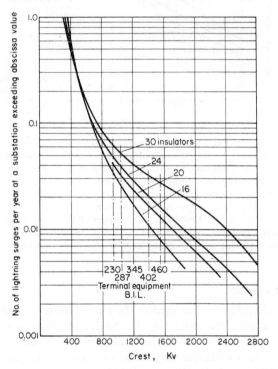

FIG. 2. Probability of crest voltage appearing at substation for different line insulation levels. Broken lines are suggested BIL for terminal equipment for different rated voltages.

present knowledge of the magnitude and wave shape of lightning-stroke current and the assumption of 100 strokes per 100 miles of line per year. They take into account voltages that might be induced by indirect strokes, by strokes contacting the ground wire, and by strokes that contact a phase-conductor.

Insulation Co-ordination

It is of interest to consider the probability of experiencing a surge at a station in excess of the strength of its apparatus. The vertical lines plotted in Fig. 2 are suggested basic impulse levels (BIL) for the indicated operating voltages. These are somewhat different values than would be derived from the following present standards:

operating voltage (kv)	Basic impulse level (kv)
230	1050
287	1300
345	1150

Present standard basic impulse levels are based on system operation with a fully rated lightning arrester. A large number of high-voltage solidly grounded systems are in successful operation with equipment insulation levels one step below standard values. If the extra-high-voltage range is approached on the basis of solidly grounding and using a reduced-voltage rating lightning arester, then, considering the higher impulse value of equipment dealt with, one lower class insulation is not only sound practice but may be improved upon by further lowering the insulation. This seems entirely practicable for several reasons. By reliance on solid grounding of the system's transformation point and controlling the voltage, it is believed practicable to use a lightning arrester having a rating of about 75% of normal voltage rating. There do not appear to be any economic difficulties to shielding adequately the line immediately adjacent to the substation so that there is practically no probability of ever getting in excess of 5000 amp through the lightning arrester. It appears possible, without appreciably affecting the overall economics, to resort to the use of diverter wires for a short distance from the terminal to give 100% shielding in this zone.

Switching Surges

Modern high-voltage breakers are designed for no more than one restrike, for which the maximum line-to-ground switching-surge voltage that should appear at the substation is of the order of 3.0 times normal line-to-ground operating voltage on a solidly grounded system. Table 2 shows that 24 insulators for operating voltages up to 460 kv provide a minimum 60-cycle dry-flashover ratio of 5.1 or a wet-flashover ratio of 3.6, which is considered adequate for these switching conditions.

Effect of Construction on Line Impedance

Shorter insulation strings make possible smaller line spacings, which have the beneficial effect of reducing line capacitive and inductive reactance.

Calculations of the relative amounts of power that could be transmitted over a 200-mile line have been made with the arbitrary assumption that stability considerations limit the line-reactance angle to 30 degrees. The results are presented in Table 3. A rather arbitrary conductor size was chosen with two levels of insulation, one taken from the suggested curve in Fig. 1, and the other from straight-line extrapolation of the low-voltage data in the same figure. This table shows that the increased power limit and charging kilovolt-amperes obtained with the smaller spacing are appreciable. Bundle conductors may offer even greater savings as a result of decrease in line capacitive and inductive reactances.

TABLE 3. ECONOMIC BENEFITS OF LOWER LINE INSULATION ON REDUCING

operating voltages, kv	Insulator units	Conductor spacing, ft	Ohms per Mi, 25 C	X_1, Ohms per Mi	X_1, Megohms per Mi
345 (1.61-in. conductor)	20	33	0.0887	0.789	0.1908
	24	37	0.0887	0.789	0.1908
400 (2.035-in. conductor)	20	33	0.0577	0.758	0.1838
	27	39.5	0.0577	0.780	0.1891
460 2.5-in. conductor)	24	37	0.047 †	0.750 †	0.181 †
	30	42	0.047 †	0.766 †	0.184 †

† Estimated value.
‡ Saving on terminal equipment due to reduced reactance.
†† Based on $8.00 per kva of increased reactive kilovolt-amperes available at receiving end.
‡‡ Based on 30-degree angle with $E_s = 1.05 E_r$.

Corona

The disruptive corona voltage and the corona power loss are very much affected by size of conductor and spacing. The greater the insulation level and the larger the spacing, the higher will be the disruptive corona voltage and the lower the corona power loss. As far as lightning and switching surges are concerned, 16 insulators with normal spacing for 345 kv would appear adequate. However, an abnormal conductor size might have to be used or the corona loss would be too high. This may be the factor that would set the lower limit on the dimensions of an extra-high-voltage transmission line. Since this factor has considerable effect on the cost of building a line, it deserves critical study.

To obtain a basis for discussing the corona-loss performance of various line constructions being considered here, fair-weathered corona-loss calculations have been made using Peterson's formula. These are based upon the assumption of a smooth conductor, an altitude of 1000 ft, a temperature of 25°C, and a surface factor of 0.9, which Peterson found to be applicable to general operating conditions for type HH cable at 220 to 287 kv.

If the required conductor diameter were chosen to give equal power loss for different lengths of insulator strings at a given operating voltage, the range of conductor diameter would be considerable. For example, an operating voltage of 345 kv, limiting the corona loss to 0.65 kw per mile, would require about a 1.75-in. diameter conductor for a line with 20 insulators, a 1.6-in. conductor with 24 insulators, and only a 1.45-in. conductor with 30 insulators. If, however, a 1.5-in. diameter conductor were used for all three line designs, the range of power loss would be 0.6 to 0.9 kw per mile.

Probably it would be impractical to limit the power loss to a constant value independent of operating voltage if a level as low as 0.5 or 0.6 kw per mile is desired, since this would call for an extremely large and impractical conductor spacing.

LINE REACTANCE AND INCREASING LINE CAPACITANCE ON A 200-MILE CIRCUIT

Power delivered Mw	Reactive power delivered Mva	Per cent X_1 reduction load base	Economic Benefits		
			Reduced reactance ‡	Increased reactive ††	Reduced line cost ‡‡
396	48.5	0.80	$48,000	$10,000	$480,000
390	47.2				
561	49.2	1.54	$130.000	$21,000	$880,000
545	46.6				
752	55.2	1.14	$129,000	$18,000	$940,000
737	52.9				

Satisfactory operation should be obtained with operating voltages up to 400 kv with transmission-line insulation corresponding to 20 insulators. At 460 kv, the minimum number of insulators probably would be 24. A more accurate determination of the best combination of line construction and conductor diameter requires study of the cost of the various combinations, taking into account loss evaluation; but no final conclusions can be drawn until more is known about corona and radio influence. Since the decision on this point alone might affect the cost of extra-high-voltage transmission very appreciably, tests are justified to give the line designers better data.

Corona loss, determined on present information and design practice, is a small factor in the overall operating cost of a high-voltage line, but if sufficient data were available on corona loss under all conditions, a closer design would be permissible with a possible large saving in the high-voltage system.

Rough estimates show an increase of about 10% in capital cost when spacing is increased from 33 ft to 42 ft, and about 25% when conductor size is increased from 1.558 in. to 1.901 in. for a given spacing. It better date will allow a smaller conductor diameter or smaller spacing, or both, capital cost can be decreased by a sizeable factor.

Conclusions

The need for developing higher-voltage transmission and the fact that costs increase very rapidly with increased voltage make it increasingly important to evaluate carefully and precisely all engineering factors entering into the design of higher-voltage transmission systems. This necessitates precise and reliable data on the characteristics and performance of constituent materials and equipment.

The two factors that influence design and costs of extra-high-voltage transmission are corona and basic impulse level, and are interrelated. There is good engineering reason for believing that materially lower insulation levels than heretofore attempted on extra-high-voltage transmission can be used successfully and when all the unknown engineering questions that this raises can be answered.

Fairly good engineering data are now available on lightning and switching surges, insulation coordination, and the effects of line spacing on reactance and capacitance. But line spacing also affects corona. Some data are available on corona and radio influence, particularly on fair-weather corona losses for horizontal configuration of single conductor per phase, and on how the surface affects these losses. There is need for corona-loss and radio-influence data showing the effect of ground wires on various combinations with single conductors, and the effect of rain, fog, clouds, and other natural elements that make up the yearly weather conditions. Bundle conductors offer some advantages that need to be investigated carefully to weigh advantages versus disadvantages.

Such data would allow a closer estimate of corona losses so that they could be considered on an average loading basis, the same as other variable losses. All of this should lead to more precise design and the development of transmission systems for extra-high voltages that would yield the maximum possible economies.

The investigation and test program now in progress at the Tidd plant is planned to obtain the necessary data to answer these and other pertinent questions and thus make possible the economical design of extra-high-voltage transmission systems.

5. 500-kv TESTS FOR CORONA AND RADIO INTERFERENCE †

DURING the past two years a great deal of valuable information on the technical features of high-voltage line design has been obtained from the field research which is being carried out near the Tidd plant of the Ohio Power Company, by the American Gas and Electric Service Corporation in co-operation with eight manufacturers of high-voltage equipment.

The broad objective is to obtain more extensive, and more precise information on some of the technical problems involved in extra-high-voltage transmission and, if possible, set at rest the fear that a satisfactory solution of some of these problems might put a limit on the voltage one might otherwise choose to improve the economics of transmitting the large blocks of electric energy which are certain to be needed in the relatively near future.

The test facilities include a 5000-kva bank of three single-phase transformers which can vary the 3-phase line-to-line voltage in steps from 265 to 532 kv. Three test lines, two 1.4 miles long and one 800 ft long, are available. Supersensitive metering equipment is used to indicate and record corona loss. The metering also includes indicating and graphic radio-influence meters, commercial amplitude-modulated and frequency-modulated receivers, a television set, a voltage-gradient meter, and graphic meters to record temperature, barometric pressure, precipitation, and humidity. In addition, an automatic camera, set to photograph sample conductors and insulators every 20 minutes, is used to record the effects of weather conditions, such as rain, snow, sleet, and fog. A setup is also provided to measure the loss on a string of insulators. Line conductors from 0.92 in. to 2.0 in. in diameter have been tested.

The data presented here must be considered as preliminary; complete analysis of data already obtained and further work now in progress may modify the results.

Corona-loss Test Program

In the corona-loss studies emphasis has been placed on establishing a method of estimating the magnitude of the annual corona loss on various conductor and line designs, and the manner in which the corona loss might be expected to vary with voltage and weather. Information upon which to base estimating procedures can be obtained by recording the corona loss on a full-scale line at different voltages over long periods that include all kinds of weather. Because of the many variables involved, and the fact that weather conditions cannot be controlled, the procedure was adopted of recording for later analysis those factors which are considered important, rather than depending upon making individual tests under a variety of weather conditions.

To provide information on the full voltage range over which the conductors might be expected to operate, the lines were energized continuously, but the applied voltage was changed every few days. Voltages of 280, 345, 396, and

† AIEE Summer General Meeting (with A. C. Monteith), Pasadena, Calif., June 12, 1950.

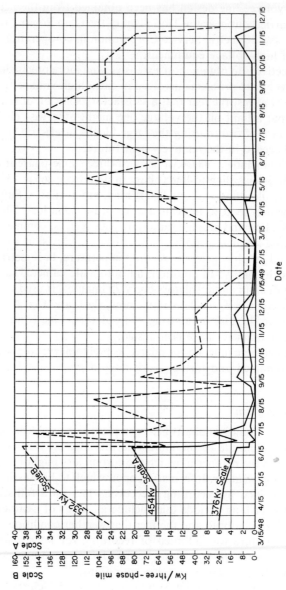

FIG. 1. Change with time in corona loss at different voltages during fair weather. (Conductor spacing changed July 18, 1949 from 45 to 32 ft.)

452 kv phase-to-phase have been used. In addition to obtaining graphic records of corona loss at these four voltages, voltage-versus-corona-loss runs were made every 4 weeks using indicating instruments. These tests were made under both single- and 3-phase conditions. Most voltage runs were made under fair-weather conditions so that the results of successive tests could be compared. Some tests under adverse weather were made, but the results are not readily comparable as rainfall is rarely constant for a sufficient time to obtain a complete test under constant conditions.

Factors Producing High Corona Loss

Great variance was observed in corona-loss measurements as noted by previous investigators. Two factors influence this variance—the conductor surface and weather conditions. Different conductors might have comparable corona loss under fair-weather conditions but different losses after aging or during foul weather.

Surface Factors.—The surface of an energized conductor undergoes a change with time such that the loss usually decreases, the magnitude varying with the type of conductor. From fair-weather runs made periodically during almost two years, the loss values corresponding to the 532-, 454-, and 376-kv points are plotted in Fig. 1 for a 1.65-in Heddernheim conductor. This graph shows how much loss might be expected on a new conductor when it is first placed in service. The wide range of loss readings obtained at a given voltage on successive fair-weather voltage-runs clearly indicates that comparisons of conductors should not be based on single observations.

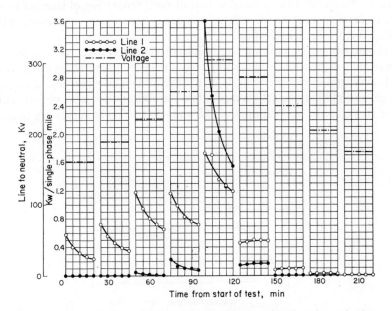

FIG. 2. Change in corona loss on two lines as affected by duration of energization.

Another surface effect is the temporary increase in the corona loss on a conductor after it has been de-energized for a day or more. If the conductor is energized again, the initial value of the loss is considerably higher than normal, but immediately starts to decrease, requiring several hours or more to become constant. Because of this effect, the corona-starting voltage of a given conductor might be 100 kv or more below its normal value. Conversely, although to a much lesser extent, if a conductor is energized at a voltage high enough to produce profuse corona, the loss at a lower voltage is temporarily reduced.

Figure 2 shows readings taken with increasing and then decreasing voltages, each voltage being applied for 20 minutes. Each time the applied voltage was increased the conductor on Line 1 showed a marked decrease in corona loss from its initial value for the interval, but when the voltage was reduced, it showed a slight increase in loss above the initial value for the interval. At the lower voltages the conductor on Line 2 was not in corona, but at higher voltages a loss decrement with time can be observed which is similar in character to the conductor on Line 1 but different in magnitude. To obtain consistent results it was found necessary to precondition the conductors by energizing them at maximum voltage for about 15 minutes prior to testing. This procedure unfortunately yields values which are lower than the loss which would exist if the conductors were energized continuously at each test voltage.

Weather Factors.—Of the weather factors causing high corona loss, the most important are rain, sleet and snow. High corona loss has been recorded during fog, but it is believed that this condition was caused by condensation of moisture on the surface of the conductor which is not likely on a line carrying load. Temperature, humidity, barometric pressure, and atmospheric voltage-gradient have an effect which is small compared to that of rain. Separation of weather factors requires detailed analysis of a large number of records because several of the weather factors change together.

Preliminary analysis shows a consistent correlation between corona loss and average rate of rainfall. Remarkably smooth curves are obtained if the corona loss is averaged over periods of 48 hr to several days of operation at constant voltage, and plotted against average rate of rainfall for the same period. This is shown in Fig. 3 for line voltages of 398 and 454 kv. These curves are based on

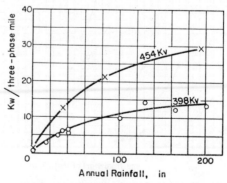

FIG. 3. Correlation between corona loss and rate of rainfall at constant voltage.

data taken under a variety of weather conditions and during different seasons. Except for a few points at high rates of rainfall, no attempt has been made to select rainy periods. The remarkable consistency of results indicates that such curves might be an acceptable basis for estimating annual corona loss.

Average All-weather Loss.—More significant information is obtained from the graphic records of all-weather corona loss than from individual fair-weather voltage-runs. From graphic records, Fig. 4, the annual corona loss can be estimated and its economic importance determined.

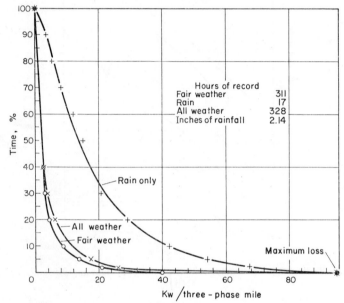

FIG. 4. All weather corona loss for one month on 395-kv line.

The average corona loss of different types of conductors depends not only on the applied voltage but also on the relative duration of fair and foul weather.

The average values of corona loss can be expected to be the same on operating lines as they are on the test lines. However, for lines of practical length the maximum value of corona loss will be much less than on test lines because extremes of weather are not likely to exist all along a line simultaneously. An evaluation of the weather conditions along a proposed right-of-way is necessary before an attempt to estimate the average and maximum values of corona loss can be made.

Radio Influence

When this investigation was started, published information available on radio-influence of transmission lines was believed insufficient for prediction of this characteristic for a line designed to operate at voltages appreciably higher than used at present. Because of the clearly indicated need for the consideration of radio influence a major study was made using the Tidd test lines.

20*

The radio-influence problem consists essentially of two parts: Determination of the radio-influence factors of transmission lines, and evaluation of these factors in terms of their effects on the communication services.

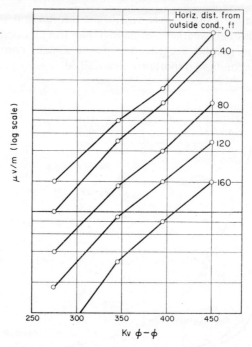

Fig. 5. Radio influence variation with line voltage at right angle distance up to 160 ft.

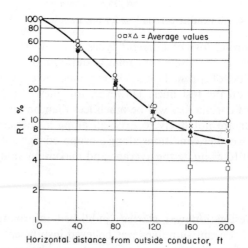

Fig. 6. Attenuation of radio influence in per cent of values measured directly below outer phase conductor for line voltages from 138 to 450 kv phase-to-phase.

Rapid rise of radio influence with increase of line voltage and drop with increase in horizontal distance from the conductor are apparent in Fig. 5. The slope of the radio-influence curves with voltage is approximately the same as obtained by other investigators. The lateral attenuation is practically independent of line voltage. Variations of radio influence, Fig. 6, along the line are believed due to shielding by towers, change in conductor height, and reflections.

Radio Influence and Frequency.—Measurements over the frequency range show that radio influence values vary somewhat inversely as the frequency. Tests with amplitude-modulated, frequency-modulated, and television receivers indicate that radio influence may affect amplitude-modulated (am) broadcast reception without appreciably affecting frequency-modulated (fm) and television reception.

Factors Affecting Tolerable Radio-influence Levels.—The evaluation of radio influence on communication services and tolerable limits requires consideration of several factors: Characteristics of radio influence obtained and its effect on the various communication receivers; available signal intensities along the line; satisfactory signal-to-noise ratios; effects of weather, number and type of receivers in vicinity of line; distances of receivers from the line; transfer of radio influence to lower-voltage circuits; and means for improvement of reception at a particular receiver.

It seems reasonable to permit higher levels of radio influence on extra-high-voltage transmission lines than on lower voltage lines for several significant reasons: The direct coupling to receiver antennas will generally be less; there will be fewer receivers near the line; and there will be fewer parallels and crossovers.

Tentative Conclusions

1. Voltage-corona-loss runs are less significant for comparing conductors than voltage tests of long duration. The all-weather corona loss of conductors is believed most significant in studying the economic value of corona loss on a high-voltage line.

2. Corona loss is a function of the combined effect of surface and weather factors. New conductors generally produce more corona than aged conductors.

3. The radio-interference problem involves the determination of radio-influence factors to be expected on a line of specific design, and the evaluation of their effects on communication services. Much work has been done on both phases of the problem. Serious consideration must be given to this problem in the design of extra-high-voltage lines.

4. While some definite values on corona loss and relative figures on radio influence are presented in this paper, they must be considered tentative. A series of reports to be presented in the near future will give complete information on both corona loss and radio influence, and on the application of these data to extra-high-voltage transmission line design.

6. PLANNING EHV TRANSMISSION†

ALTHOUGH the basic need for higher-voltage transmission has been discussed previously in general, it may be pertinent to summarize the particular case presented by a power system such as that of American Gas and Electric Company. Because of the availability of coal and water resources and unusual flexibility in location of its generation facilities, its problem involved the transmission of very large blocks of power over only moderate distances.

Growth in system peak from approximately 1,000,000 kw in 1940 to 2,250,000 kw by 1950, with an expected doubling of that peak to 4,500,000 kw within the next 12 years, has brought sharply into focus the need of larger and technically improved transmission facilities to keep unit costs down. This need is sharpened by the introduction of larger and larger units at steam-electric generating stations to the point where units of 200 Mw net capability are in process of being installed on the system.

The need for greater transmission capacity is brought about by the following factors:

1. The necessity of carrying larger blocks of power to load centers even though the short distances involved preclude classing the lines as long-distance transmission.

2. The necessity of shifting generation from one plant to another for emergency and economy reasons.

3. The required ability to transmit power from points of overbalance between generation and load to points of deficiency during relatively brief but frequently occuring periods as a result of expansion and construction in relatively large blocks at successively different points on the system.

4. The need to restore flexibility in the basic transmission system which to a considerable degree has been lost due to system loads and generating units outgrowing the present 138-kv system.

5. The fact that expansion of transmission capability at the present 138-kv level is becoming more difficult because of the multiplicity of rights-of-way required. These difficulties, almost certainly, will grow into almost insurmountable obstacles as densities of load and population increase, and particularly as relatively sparsely settled areas in which centers of transmission were originally erected become more thickly populated.

6. The obligation to provide for the needs of national defense.

7. The pressing, almost imperative, need to counterbalance in part or in whole the natural increase in cost of transmission at a given voltage as the system and areas served grow and develop; also because of the abnormal increases in cost of basic materials, labor, and equipment.

† AIEE Winter General Meeting (with E. L. Peterson, I. W. Gross and H. P. St. Clair), New York, N.Y., January 22, 1951.

Selection of Voltage Level

With the superposition of higher-voltage transmission inescapable, the central question for research for the past several years, has been what level to use, since the final selection of this voltage represents an approximate balance between such opposing factors as technical difficulties, economic considerations, and practical performance.

Since the present voltage (138 kv) has taken care of system growth for more than 30 years, it seems reasonable to expect that the new voltage level should be capable of meeting internal system requirements for at least as long, based upon the best load predictions possible at the present time.

In addition to adequacy for internal system requirements, consideration was also given to interconnection capacity between systems and areas. As neighboring systems grow in total load and size of individual generating units, it is logical that capacity of interconnections between systems should be increased on a similar scale. Adequate capacity in this respect is not only important for emergency arrangements and staggering of generating capacity but also for national defense.

Sleet Melting.—Other considerations in selecting the voltage level had to do with keeping lines in service. One is the sleet problem which is very serious in a large part of the territory covered by the system. The solution of this problem by outright melting of sleet on 138-kv circuits has been developed and used successfully on repeated occasions. Based upon this experience and weather records, it has been concluded that the design of the new high-voltage system must be such as to permit sleet melting. Obviously, since both the conductor diameter and conductivity necessarily go up as the voltage level is increased, the sleet-melting problem will become more and more difficult and may constitute a practical limit to the voltage level selected.

Hot Line Maintenance.—Another important consideration in keeping lines in service is the use of hot-line maintenance procedure. On the present 138-kv system this is considered a necessary routine. While it may not be feasible to retain this practice if and when much higher voltages are used, it is believed entirely practicable to use hot-line maintenance in the range of voltages under consideration for this project.

Double vs. Single Circuit.—In addition are the practical questions and economic implications of double-circuit versus single-circuit lines. While it is not necessarily a foregone conclusion that 345-kv or even 360-kv double-circuit lines could not be practical, a double-circuit line at these voltages would, in effect, tie up excessively large transmission capability on one right-of-way. For example, on 100- to 150-mile sections this capacity would approach 1,000,000 kw at 360 kv. On the other hand, a 315-kv double-circuit line appears to offer feasibility and advantages of sleet-melting and hot-line maintenance, and does not concentrate so much capacity in any one route that the voltage would have to be rejected for that reason.

Corona and radio influence were early recognized as factors to by considered in choosing a satisfactory transmission voltage; steps were taken over three years ago to investigate these factors by extensive field research. Basically, corona produces two adverse effects: energy loss which must be kept within economic limits; and radio influence in areas adjacent to the line which must be held within tolerable limits.

Both corona and radio influence increase rapidly as line voltage is raised for any size, type, and arrangement of conductors within practical limits.

In determining a final voltage level for the new transmission system, it was necessary to follow through each of the general considerations mentioned with quantitive evaluation of all important factors. The entire project hinged on these evaluations.

Power-flow Studied.—While estimated future loads constitute a quantitative factor in the determination of the best transmission voltage, evaluation of this factor does not permit a simple direct solution on an interconnected system where line-loading requirements arise from a variety of conditions. Rather extensive network-analyzer studies of the performance of the future system were made as plans were developed, using voltage levels from 287 kv to 360 kv and higher. For each voltage, critical system conditions such as generator unit outages and interconnection loadings were varied to test the capability of various transmission elements under the heaviest assumed loads. Since these studies were carried out over a period of several years, important economic developments and concepts occurred which required basic changes from time to time in the future system plan—changes in plant locations, size of units, and the transmission and switching layout. From all of these studies it was possible to judge fairly well the adequacy of a given voltage to handle future load requirements, even though quantitative exactness was not obtainable.

Although the overall length of the high-voltage system will be quite large, the effective transmission distance involved in any one line will be moderate. Distances vary from a maximum of about 175 miles down to 50 miles or less. The load-carrying capability per circuit of these lines will go from around 1.5 times surge-impedance loading for the longest line, up to at least 2.5 times for the shortest. At 287 kv this gives 300 to 500 Mw whereas at 345 kv it would be 450 to 750 Mw. Hence the capability limit of the 175-mile line would be stability and the ability to hold voltage and supply the reactive loss in the line (approximately 125,000 reactive kva at 345 kv and 450 Mw). On the other hand the limit of the 50-mile line may be a matter of thermal capacity of conductors and reactive loss. For example, the line currents for the 50-mile loading with those limits would be about 1000 amp and 1250 amp, respectively, for 287 kv and 345 kv.

The conclusion drawn from the power-flow studies was that adequate transmission capacity was provided for the loads assumed by any of the voltages tried, ranging from 287 kv to 360 kv. The main difference was that 287 kv required double circuits on some lines, whereas single circuits were adequate throughout with 345 kv or 360 kv. This brought out the fact that the capability of a 287-kv double-circuit line exceeds that of a 345-kv single circuit by 38% and 360 kv by 27%. As the studies progressed with ever-increasing goals of future loads, the possibilities of double-circuit lines became more and more attractive.

Concurrent with power-flow studies of the future system, economic studies and investigations of corona and radio influence were carried on intensively; the observations particularly on radio influence, played a very important if not decisive part in the final selection of voltage level. Corona losses, as shown later, definitely were not a determining factor in selecting the system voltage.

Radio Interference.—To facilitate evaluation of the radio-influence factor, a summary of representative data accumulated from the 500-kv test project, from field tests on operating lines, and from laboratory tests, is shown in Fig. 1.

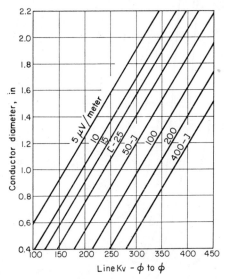

FIG. 1. Typical radio influence in fair weather for various line-to-line voltages and conductor diameters.

It was immediately evident that an excessively large and unwieldy single conductor would be necessary with a voltage level as high as 360 kv, or even 345 kv, if radio influence were to be maintained at a reasonably low level. Bundle conductors were considered, but ruled out rather early because of complications and expense introduced in stringing, sleet melting, hot-line maintenance, high tower costs, and so forth.

The result tended to discourage the adoption of the higher voltage single-circuit idea and to favor more than ever the use of double-circuit lines at a more moderate voltage.

With the upper limit of voltage fairly well established further examination of sleet melting, hot-line maintenance, and other factors led to the conclusion that the new transmission lines should be double-circuit construction rated tentatively at 300 kv (315 kv maximum), and using a 1.6-in. diameter conduc-

20a   VEP

tor; all other design features were to be suitable for a 315-kv nominal, 330-kv maximum rating. All insulation requirements, including line units and transformer basic impulse insulation levels have been based on the 315–330-kv level.

Corona Loss Not Determining.—While the final decision on voltage level was based largely on radio influence and other considerations previously mentioned, corona losses and their economics also were carefully analyzed. Obviously, any economic evaluation to be significant must be based on annual integrated energy losses under all types of weather, since stormy weather losses are many times greater than fair weather. Test site data, Fig. 2, show typical

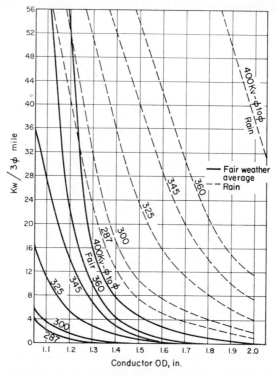

FIG. 2. Typical corona loss at various conductor diameters and line voltages in fair weather and rain.

corona losses to be expected for conductor sizes from 1.1 to 2.0 in. in diameter, and for line voltages from 287 to 400 kv line-to-line, under fair weather and rain. Similar typical data under apparent fair weather, all weather, stormy, and maximum stormy weather are given in Fig. 3 for the voltage range 287 to 315 kv, and for conductor diameters from 1.3 to 2.0 in. The large increase in corona loss as the conductor diameter is decreased and the voltage increased is evident.

With a conductor diameter of 1.6 in. and a line voltage of 315 kv, the maximum apparent fair-weather loss is 0.2 kv, the all weather loss 1.8 kv, the stormy weather loss 10 kv, and the maximum stormy weather loss 69 kw per 3-phase mile of line.

While widespread storm conditions involving considerable line mileage could produce corona loss that would increase the system demand, it does not appear that the increase would be significant or burdensome with a 1.6-in., or even a somewhat smaller conductor.

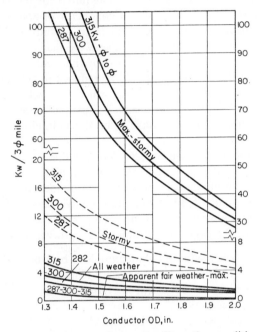

FIG. 3. Typical corona loss under all weather conditions.

On a 315-kv line an increase in conductor diameter from 1.6 to 1.65 in. shows the annual savings in corona loss (using energy at $0.0025 per kwhr and annual charges at 16.25%) would justify an increased investment of only $94 per mile. Since this amount would be far less than the actual cost of installing the increased conductor diameter, it is obvious that the annual integrated corona loss is not a determining economic factor.

Various System Features

From the very beginning of the project a great deal of study was given to all features of the high-voltage system, including line and transformer insulation, tower design, switching layouts, conductor conductivity and relaying. However, it was impossible to bring most of these to final decision until the voltage level had been decided.

Conductors.—While the conductor diameter of 1.6 in. was determined principally by radio-influence requirements, the conductivity required is based upon at least three considerations: thermal requirements to carry the maximum load;

20a*

economics of resistance losses, considering both capacity and annual energy losses; and feasibility of melting sleet.

As pointed out previously, the shorter line sections should be capable of carrying not less than $2\frac{1}{2}$ times surge-impedance loading, which at 315 kv would be 625,000 kw, or a current of approximately 1150 amp. However a study of loss economics, based upon reasonable line-loading assumptions, indicated that substantially higher conductivity than that required for the 1150-amp thermal limit would be justified. At this point, a conductor of 850,000-cir mil copper equivalent was seriously considered, subject to a study of the sleet-melting problem. Extensive network-analyzer studies were made in addition to actual ice-melting tests on different-size conductors by the conductor manufacturer. It was concluded that the conductor should be reduced to 800,000-cir mil copper equivalent to aid successful sleet melting, which it is believed can be accomplished in spite of some difficult problems. It is planned to apply 138-kv sources directly to 315-kv lines which will have to be short circuited at the far ends of 60 to 75-mile sections.

The final conductor design will be 1,269,300-cir mil steel-reinforced aluminum cable, or 798,300-cir mil copper equivalent, 1.6-in. diameter, and 1883.6 lb per thousand feet. The 1.6-in. diameter was obtained by incorporating between the aluminum and steel core a filler consisting of two layers of twisted, treated paper twine and two aluminum wires in each paper layer to increase the compression strength of the cable. The ultimate strength will be 41,970 lb.

Overvoltages.—A knowledge of the magnitude of both 60-cycle and switching-surge overvoltages on the system is necessary if insulation is to be applied on the most economical basis. Transient-analyzer studies and calculations were made to determine these values.

The interconnected network involved in the system does not present the severe 60-cycle over voltage possibilities inherent in systems with long one-way bulk-power transmission lines, particularly where overspeeding of hydro-generators must be added to other causes of overvoltage. Studies on the proposed high-voltage system indicated no more than 18% 60-cycle overvoltage above a maximum operating level of 300 kv.

Switching surges, or transient overvoltages accompanying one or more restrikes in a circuit breaker during the interruption of line charging current, were a matter of some concern. Transient-analyzer studies were made to simulate the most severe switching operations that can take place, including line sections of considerable length switched on high or low side of transformers, and with simulated restrikes from 0 to 3.

Although an essentially non-restriking type of circuit breaker may be used, it was felt that it would be wise to allow for one restrike, which would mean a switching surge transient of 2.8 times normal line-to-ground voltage.

Line Insulation.—Three types of overvoltages must be considered in determining line-insulation requirements: nominal frequency or 60-cycle voltages under both normal operating and line-fault conditions; overvoltages of short duration due to switching surges; and overvoltages due to lightning against which it may not always be possible to insulate completely even with a well shielded line and tower-footing resistances. A fundamental engineering approach was made to the 60-cycle and switching-surge problems rather than the somewhat arbitrary and rule-of-thumb method of supplying insulation on a three to five times normal line-to-ground voltage basis.

60-cycle Requirements.—As the 300/315-kv system will be solidly grounded the 60-cycle requirements will be based on the rated line-to-ground voltage. During system faults this voltage will increase momentarily until the fault is cleared. Calculations on the ultimate system indicated this might increase to 1.18 times normal. A rounded figure of 1.20 was used in this study.

Flashover Frequency.—Assuming eighteen $5\frac{3}{4}$-in. insulator units, tower-footing resistances not to exceed 20 ohms, and a ground-wire separation in midspan of 48 ft, the line lightning flashover per 100 miles of line per year has been determined for an isoceraunic level of 45 as 0.3, or one in three years. Most of such lightning flashovers, although infrequent, would not be expected to cause a service interruption since high-speed reclosing circuit breakers will be used, and the timing of the lightning on the 60-cycle wave will not always produce dynamic current.

Tower Design.—Tower design is based on a number of factors such as conductor-loading assumptions, height of conductors above the ground, angle of sideswing of the insulator strings, and clearance of the conductors to the tower members under maximum assumed insulator-string deflection.

Studies on system load requirements resulted in a decision to utilize double-circuit towers on the first link of the high-voltage network.

One of the prime considerations in mechanical line design is the choosing of the conductor tension. On our first line, which traverses rugged country, a value of 13,500 lb was chosen for a loading condition of $\frac{1}{2}$-in. ice and 8-lb wind. This load is only 32.5% of the tensile strength of the cable, but was chosen to give the most economical tower weight per mile of line for this particular topography where advantage can be taken of the rolling nature of the profile without resorting to high conductor tension. In flat country, on the other hand, higher tensions are indicated.

Only a single ground wire will be used. This will be installed at the center of the tower and will be a standard steel-reinforced aluminum cable of 159,000 cir mils.

The minimum clearance from the bottom conductor to ground will be 35 ft. The minimum length of crossarms was determined by the assumed clearance of conductors to steel of 7 ft 9 in. under a sideswing deflection of 35 degrees. The vertical separation of the crossarms was determined by this same condition plus the clearance required for off-setting the crossarms. The middle arm extends 6 ft 6 in. beyond the top arm and 5 ft beyond the bottom arm. The offset is to protect against conductor contacts with dancing conductors or during sleet unloading.

The basic tower height is 148 ft over-all although it can be shortened 20 ft to take advantage of favorable topographical conditions in hilly country.

To reduce the weight of the towers it was decided to utilize silicon steel for the leg members. Silicon steel has a minimum yield point of 45,000 lb per sq in. against 33,000 for carbon steel; the consequent weight reduction more than offsets the higher cost of material. The suspension tower is designed for a normal span of 1100 ft with a 5-degree angle in the line or a tangent span of 1700 ft. The angle tower is designed for a normal span of 1100 ft with an angle in the line of 11 degree. The strain tower is designed for a normal span of 1100 ft with an angle in the line of 30 degree.

Line Hardware.—As cable diameters become larger vibration troubles tend to increase. On the first line of the high-voltage system both armor rods and vibration dampers will be used. Although our experience with Stockbridge dampers on some 5000 circuit miles of 132-kv line has shown that they are effective in preventing conductor breaks, the added precaution is taken of installing armor rods. These are desired partly to prevent conductor damage from flashovers of the insulator strings. The dampers installed at the ends of the armor rods are designed to form corona shields for the clamps holding the ends of the armor rods. The dampers are 29 in. long and weigh 25 lb.

Shields will be installed at the line end of the insulator strings to grade the voltage distribution over the insulators and also to shield the conductor clamps from corona.

Conventional-type suspension clamps will be used. The dead-end clamps will be of the hydraulically compressed type.

Insulation for Transformers.—As large savings in transformer costs are possible in the high-voltage range if insulation can be reduced safely below present-day standards, an intensive study of this possibility was made. As with line insulators, three types of voltage have to be considered: impulse or lightning, switching surge, and 60-cycle.

The impulse (or basic insulation level—BIL) requirements were studied first. As lightning arresters will be installed close to the transformers to limit lightning and surge voltages, the 60-cycle rating of the lightning arresters which would seal against dynamic voltage was determined, based on the maximum line-to-ground overvoltage under fault conditions, and 10% added as a factor of safety. The protective level of this lightning arrester (maximum voltage drop when discharging 5000 amp) was increased by 20%. This was the required BIL of the transformer. For the 315-kv rated system, the BIL so arrived at was 1100, based on present-day station-type valve lightning arresters. A reduction of about 5% in this BIL would seem justified if advantage were taken of the diversity factor in stacking a large number of separate lightning arrester units, all of which would not have the maximum protective level. This would give a 1050 BIL, the value now known as the 230-kv insulation class.

However, this 1050 BIL was not selected as further study indicated that the 60-cycle voltage test requirements inherently produced a higher BIL, namely 1125.

A transformer, as well as other insulation, must have a 60-cycle insulation strength suitable for continuous normal operation. On the basis of present-day transformer insulation characteristics, experience, and risks in setting the 60-cycle test at an impractical low figure, a 60-cycle 1-minute test of 500 kv was agreed upon as the lowest permissible. Since there is a pretty well defined ratio between 60-cycle test and impulse test in high voltage transformers today, this 500-kv 60-cycle test resulted in an inherent transformer impulse strength of 1125. Thus the required 60-cycle test, in effect, set the BIL at a value higher than required for protection by the lightning arrester. This in effect results in a protective margin of nearly 30%, instead of 20% between the lightning arrester level and the transformer BIL. This is in the right direction, since the lightning arrester cannot be located at the terminals of the transformers on a high-voltage system of this type because of physical dimensions of the lightning arrester and required electrical clearance; the added impulse margin may well be partially consumed in this "distance factor".

Insulation for the transformers has, therefore, been set at 500-kv (60-cycle 1-min tests) and BIL of 1125. These values were found to provide adequate insulation strength against switching-surge voltages permitted by the arrester.

Insulation of Station Equipment.—Insulation requirements within the station (except the transformers) have not yet been definitely decided. However, some of the preliminary thinking may be of interest. It is planned to continue the philosophy of having the main bus insulated higher than other equipment in the station. Clearances to ground and between phases will be influenced by studied electrical requirements, the expected use of grading or corona shields in many locations, and the provision for hot-line maintenance.

Relaying.—Preliminary studies indicate that the electrical characteristics of the high-voltage system approach sufficiently close to those of systems now in operation so that existing relaying methods and equipment can be applied with the same degree of success now being obtained on our lower-voltage systems. There will be some form of carrier relaying, which probably will be controlled by distance relays having phase-angle discrimination, and by conventional ground directional relays. Some other possibilities of new types of relaying also are being considered, but these investigations have not yet been carried far enough to indicate their practical value.

7. ECONOMICS OF BULK POWER TRANSMISSION†

UNDERLYING all other considerations in the design of a high-voltage transmission system to be superimposed upon an existing system, are the basic economics of power transmission at various voltages and distances. Long-distance bulk-power transmission is not one of the immediate requirements for the American Gas and Electric Company system. Nevertheless, with fairly heavy loadings over distances approaching 200 miles in the immediate future and undoubtedly still heavier transmission and perhaps longer distances later on, it was felt that a basic study of bulk-power transmission at various distances, voltages, and load factors would yield data of importance in our voltage-selection problem.

To cover the entire range of possible transmission distances, these studies were carried out for line lengths of 100, 200, 300, 400, and 600 miles at voltage levels of 230 kv, 287 kv, 315 kv, and 345 kv.

Essential to a consideration of transmission costs are the installed costs of lines and associated terminal equipment, such as transformers and switchgear; and the inherent load-carrying capability of transmission circuits in terms of voltage and distance. Conductor size has not been included as a factor since it is assumed that an adequate size will be used to take advantage of the inherent capability of any circuit of a given length. In other words, it is assumed that under given terminal conditions basic capability can be expressed in terms of voltage and distance alone.

The first step in our economic study, therefore, was to set up practical limits of power-carrying capability of transmission lines and to develop the basic

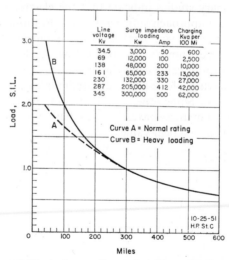

FIG. 1. Basic capability in terms of surge-impedance loading (2.5 kv²).

† Conférence Internationale des Grands Réseaux Electriques à Haute Tension (CIGRE). (with H. P. St. Clair and E. L. Peterson), Paris, France, May 28, 1952.

capability curves shown in Fig. 1. These loading curves, which are expressed
in terms of surge-impedance loading, are not the result of any single formula;
they are based on empirical determination, including extensive network-analyzer studies and a review of actual operating experience on our own and other
systems. In the upper range of distances, 300 miles and above, the load values
shown are limited by stability considerations, assuming practical values of
terminal impedance at the two ends of the lines without the use of series-
capacitor or other compensation. Below 300 miles it will be noted that the
curve is divided into two parts, the lower part (Curve A) representing normal
or conservative loading, and the upper part (Curve B) representing a heavier
degree of loading, which is entirely permissible if conductor sizes are adequate
and if sufficient reactive-kva sources are available to supply the higher reactive
losses. Curve A loadings were considered normal for design purposes, and the
economic studies described later were based mainly upon these values. The
entire curve, omitting the "B" portion, follows quite closely the conclusions
worked out and published by Crary.

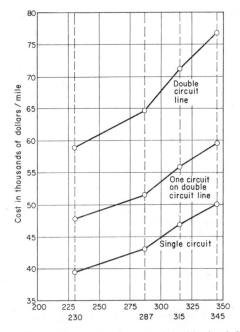

FIG. 2. Estimated costs for single- and double-circuit lines.

Installed costs of transmission lines, transformers, and switchgear were
based upon general price levels as of September 1950. Transmission-line costs
were based upon a type of construction suitable for the American Gas and
Electric system territory, and upon conductor diameters selected in the light
of corona and radio-influence requirements. Conductivity of conductors was
determined by both thermal limits and economics of losses, tempered to some

extent by sleet-melting considerations. The principal design characteristics, including conductor sizes, insulator units and spacing used in the cost estimates for line construction at each of the four voltages levels used in this study are shown in the accompanying table. For the three lower voltages, both single and double-circuit construction was considered; only single circuit was believed practicable at 345 kv. Estimated costs for these lines are shown in Fig. 2.

Installed costs for transformers in banks of three single-phase units and for circuit breakers are shown in Figs. 3 and 4, respectively.

Cost of intermediate sectionalizing stations, Fig. 5, was determined by a separate economic study based upon the capability remaining in any combination of two or more circuits in parallel following an outage of one circuit, or of any one line section if sectionalizing stations are installed.

These studies were carried out in terms of investment per kw/mile for each distance and voltage and for combinations of 2, 4 and 6 circuits. For the three lower voltages, including 315 kv, double-circuit construction was assumed, whereas for 345 kv the studies were based on single-circuit construction only. The reason for excluding double-circuit construction at 345 kv was not so much a belief that double-circuit towers at this voltage would be impract-

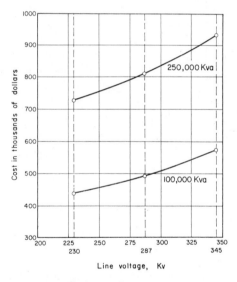

FIG. 3. Installed cost of stepup transformers.

icable, as that the amount of capacity tied up would be excessive. For example, at distances of 100 miles or less a 345-kv double-circuit line would have a capability of the order of 1 million kw, which was judged to be excessive as compared with 600,000 to 750,000 kw at 315 kv.

Typical results of this study showing transmission cost in dollars per kw/mile are shown in Fig. 6 for a two-circuit line. As would be expected, a definitely decreasing cost is indicated with rising voltage up to 315 kv but a pronounced

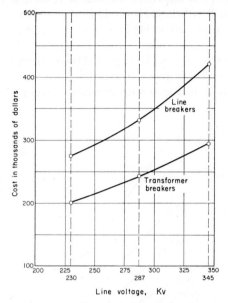

FIG. 4. Installed cost of circuit breakers.

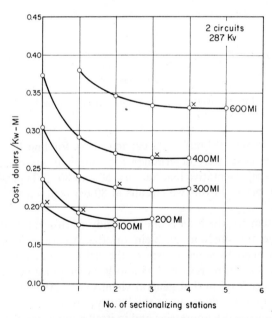

FIG. 5. Installed cost of intermediate sectionalizing station (points chosen for study marked "x").

rise in cost at 345 kv, due largely to the change from double-circuit to single-circuit construction. It is believed entirely probable that the downward trend in cost would have continued to 345 kv if double-circuit construction had been used there also.

Transmission costs in miles per kwhr probably constitute the ultimate measure of the economics of electric power transmission. Basing investment costs on results similar to those in Fig. 6 with carrying charges on the investment at

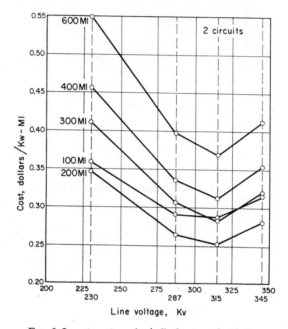

FIG. 6. Investment per kw/mile for two-circuit line.

13.75% energy losses at 3 mills and capacity cost for line losses at $125 per kw, transmission costs were determined for combinations of 2, 4, and 6 circuits and at two representative load factors, 50 and 80%. Typical results showing costs in mills per kwhr for two circuits at 80 per cent load factor are shown in Fig. 7.

While the extent to which an economic study of this kind can be applied to the high-voltage transmission problem on the American Gas and Electric system may be open to some question, the definite indication of maximum economy at 315 kv can be regarded as another consideration favorable to the choice of a voltage level in that general vicinity. Also, the economic advantage of double-circuit construction up to the limit where such construction may be considered feasible is very impressive and in itself constitutes a strong argument for adopting a voltage level at which double-circuit construction can be used safely.

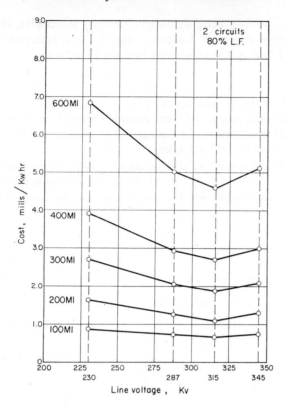

FIG. 7. Transmission cost per kwhr for two circuits at 80% load factor.

Final Selection of Voltage.—The various factors bearing upon the selection of a system voltage level for AGE's high-voltage system were: corona loss, radio influence, sleet-melting, hot-line maintenance and economics.

Extensive study of each of these factors indicated a very remarkable convergence upon the selection of a voltage level in the vicinity of 315 kv. For this reason it was felt that detailed design of the various elements of the system,

TRANSMISSION-LINE PHYSICAL DESIGN CHARACTERISTICS

Voltage	230 kv	287 kv	315 kv	345 kv
Conductor, steel-reinforced aluminum cable	874,500	1,058,900	1,272,000	1,351,500
Conductor, copper equivalent	550,000	666,000	800,000	850,000
Conductor diameter (in.)	1.1	1.4	1.6	1.75
Insulator units	14	16	18	20
Single-circuit phase spacing (ft)	25.5	27.5	29.5	30.5
Double-circuit phase spacing (ft)	20.5	21.5	23.8	

including lines, transformers, lightning arresters, switchgear, etc., could be undertaken with a degree of assurance that the final selection would be as nearly a correct one as it is reasonably possible to achieve.

Conclusions

A basic study of the economics of high-capacity, high-voltage transmission, coupled with considerations of sleet melting, hot-line maintenance, corona and radio influence indicated the desirability of limiting the new transmission system voltage level to 315 or 330 kv.

The favorable economics of double-circuit construction, coupled with the limitation of voltage level at 315 or 330 kv, led to a definite decision to adopt double-circuit construction for the major links in the proposed extra-high-voltage system.

8. TWIN 345-kv LINE CARRIES MILLION kw†

A RECENT experience in the operation of Ohio Valley Electric Corp (OVEC) during which a 1,000,000 kw load was carried on one double-circuit 345-kv line for close to a week marks a new milestone in magnitude and effectiveness of high-voltage, high-power transmission.

OVEC was sponsored and organized by 15 private electric companies in Indiana, Kentucky, Ohio, Pennsylvania, and West Virginia for the express purpose of supplying the power requirements of the Portsmouth, Ohio, uranium diffusion plant of the Atomic Energy Commission. Designed to carry the 1.8 million-kw initial contract load of the project the OVEC system comprises the Clifty Creek and Kyger Creek generation stations together with some 390 line-miles of double-circuit, extra-high-voltage transmission lines, as shown.

Four double-circuit lines, two from each generating station situated west and east of the project, are capable of handling the entire 1.8 million-kw load with any two single-circuit line sections out of service simultaneously. The actual load has been running in excess of 1.8 million kw.

During the recent tornado in Ohio, two towers on one double-circuit line from Kyger Creek were wrecked, leaving only one line in service. Automatic relays dropped a small portion of AEC load until conditions were stabilized. Then full load was recovered and carried.

During the close-to-a-week's outage of the damaged line, the project was maintained at full load without any curtailment, and for the greater portion of this time the remaining line from Kyger Creek to the project carried in excess of one million kw.

Although this loading is well within the predicted capability of the circuits and the outage of any two circuits simultaneously was a specific design criterion, nevertheless this is believed to be the first time in the history of electric power transmission that more than one million kw has been carried on one double-circuit line. Carrying of this load was facilitated in part by raising the voltage level from about 350 kv to 356 kv at the Clifty Creek bus and from 340 to 346 kv at Kyger Creek. While these adjustments were sufficient to hold voltage at the load appreciably above minimum operating levels, an additional boost, or safety margin, was provided by utilizing the reactive generating capacity of synchronous condensers at the load. No difficulty was experienced in operating at the higher voltage levels at either plant. In fact these higher levels have remained in effect since restoration of the damaged line.

From a single-circuit standpoint an even more spectacular loading was carried for a short time only a few hours after the damaged line was returned to service. During about two minutes when three of the four circuits feeding the AEC project from Pierce switching station to the west were out of service simultaneously because of storm conditions, the one remaining circuit carried approximately 725,000 kw.

† *Electrical World*, April 30, 1956.

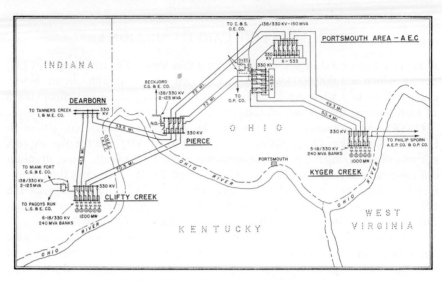

Double-circuit 345-kv system serving 1800 Mw atomic energy load.

In both cases high-speed relaying of these extra-high-voltage circuits in conjunction with 3-cycle circuit breakers provided a fault-clearing time of only four cycles, and proved its effectiveness in maintaining continuity of a normally rather critical load.

In the first case, even with the extreme severity of a double-circuit, 3-phase fault fairly close to the project, the tripping of the faulted line was fast enough; coupled with automatic tripping of a small portion of load, to prevent the loss of the main project load. For a line length in this case of approximately 50 miles, the loading of 1,000,000 kw on one line or 500,000 kw per circuit, representing about 1.7 times surge-impedance loading at 330 kv, is considerably below stability limits. The extreme severity of this occurrence lay in the double-circuit solid 3-phase fault which initiated the trip-out.

On the other hand, the 725,000-kw load carried by the single-circuit line in the second instance at a delivered voltage of approximately 320 kv over a 70-mile line is equivalent to about 2.6 times surge-impedance loading at that voltage and is not far below practical stability limits for that distance. It may also be of interest to learn that this load represented an actual line current of approximately 1300 amp.

These demonstrated capabilities bring sharply into focus the capacity, flexibility, and inherent ability of extra-high-voltage networks to reduce transmission costs for the clearly indicated coming of mass generation of electric energy. If in some cases this will be delayed a bit, no one can doubt that its eventual coming is sure wherever there is a dynamic complex consisting of population, industry, growth, and belief in the future.

9. SOME OPERATING ASPECTS OF 330-kv LINES

WHEN the first 330-kv line on the American Gas and Electric system was energized in October 1953 the radio-influence measurements were somewhat higher than had been predicted by the Tidd tests. Examination of the line indicated that the corona bursts on the conductor surfaces were due to several causes. Some of the saturant used in the paper filler had seeped to the cable surface, producing areas of high-voltage stress. Also, during stringing operations, the conductors had picked up some dirt and abrasions.

On lines where conductor stringing was not yet under way, the following steps were taken to remedy this situation:

The amount of saturant in the paper filler of the conductor was reduced;

The conductor diameter was increased from 1.6 in. to 1.75 in., thereby reducing the average voltage stress at the conductor surfaces;

To minimize the amount of dirt and abrasions picked up during erection, stringing methods were modified to keep the conductors off the ground at all times by using wooden conductor supports at proper intervals, depending upon the terrain, and by tension stringing without the use of intermediate conductor supports;

Motor-driven nylon brushes were constructed and used at the erection sites to remove any loose particles of metal or dirt which might have collected on the conductor during the shipping operation;

Special attention was given to the various items of hardware to eliminate sharp corners and projections which would induce corona.

While it has been extremely difficult to determine the individual effect of these corrective measures on line radio-influence levels, the cumulative effect has made it possible to operate sections of the system at voltages up to 345 kv without encountering disturbing radio-influence levels.

Radio-influence Studies

After the initial 330-kv lines had been placed in service, it became possible, for the first time, to make detailed investigations on operating lines of the various factors in the radio-influence problem. Brief discussions of these investigations follow:

Basic Problems.—The level of radio influence produced by an operating line is roughly inversely proportional to frequency, as determined by the characteristics of the surges produced by corona bursts. Therefore, the broadcast band of 540 to 1600 kc is of major interest in the study of radio influence in the United States. Interference to the reception of frequency-modulation broadcasts in the range between 88 and 108 Mc is negligible, due to the low line level at these frequencies and to the inherent noise-limiting characteristics of the FM system. Television transmitted in the bands between 54 to 88 Mc, and 174 to 216 Mc is not a serious problem.

† Conférence Internationale des Grands Réseaux Electriques à Haute Tension (CIGRE (with W. S. Price), Paris, France, May 30, 1956 .

Satisfactory reception by radio users along transmission lines depends upon the ratio of broadcast signal strength to line noise at the antenna location. Thus, the problems arise in areas which are remote from the transmitters or where the user's antenna is unusually close to the line.

Instruments.—It is not too difficult to measure the field strength of a broadcast-station carrier to the order of 10% accuracy but the measurement of noise is much more complicated since it involves integration over a specified pass band. For accurate noise response, the selectivity characteristic of the receiver must conform to a pre-determined shape, and the time constants of the detector and indicating instruments must be accurately controlled. Thus, with the greatest possible care to maintain and calibrate instruments, the accuracy is probably not better than 20%.

All RI measurements were made with Ferris 32A meters, with a nominal band width of 10 kc and 1 Mc, and detector circuit charge and discharge time constants of 10 and 600 ms, respectively. Since the 540 to 1600-kc broadcast band is of principal interest in studying radio interference, most of the line level readings were made at 1 Mc.

Line Level Measurements.—Determination of a representative RI level of an operating line is made very difficult by the many factors which influence the reading of a noise meter near a line, such as RI variations with line height,

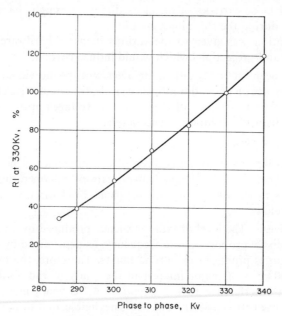

FIG. 1. Voltage correction factor.

line voltage, background ambient noise levels, and weather. Line levels were usually measured by a combination of under-the-line readings, corrected to standard line height, and by lateral profiles showing the attenuation of field strength along a path perpendicular to the line. All readings were normally corrected to a standard line voltage for comparison with other readings taken at different line voltages.

A voltage correction factor, shown in Fig. 1, was obtained by connecting a section of line to a separate generator and reading RI level under the line for voltages between 285 and 340 kv. This curve represents the average of runs made on three separate days, all in fair weather. For this line, the RI varies as the 7th power of the voltage making the RI level double with each 10% increase in voltage.

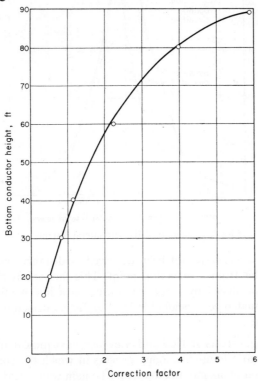

Fig. 2. Height correction factor.

The height correction factor, shown in Fig. 2, was obtained by lowering the bottom conductor on one span of line to within 15 ft of the ground, and noting the increase in meter reading under the line. Although the height of the middle and top conductors was not varied on this test, the resultant correction factor was reasonably good.

The reduction in RI with lateral distance from a 330-kv line is indicated in Fig. 3 which shows that only 9% of the RI under the lines is obtained 150 ft from it.

Standing Waves.—Standing waves of RF voltage, appearing near reflection points such as substations, caused variations in line level of up to 6 to 1 between two points on the line one-quarter wave-length apart (250 ft at 1 Mc). In such areas it was necessary to average readings along the line to obtain a representative line level.

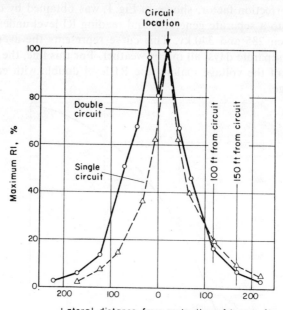

Fig. 3. Reduction of RI with lateral distance.

Such standing waves have not been observed at distances greater than approximately 5 miles from a reflection point. This fact seems to indicate that the noise signal produced by a corona source on a line would, in traveling 10 miles, be attenuated to a negligible level as far as detection and measurement are concerned.

Conductor Aging.—Tests at Tidd and elsewhere have indicated that energized conductors undergo a long-term aging process in which the corona losses decrease over a period of six months to a year. To gain some ideas as to the effect of this phenomenon on radio-influence levels, a series of measurements have been made on each of the 330-kv circuits of the Ohio Valley Electric Corporation, beginning at the time the lines were first energized. Periodic readings were taken at three locations at least 10 miles from the nearest substation to avoid standing waves. The curves resulting from these observations are not smooth, even for the fair weather values. However, by averaging a great many observations, and correcting for line-voltage variation, an indication has been obtained that the RI levels diminish with age. These measurements will be continued until the extent of the aging effect is definitely verified.

Lightning Performance

Eighteen 5¾-in. insulators in the 330-kv lines provide a critical impulse flashover level of approximately 1600 kv. The single ground wire provides a shielding angle of 35 degrees. It was expected that, with such insulation and shielding, the 330-kv lines would be relatively immune to lightning flashovers, provided the tower-footing resistances were reasonably low.

However, the number of flashovers due to lightning has exceeded expectations. These 330-kv flashovers have not caused any operating difficulties such as loss of load or instability. Furthermore, no such difficulties are expected, since the use of ultra-high-speed reclosing circuit breakers, first developed for the AGE system some 20 years ago, has been extended to the 330-kv system.

The discrepancy between predicted and actual lightning performance of the 330-kv lines has, however, prompted a re-examination of the entire mechanism of transmission line flashovers. A critical evaluation of the performance data has shown that the concept of voltage drop in the tower-footing resistance does not account for the increased flashovers since most of them have occurred at towers with footing resistance of only a few ohms. Furthermore, it does not appear that a large percentage of the flashovers have been caused by shielding failures. One concept which is being examined in this connection is the idea that lightning current flowing through the surge impedance of the tower will produce flashover voltages in the initial fraction of a microsecond before multiple reflections in the tower eliminate this voltage.

Line Maintenance

Since 1928, it has been the practice to do much maintenance work in 132-kv and lower-voltage lines with the lines in service. This work includes replacement of line hardware and insulators damaged by lightning or rifle fire.

With the development of the 330-kv system it became apparent that these maintenance techniques would have to be extended to this system, since the great importance of these lines to the operation of the system would preclude their removal from service for routine maintenance. Accordingly, line hardware and special tools and techniques have been developed to permit maintenance of 330 kv energized lines. Insulators have been successfully replaced using these tools and techniques.

Sleet Control

Sleet melting was necessarily a factor in the selection of conductors for 330 kv. Calculations and tests indicate that, for the conductors chosen (1.6-in. dia ACSR), currents of 1200 to 1800 amp are required to remove sleet from the lines in a reasonable length of time. The currents are obtained by applying 132 kv to 330-kv line sections, 55 to 85 miles in length, short circuited at the remote terminal. A sleet melting current of 1200 amp at 132 kv required approximately 26,000 kw and 275,000 kvar; 1800 amp required 39,000 kw and 410,000 kvar. Any 132-kv bus used as a source for melting sleet must be capable of supplying this reactive excessive voltage drop.

The reactive supply necessary for melting sleet may not be available at all locations on the system, making it necessary to apply lower currents, between 750 and 1,200 amp prior to the formation of sleet. These lower currents are obtained by applying 132 kv to short-circuited line sections between 85 and 140 miles in length.

Both sleet-melting and sleet-prevention procedures were successfully applied to the 330-kv systems during the winter months of 1954–1955.

Summary

Somewhat higher RI level of the first 330-kv line to be energized than predicted was attributed to corona produced by dirt, abrasions, excess conductor saturant on the conductor and hardware surfaces which was improved by modified stringing methods and less conductor saturant.

On 330-kv double-circuit lines, a phase sequence of 1, 2, 3, top to bottom, on both circuits, produced an appreciably lower RI level than other phase arrangements.

Extensive tests indicated that radio influence coupled to adjacent telephone and distribution circuits would not cause significant interference to radio reception.

RI levels on 330-kv lines have decreased over a period of 6 to 12 months, apparently due to aging of conductors.

Lightning outage rates have been somewhat greater than expected, considering the relatively higher insulation levels of these lines, but ultra-high-speed reclosing circuit breakers have prevented any adverse system effects.

Both the radio-influence and lightning performance problems are now being studied further by theoretical, laboratory, and field investigations. It is expected that these investigations will provide the basis for material advances toward the solution of these problems.

10. EXTRA-HIGH VOLTAGE SLOW IN GETTING STARTED†

EXTRA-HIGH VOLTAGE transmission at 230 kv and above, either in actual operation or advanced planning stages, has become a reality in many parts of the world. A roster of countries where this is true would include the United States of American, Sweden, France, England, Germany, Australia, Japan, Finland and Russia.

While this shows a substantial development, extra-high voltage has been slower in getting started than one would have expected, considering the tremendous expansion that has taken place over the past two decades in growth of power systems and their loads, and in size of generation plants and their units.

The reason may be that transmission systems generally have greater flexibility and greater reserve capacity than generating systems. An even more important probable reason is the fact that at a given transmission voltage, particularly the higher levels up to 230 kv, total transmission costs may remain rather flat over a considerable range of loads. As a result, there is a natural tendency, which has not been fought too successfully, to postpone the large initial capital expenses involved in undertaking a new and higher transmission voltage.

However, if the growth curve of a system is projected far enough ahead and thoroughly enough so the results are really believed, the economic necessity for extra-high voltages becomes unmistakably clear because of eventual breakdown, economically, of existing voltage levels. Wherever this forward-looking and optimistic but sober projection has been made, I believe that records almost always show a decision to proceed with extra-high voltage.

Combined circuit mileage of 345-kv and 400-kv systems today is well over 3000 miles and many more hundreds of miles are under construction to be in service within a matter of months.

† American Power Conference, Chicago, Ill., March 27, 1957.

11. A PROPOSAL FOR INTERNATIONAL EHV STANDARDIZATION†

THROUGHOUT the world increasing needs for long-distance, high-capacity electric-power transmission are focusing much interest and study on extra-high-voltage problems, not the least of which is the choice of future transmission voltage levels. Much emphasis has been given to the desirability of limiting the number of future EHV levels and adopting internationally-standardized voltage steps. There is little doubt that such standardization would create positive gains in effective research and development, improved reliability and economy in manufacture of equipment, more extensive use of EHV transmission in general, and more extensive interconnection of EHV systems in particular.

The outstanding example of international standardization in high-voltage transmission is the 230-kv level which has had world-wide usage for many years. Above this level, however, standardization has not fared so well, particularly since 380/400 kv prevails in Europe, and 345 kv is established in the United States, Canada, and elsewhere. Present IEC Standards, now under revision by Technical Committee No. 30, include only the 230-kv, 287-kv and 400-kv levels, with the addition of 345 kv still under consideration.

This paper will discuss the basis for the 345-kv development and also will propose two additional EHV steps beyond 345 kv and 400 kv to serve both near and long-term future transmission requirements. The first step would meet early requirements for long-distance, high-capacity transmission in this country and Canada, while the second step would look ahead to long-term future needs. Together, these steps should provide a solid basis for international agreement on EHV standardization above the 345-kv and 400-kv levels.

Historical Consideration

The need for long-distance, high capacity transmission became urgent during the past fifteen years, particularly in Sweden where internal fuel resources were nearly non-existent and the largest undeveloped hydro sites were 500 to 600 miles from principal load centers. Extensive 230-kv transmission was in use there as well as in other countries in Europe and 380/420-kv transmission was developed primarily for superposition on 230-kv systems. In comparison with existing 230-kv lines, most of which were of single-conductor design, the 380-kv lines with twin conductors provided a capacity increase per circuit of a little less than 4 to 1. While this is substantial, an even greater increase over 230 kv would perhaps have been desirable as later studies have indicated.

In this country, however, 345 kv was developed initially, not for particularly long-distance transmission where 230 kv was already available, but for medium-distance requirements to reinforce an existing 138-kv system network.‡ For this purpose, 345 kv, in providing an increase of more than 6-to-1 in circuit capability above 138 kv, has proved to be a logical and economical choice with ample capacity for long-term growth. Furthermore, while 345-kv con-

† *AIEE Transactions* (with H. P. St. Clair), January 8, 1960.
‡ The American Gas and Electric system, now called American Electric Power system.

struction on this system to date has been principally on a double-circuit, single-conductor-per-phase basis, immediate expansion and probably much of the future construction will be on a single-circuit, twin-conductor basis for which the increase in capability over 138 kv will be nearly 8 to 1. Both single and twin-conductor 345-kv lines have been adopted by a number of other systems within the past few years and should be generally applicable to many more power systems now using 115 kv, 138 kv, or even 161 kv.

It is true that transmission at 287/300 kv is being used to some extent in this country and in Canada. But it does not appear likely that it will be adopted extensively for new projects in competition with either 230 kv or 345 kv. Sharply in contrast with this situation are the more than 2500 circuit-miles of 345-kv lines now in service in this country on three different power systems plus a number of new projects underway or under consideration at this same voltage on other systems. With this acceptance and usage, 345 kv undoubtedly will be included as an alternative voltage in the present IEC Standards. In the meantime, for this country and probably for Canada as well, 230 kv and 345 kv may be considered basic steps or starting points in any proposals for long-range EHV Standards.

Future EHV Standard Voltages

One of the first considerations in selecting a new and higher transmission voltage to superpose on an existing system is the necessity for a substantial increase in capability per circuit. The problem of securing right-of-way, capacity for long-term growth and maximum economy in unit transmission costs, all underscore this consideration.

A second consideration is the economic importance of limiting the number of voltage steps to be adopted and standardized. These should be limited to the maximum extent possible and still provide the necessary flexibility for existing and projected systems. Voltages now in standard use are separated by ratios considerably less than 2 to 1. In view of this it can be stated, as a general rule, that a one-step increase in voltage will be inadequate for superposition on an existing system. In other words, superposition should generally be on a "skip-step" basis, using a new voltage at least two steps higher than the existing voltage. Fig. 1 illustrates the application of this principle by the specific example of the 345-kv development in this country which for the most part has been superposed on 138-kv systems.

There will be exceptions where a one-step-higher voltage may be economically justified to supplement existing transmission. An example is the addition of 345-kv lines on the Bonneville 230-kv system. In general, however, the "skip-step" procedure must be applied to satisfy the first consideration of obtaining a substantial increase in capability per circuit. For most 230-kv systems, particularly where high-capacity, long-distance transmission is involved, the increase of less than 3 to 1 obtained with 345 kv on a twin-conductor basis would be considered inadequate.

500 kv. — While this question of selecting the next EHV step above 345 kv has been under general discussion for some time, the necessity for a firm decision has now become a more urgent matter in view of large transmission projects already in the active planning stage. Up to the present time, both power-system engineers and manufacturers have given a great deal of consideration to the adoption of 460 kv following the historical pattern of even multiples of 115 kv. From the standpoint of equipment manufacture, it has been assumed that this pattern of even multiples would offer the advantages of uniform "building-block" techniques. However, as discussed later in this paper, it is believed that the relative importance of this consideration may have been over-emphasized.

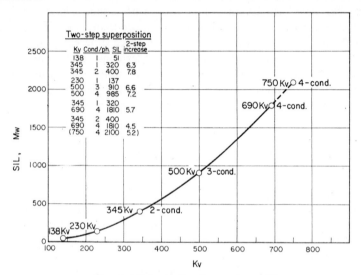

FIG. 1. "Skip-step" voltage superposition.

The important question is whether 460 kv goes far enough in providing an adequate increase in capability above preceding steps, particularly 230 kv. The gain in capacity over 230-kv single-conductor lines is 5 to 1 on a twin-conductor basis, and this can be increased to 6 to 1 by using a 4-conductor bundle. This is a substantial gain over the European 400-kv systems where the increase above 230-kv single-conductor lines is about 3.8 to 1 on a twin-conductor basis and better than 4 to 1 with triple conductors.

In making this comparison, however, it should be pointed out that while 380/400 kv was a remarkable pioneering undertaking when first placed in operation in Sweden in 1952, it is extremely doubtful in the light of today's knowledge and operating experience whether the same voltage would be adopted today for superposition on a 230-kv system. This conclusion is strengthened by subsequent studies made by Swedish engineers on the possibilities of converting 400-kv lines to 500 kv and the decision of the Russians to carry through just such a conversion. In England original plans for converting portions of their 275-kv lines to 380/400 kv are being questioned in favor of a 500-kv superposition.

For these reasons it would seem logical to regard the European 380/400-kv class and our 345-kv class as belonging in the same category and to consider that these two systems should converge on a single level at the next higher voltage. In determining what this level should be, we are fortunate that notwithstanding considerable interest in 460 kv, final commitments on actual installations at that level have not been made. There is still time to examine critically the relative merits of 460 kv as against a higher level such as 500 kv which we believe would be a wiser choice.

In growing interconnected power systems, right-of-way acquisition is becoming increasingly difficult and expensive. Therefore it is important to plan for maximum utilization of such rights-of-way to take care of long-term growth. While there is no magic in the 6-to-1 increase of capability which was obtained in superposing 345 kv over 138 kv, this increase appeared adequate to take care of long-term growth. In current planning we are now drawn most strongly toward twin-conductor, 345-kv, single-circuit construction which gives the still higher increase of 7.8 to 1. Following a similar pattern, 500 kv with three conductors per phase superposed on 230-kv systems provides capability increase of 6.6-to-1 and a 7.2-to-1 increase with four conductors per phase.

Looking to the future, the continuation of the historical pattern of using fixed 115-kv intervals offers the serious objection of a declining percentage change between steps, thereby increasing the number of steps. A departure from this pattern by adopting 500 kv not only would preserve a more appropriate ratio above the underlying 345-kv level, but would also help to minimize the number of EHV steps in the future by pushing the next step to 690 kv or possibly even higher.

Without question, the technical feasibility and economy of the design, construction and operation of 500-kv lines have been greatly augmented by the accumulation of operating experience, results of field and laboratory investigations, improvements in equipment and advances in the art that have been realized in the EHV field over the past several years. Some of the more recent results would suggest that previously considered 460-kv designs would be fully adequate, if not more than adequate, for 500-kv operation.

Benefits.—Because of the important bearing that this accumulated experience should have in the resolution of the problem of standardization of EHV transmission voltages, some of the benefits are listed below:

1. A more realistic evaluation of the radio-influence problem based upon actual experience has given increased confidence in our ability to cope with the problem successfully.

2. Substantial improvements in lightning-arrester protective margins have made possible further reduction in transformer insulation. Previous 1175-kv BIL for 345-kv transformers has now been further reduced to 1050 kv, representing a full 2-step reduction instead of the previous 1½-step, and resulting in further economies in transformer costs.

3. Completely satisfactory operation has been experienced during one lightning season on a 1-mile, 345-kv line section with insulation reduced from 18 to 14 units.

21*

4. Highly sucessful operating experience has been obtained with ultra-high-speed reclosing giving faster-than-expected over-all reclosing speeds.

5. The continuing accumulation of field measurements and laboratory research investigations is yielding an increased understanding of lightning-flashover phenomena.

Referring again to the possible advantages from the manufacturing stand-point of following the historical pattern of 115-kv intervals and staying with 460 kv, it is obvious that this consideration would not apply particularly to transformers since they are custom designed to the required ratios. As far as circuit breakers are concerned, it is believed that new developments and designs are sufficiently flexible to be adapted to 500 kv as well as to 460 kv. Referring to "building-block" techniques, it would appear that four 138-kv units could be used for 500 or even 525 kv in the same manner that four 115-kv units can be used for 460 kv. As for other equipment—lightning arresters, disconnecting switches, etc.—there would seem to be no particular advantage in the adop-tion of even multiples of 115 kv.

We may conclude, therefore, that the next EHV step, whether 460 kv or 500 kv, should be chosen primarily on the basis of such important considera-tions as circuit capability for long-term growth, simplification and limitation of future EVH standards; furthering the potential gains from international standardization and, of course, an over-all economic evaluation.

Specific Proposal for Standardization

Typical design characteristics, including phase configuration, conductor sizes for single and multiple-conductor lines, and resulting surge-impedance or "natural" loadings, are listed in the accompanying Table for line voltages 138 kv and above. In this table, phase spacings are perhaps lower than most current practice and thinking, but they are based on the assumption that some-what reduced insulation will become feasible in future designs. Conductor sizes, of course, will vary with individual requirements but the sizes shown are at least consistent with respective surge-impedance loadings and are be-lieved to be generally adequate for RI requirements.

A graphic representation of surge-impedance loadings over the entire range of voltages from 138 kv to 750 kv is shown in Fig. 1. As noted on this curve, two conductors were selected for 345 kv, three conductors for 500 kv, and four conductors per phase for 690 and 750 kv. Included in Fig. 1 is a table illustrating the "skip-step" superposition principle for combinations of vol-tages over the range shown on the curve.

The consistent relationship between the proposed voltage levels is even more clearly illustrated in Fig. 2, where loading capabilities per circuit over various distances from 50 to 500 miles are plotted on semi-logarithmic coordinates. It seems clear from the uniform spacings between these capability curves that two steps above 345 kv should be ample for presently conceivable require-ments at least for a very long period of time. While this graph shows 500 kv to be a logical choice for the next step above 345 kv, it also indicates, as pre-

TYPICAL DESIGN CHARACTERISTICS FOR EHV TRANSMISSION

Line voltage kv	Config-uration	Phase conductors					Surge-impedance loading	
		No.	Spacing	ACSR Mcm	Diam. inches	Cu equiv. Mcm	Mw	Amps
138	14' Vert.	1	—	556.5	0.953	350	51	212
230	22' Flat	1	—	954	1.196	600	137	344
345	25' Vert.	1	—	1414†	1.75	890	320	535
345	28' Flat	2	16'	954	1.196	1200	400	670
400	32' Flat	2	18"	1033.5	1.246	1300‡	530	765
400	32' Flat	3	18"	874.5	1.146	1350	595	860
400	32' Flat	4	18"	556.5	0.927	1400	640	925
500	36' Flat	2	18"	1272	1.382	1600	820	945
500	36' Flat	3	18"	954	1.196	1800	910	1050
500	36' Flat	4	18"	795	1.108	2000	985	1140
690	44' Flat	4	18"	1033.5	1.246	2600	1810	1520
750	46' Flat	4	18"	1192.5	1.338	3000	2100	1760

† Expanded conductor
‡ European practice leans to smaller conductors.

viously suggested, that 690 kv may not be quite enough for the next succeeding step, which for this reason might be pushed higher, possibly to 750 kv.

Over the entire range covered by Fig. 1 the chances are excellent that international standardization could be obtained and would be missing at only one

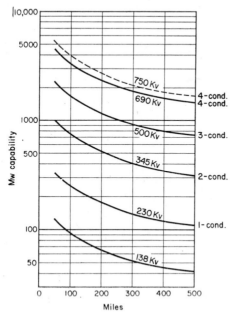

FIG. 2. Loading capabilities per circuit over various distances.

point, that between 345 kv and 400 kv. In addition to the interest in 500 kv on the part of Sweden, Great Britain, and others mentioned, Russia, as previously pointed out, is converting 400-kv lines to 500 kv and is undertaking 500-kv construction for future lines. One or two countries have expressed a preference for 500 kv as a future standard.

As far as foreseeable transmission requirements in this country and Canada are concerned it probably will be a long time before we need to go beyond 500 kv with its single-circuit capability of more than 900 Mw at 300 miles. With series-capacitor compensation, this voltage could probably be suitable for distances up to 500 miles. Likewise, recently developed 345-kv transmission will have a long period of adequacy and usefulness before superposition will be required. When this occurs, it should follow the "skip-step" pattern and go to the 690-kv (or higher) standard level, two steps above 345 kv.

In conclusion the authors firmly believe that all of the significant considerations which they have tried to emphasize clearly indicate that the next EHV step should be 500 kv rather than 460 kv. Equally clear are the indications that the step beyond that should be 690 kv or higher, possibly as high as 750 kv. There is ample time to shake down the exact value of that step. A decision on 500 kv not only will take care of our own immediate future requirements but will also provide a basis for international standardization at that level and at the level beyond that.

CHAPTER 6

TRANSMISSION PROTECTION AND OPERATION

1. SUBSTATION STANDARDIZATION†

REDUCTION of cost is the primary object of standardization.

Cost in this connection, however, must not be construed in any narrow sense; it must be understood as total cost. In substation construction, for example, an analysis of reduction of cost should embrace the factors of planning, design, construction and operation. Any attempt at standardization that does not take all these factors into account is bound to fail.

It is quite possible, for instance, that a standardized substation may contain more steel than one specially designed for the same work. Yet the standardized unit may cost less for no other reason than because in all probability, it took a few weeks less time to design it and get it on the job than the specially designed station — that and the additional revenue obtained in getting service to the consumer that much soon could easily more than offset the cost of the additional steel.

An engineer who would not dream of designing every distribution installation on an individual basis will very often feel constrained to exercise his individuality on substations. However, where he is faced with quantity production, as for large holding companies, standardization becomes an absolute necessity. The only alternative is a limitless engineering force that would move with such ponderous slowness that the cost of engineering would become prohibitive and nullify its value.

Feasibleness

Can standardization be applied to substation design? The answer is a decided affirmative, although it may involve a greater variety of standards than would be required in some other lines of work. Visualizing the substation, however, is very much simplified if a group of standards exists.

After engineering has been completed material must be ordered. If it is to be a standard substation delivery can be very much expedited, since it is possible to order the material before the drawings are put together. Likewise standardization makes it possible to order certain materials in larger quantities and even to stock other items with resulting lower costs. A further important advantage is interchangeability of material. If, for example, a substation is planned for one locality and later found to be needed less than one at another place it is possible where standardization exists to remove available material temporarily from one site to the other to gain time.

From the construction standpoint standardization again has many advantages. Bills of material for many substations, for instance, are bound to have great similarity and in some cases will be exact duplicates. After a certain number of substations have been built any inaccuracies or deficiencies in bills of material can be cleared up quickly if the standard has been established. Material and parts are bound to fit readily together unless workmanship is far below

† *Electrical World* (with M. L. Sindeband), November 20, 1926.

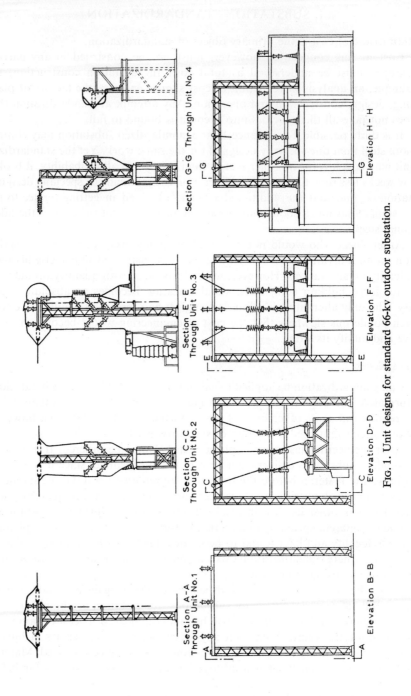

FIG. 1. Unit designs for standard 66-kv outdoor substation.

standard. The construction men after handling a number of standard substations become familiar with the design and therefore can go ahead on repeat jobs with greater speed and precision.

Standardization Procedure

The first most important step of standardization is the rationalization of procedure. For example, consider a 300-kva station stepping from 33,000 v to 2300 v. Transformers could be placed on a platform, or on the ground; they could be connected to the line by an ordinary disconnect switch, or by a horn-break switch; they could be tied in solidly, or connected through a high-tension fuse; the low-tension side could be connected through an expulsion fuse, oil cutout or oil switch. The high side could be left unprotected, or it could be protected by a station-type, or pole-type arrester. If the station is one of a series on an isolated transmission line, the two sides of the line could be sectionalized at the station, or carried through solid; the station could be made of wood, wood and steel, or all steel. Unless all these variables are brought down to some rational basis, standardization might as well be given up.

Standardization should first tackle those cases that are used most, then those next in frequency of use, until finally one reaches a point where it becomes necessary to weigh carefully the possible frequency of use of a standard against the labor involved in establishing it. Probably it will be found that greater effort is involved than in many other branches of electrical work, because of the great variety of combinations and different types of substations needed to cover the field. Nevertheless as the size and complexity of stations increase the number of possible combinations becomes so much greater that one ought to hesitate before making a separate design for every possible combination.

A procedure that has been found highly satisfactory for such stations is to split the design into elements or units, design each element, and then work out a sufficient number of combinations that are very common. Other combinations can very easily be put together with these elements, and will have practically all the advantages of the normal standard design and of a special design, without the disadvantages.

Limitations

What limits the scope of standardization, or at what point does it become desirable? It depends a good deal on the organization, its size, amount of work, and the type of men available to handle it. In general the limit will be an economic one. Where enough work is involved and the possible benefits of standardization overbalance the costs of carrying it out, then most surely standardization should be worked out.

21 a*

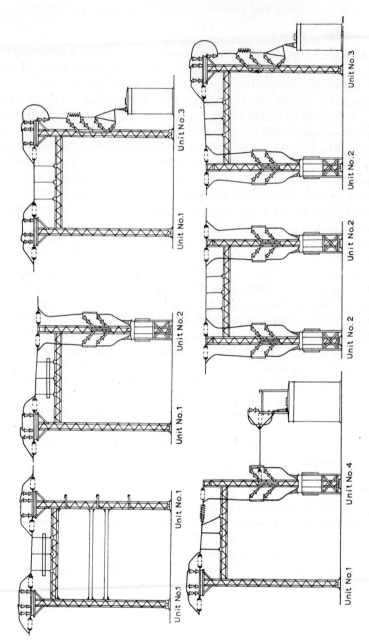

FIG. 2. Assemblies of standard unit designs for 66-kv substation.

2. OIL CIRCUIT BREAKER TESTS REVEAL ADEQUACY OF DESIGNS†

THE American Gas and Electric Company has carried out a number of tests on high- and low-voltage oil circuit breakers, falling into three principal groups.

The first group of tests was on Brown Boveri Company 150-kv and 35-kv breakers, the acceptance of which was made conditional upon the results of rupturing capacity tests. The 150-kv breaker is of the round-tank multiple-break type equipped with oil-filled 150-kv bushings. A total of 10 breaks per pole is employed using simple ball-type butt contacts. Although it was not possible to obtain sufficient short-circuit current on the company's interconnected 132-kv system to test the breaker at its full rated interrupting capacity of 1.5 million kva, it was felt that a series of tests at approximately 750,000 kva would indicate whether the breaker would be acceptable for the intended service. The 35-kv breaker was of the plain-break type, two breaks per pole, with all three poles in one rectangular tank. Several of these switches had been purchased subject to the results of tests to be made at 22 kv and carried to the full breaker rating of 250,000 kva.

The second group of tests using the regular testing equipment of the General Electric Company, was made on one unit of the Reyrolle compound-filled switch-gear, type C-1-ORD, rated at 7000 v, 400 amp, and having a guaranteed rupturing capacity of 75,000 kva. These units also had been purchased subject to satisfactory performance under short-circuit tests. These tests were made first at 2300 v beginning with less than rated duty and carried to a point considerably beyond the rating, and later at 6600 v at more than full rating.

The third group of tests was made on two General Electric Company 132-kv breakers, one an FHKO-39-B and the other an FHKO-136-B; both were explosion-chamber type. The FHKO-39-B unit had a rated rupturing capacity of 1.25 million kva, and a round tank. The FHKO-136-B unit was rated at 750,000 kva and had an oval tank.

Value of the Tests

Tests of oil circuit breakers are very expensive, and frequently considerably upset the system. Even if no actual physical damage results there is always the effect of the short circuits on the system voltage and perhaps upon apparatus susceptible to voltage changes and dips. Therefore, before tests of this sort are undertaken the benefits expected should be weighed to make certain that they will overbalance the possible harmful effects. After the tests the question should be raised again as to what of value actually has been obtained.

Brown Boveri Tests.—Reviewing the results of tests on the 150-kv Brown Boveri breakers, the following benefits were obtained:

Whether a multi-break breaker could be built to handle successfully rupturing capacity in the order of one million kva was not definitely answered but the per-

† *AIEE Transactions* (with H. P. St. Clair), February 11, 1927.

formance of the breaker when rupturing a short circuit of 75% of that value was such that no doubt arose that its limit had not been reached.

Definite information was obtained as to the ability of the breaker to go through a cycle much more severe than the standard duty cycle. It was shown for the AGE system that the breaker tested, if necessary, could interrupt a short circuit four or five times in rapid succession with perfect safety.

In all, 26 short circuits were placed on the 132-kv system of which 13 were at approximately full system capacity. So far as is known no appreciable damage of any sort resulted to the system. There were minor exceptions. One was the breaking of jewels in meters connected in secondaries of current transformers that fed heavily into the short circuit and which, through an oversight, had not been removed from the circuit. A strain choke coil on the circuit supplying the full short-circuit capacity collapsed, and half of the primary of one of the current transformers on a 132-kv circuit supplying the short circuit was short-circuited by arcing between turns. But with these exceptions no damage of any kind was experienced.

When the tests were originally contemplated, doubt was expressed as to the advisability of deliberately placing severe short circuits on a healthy system. The view that finally prevailed, however, was that if our system was not in a position to stand up under such a short circuit, the sooner that condition was found and remedied the better off the system would be. It was satisfying to find that the system was able to go through all these short circuits without any appreciable damage.

Until the tests were made no check was available as to the calculated short-circuit capacity of the systems. Many calculations and much design work and specification work had been done, however, on the basis of calculated values and it was felt that sooner or later some of these calculations ought to be subjected to test to determine whether the actual values were in agreement within reasonable limits. The tests demonstrated that the system calculations were correct to within 10%.

In the test of the 37-kv Brown Boveri breaker the principal benefit obtained was a complete demonstration that the designer had completely missed his mark in guaranteeing the breaker for 250,000 kva. Perhaps under no conditions will it be possible to build economically a single-tank breaker for that voltage and a rupturing capacity of 250,000 kva. It would be interesting if, at some future time, this point could be definitely established or disproved.

Reyrolle Test.—Regarding the Reyrolle test, the benefits can be summarized as follows:

It gave confidence to the engineers' original decision to install this equipment although, so far as was known, none of that type had then been placed in service in this country. However, the equipment is widely used on the continent and particularly in England. The system of baffling, consisting of wooden linings with wooden barriers fastened to the linings, was inherently weak and not advisable for breakers expected to rupture even such a moderate amount of power as 75,000 kva.

General Electric Tests.—Coming to the tests of General Electric breakers, the following distinct benefits resulted:

A further check was obtained on calculations of the system short-circuit capacity that had been employed in the past. In all, 64 short circuits were placed on the 132-kv system of which 35 were at approximately full system capacity. It was highly satisfying to find that the system went through all 64 tests without any appreciable damage. Full advantage, of course, was taken of the experience gained during the Brown Boveri tests, so that any troubles encountered at that time were not met with during these tests.

The tests served as a very thorough check on the explosion-chamber type of breaker and particularly on the assembly that was standard before the tests were undertaken. They showed the weaknesses of the original assembly and demonstrated the complete efficacy of the remedial measures finally applied. This information was of great value not only for future breakers but also for some breakers already in service and employing the original explosion-chamber assembly. An example of this occurred when six or seven 132-kv breakers of the KO-39-B type later failed in operation due to their inability to go back in circuit after opening a number of times under short-circuit conditions. An examination of the explosion-chamber assembly showed that the failure was exactly similar to that during the tests on the 136-B breaker. Therefore, as a direct result of the test experience, the new type of explosion-chamber assembly was substituted for the old type and no trouble of any kind has been experienced since.

A benefit that must not be overlooked is that a breaker of high rupturing capacity was subjected to a test very close to its rated values and on a cycle which might be interpreted as having subjected the breaker to a duty considerably in excess of its rating, in view of the present tentatively adopted standard for derating breakers for other than a standard duty cycle. The test clearly showed, at least for this particular type of breaker, that the guaranteed limit could be handled successfully after certain changes had been made. If it is assumed that the design had been made on a rational basis involving definite empirical and other fundamental data, then it may be safely concluded that the tests of the breaker served as a check on it and on other breakers designed on the same principle.

3. PHILO OIL CIRCUIT BREAKER TESTS†

BEFORE undertaking the Philo tests of 132-kv oil circuit breakers, which involved placing short circuits directly on the 132-kv bus of the largest generating plant on the AGE interconnected system, careful consideration was given to the benefits obtainable:

First, the proposed tests offered a means of determining positively whether a high-capacity 132-kv breaker would be able to withstand its full rated duty. This in itself was considered of great value. In addition, past experience justifies the conclusion that the carrying out of such tests almost invariably reveals one or more weak points in the breaker assembly; the remedies which are devised often constitute important advances in the art of circuit breaker design.

Second, the test promised to contribute materially to the development of high speed in the explosion-chamber type of breaker. The high-speed breaker is recognized as a very important development which promises considerable relief from the damaging effects of system disturbances, such as flashovers and insulator failures and by means of it system stability in many cases may be greatly improved. In addition, the high-speed breaker is expected to prove of great value in customer service by shortening the time of voltage surges and dips. This is particularly important to many customers using motors which either inherently or because of control apparatus have a tendency to drop off the line on voltage dips.

Third, the tests offered an opportunity to demonstrate further the feasibility of the multi-duty cycle to replace the present standard-duty cycle for circuit-breaker performance. As power systems grow and standards of service are raised, it is becoming more and more evident that the present standard-duty cycle of two breaker openings separated by a two-minute interval is hopelessly inadequate. It is well known that breakers are frequently called upon during storms to operate many times without any intervening inspection, and that the interval of two minutes is entirely too long, and cannot be tolerated in most well-operated systems. One of the aims of the Philo tests was to carry out further tests to provide additional evidence of the feasibility of changing the present duty cycle.

Fourth, it was felt that heavy-duty short-circuit tests directly on the high-tension bus of a large generating plant would give valuable information not only on the performance of a breaker located at such a point, but also on the performance of the plant itself under short-circuit conditions. Carrying out the tests at this location also made it possible to obtain experimental confirmation of a recently enunciated theory regarding recovery voltage phenomena in oil circuit breaker operation. To obtain this confirmation it was necessary to carry out short-circuit tests with various numbers and lengths of 132-kv lines connected to the test bus, a condition which was met by the Philo setup.

Almost the entire American Gas and Electric system, as well as some important external interconnections, were tied together in a practically normal

† AIEE Winter Convention (with H. P. St. Clair), New York, N.Y., January 26, 1931.

setup. No special provisions were made to take care of any portions of the system load from sources not connected with the system supplying the tests; normal interconnections with West Penn Power Company on the east, with Chicago on the west, and with North Carolina on the south, were maintained. With this setup a short circuit of over 1.5 million kva was obtained at the Philo 132-kv bus. The major portion of this total, or approximately 1 million kva, was supplied by the Philo plant itself, while the remainder was supplied by the various generating plants distributed at more distant points throughout the system.

Tests made at Canton in 1925 and 1926 involved only about one-half of the short-circuit kva obtained in the Philo tests and was supplied from several distant generating plants. Therefore, in severity of duty on any one generating plant and on the breaker the Philo tests represented a step of considerable size.

Results of Tests

The tests comprised a total of 81 shots divided into three separate groups on three different days. To carry them out at full rating, a 60-in. tank-breaker having a rating of 1.5 million kva was used.

The first group, consisting of 28 shots on a breaker equipped with standard butt-type explosion-chamber contacts, included seven shots with Philo generators alone supplying the short circuit and no transmission lines connected to the short-circuited bus. These shots, of which four were made with the full capacity of Philo, undoubtedly imposed a severe duty on the breaker as evidenced by length and duration of arc. The greatest distress was shown by the breaker on two of the shots in which approximately two-thirds of the Philo capacity was used. While the breaker succeeded in rupturing the arc on both shots before the opening of the backup breaker, there was an unusual amount of smoke and some oil discharge from the breaker vents, and the oscillographic records showed arcing continued to the end of the stroke.

Inspection at the conclusion of the 28 shots revealed one of the explosion chambers completely shattered. Burning on the contact rod indicated that the breaker had been operated a number of times in this condition. While it was thought that the explosion-chamber might have broken on the 12th or 13th shot, when maximum distress was apparent, no other evidence as to the particular shot on which the breakage occurred was available.

The second group, consisting of 40 shots on the same breaker equipped with the new "oil blast explosion chamber", included 16 shots with the maximum rate of recovery-voltage rise obtained with no lines connected to the bus. Seven of these shots were made with the full capacity of the plant behind the short circuit giving a rupturing duty on the breaker varying from a little below one million kva to 1.285 million kva. Six shots were made with maximum available system capacity, all at a duty of more than the full breaker rating of 1 million kva with a maximum of 1.735 million kva on the last shot.

The performance of the breaker was satisfactory throughout. All of scheduled shots were taken in spite of the fact that on the last test a contact rod was lost through breakage of a crosshead, due to use of a rod which was too short to be properly threaded into the crosshead. Oscillograms showed that the breaker carried its full three-phase short-circuit current on this last test, although, of course, one phase would not have closed had any further tests been made.

System disturbances due to the tests were not very severe and caused very little trouble or complaint except at nearby points such as the city of Zanesville.

Benefits Obtained

Without question the tests have increased confidence in the higher ratings of modern breakers. A few years ago 132-kv breakers of 1.5 million kva rupturing capacity were considered adequate for any location on the system; and, as a matter of fact, higher ratings were not available. Unforeseen developments, however, such as installation at Philo of three-unit turbines of much higher capacity than the single-unit machines originally planned have made it necessary to employ breakers of considerably more than 1.5 million kva rating. Developments in the art of designing and building breakers have given the manufacturers sufficient confidence to offer breakers of 2.5 million kva rating. Not only have a number of breakers of this rating with 72-in. tanks been purchased, but it has been possible with certain changes to re-rate existing 72-in. tank-breakers at 2.5 million kva.

The definite realization of the 8-cycle breaker, as demonstrated by the tests, marks a forward step of real practical value to the power system. In the past the types of relay systems used have required too much time to operate. It has become desirable and necessary to meet continually improving standards of service by reducing relay time to an absolute minimum. In such cases the time consumed in breaker operation becomes a very large factor.

Recently developed relays have helped to reduce the time required for clearing sections of single-circuit lines, but there has been little incentive to adopt new and faster relays for balanced double-circuit lines so long as the breaker operating time is so much slower than that of existing relays for this service. However, by employing the high-speed circuit breaker and using faster relays it will be possible to cut down the time required to clear one circuit of balanced double-circuit lines by at least half and probably more.

The increasing percentage of two-line outages on double-circuit balanced lines may be due to power-system growth without accompanying improvements in relay and breaker operating time, with the result that the greater concentration of power in the arc following flashover increases the tendency for the flashover to be communicated to the second circuit. If one circuit on a balanced double-circuit line could be cleared very quickly by fast relays and a high-speed breaker, the number of outages involving both circuits could be substantially reduced.

4. RATIONALIZED SWITCHING ARRANGEMENTS†

A STUDY of the switching schemes used throughout the United States discloses a great variety of arrangements. The question arises whether these schemes are as different as they would at first appear. Further study indicates that generally they use certain basic ideas; it is the special additions and trimmings that usually make them seem different. These special features undoubtedly are often due to precautions, but too frequently are caused either by predjudices arising out of experience or by an attempt to take care of all types of abnormal conditions, some so unlikely of repetition that the additional investment is not warranted. Hence, it is questionable whether there is any real justification for the rather wide variation in types of switching schemes that are commonly employed.

Quite often the difference in switching schemes are due to failure to understand thoroughly the purpose and functions of the various components. A great deal can be gained by proper definition of components as follows:

Bus.—Any switching scheme has a bus in some form or another, the real function of which is to provide a medium for pooling power sources at generating or distribution stations so that power can be sent to distribution points with optimum economy and reliability. The fact that special or abnormal conditions occur makes it necessary ordinarily to install not only a single bus but some combination of buses.

Main Station Bus.—The main bus is the one normally energized. All incoming power sources and the outgoing feeders are connected to this one bus.

Reserve Bus.—A reserve bus normally is one not energized but is a duplicate of the main bus. All circuits connected to the main bus may be switched to the reserve bus in order that the main bus may be taken out of service for inspection or repairs.

With all circuits connected to the reserve bus it functions in the same way as the main bus. There is no fundamental difference between the main and reserve busses; the names merely distinguish which bus is normally kept in service.

If power is normally taken from both the main bus and reserve bus the latter must be considered as a part of the main bus, and one then has a split main bus. The importance of maintaining the reserve bus as such under normal conditions increases with the size and importance of the station. With both busses normally energized, the problem of service reliability may become very serious.

Transfer or Inspection Bus.—A transfer bus provides means for by-passing a line or piece of equipment through a spare or transfer breaker connected to the main bus so that the line or apparatus connected thereto may be taken out of service for inspection or repair. Like the reserve bus, the transfer bus is not normally energized.

† Conférence Internationale des Grands Réseaux Electriques à Haute Tension (CIGRE), Paris, France, June 18, 1931.

Reserve and Transfer Bus.—A reserve bus may be so connected that it can serve the function of either a reserve bus or a transfer bus but cannot combine the two functions simultaneously.

Main Reserve and Transfer Bus.—In some cases a main, reserve, and transfer bus may be desirable, that is, three separate busses.

Ring Bus.—The ring bus is a means for transferring power from any section of the bus to another. In effect, it simply ties the two ends of any main bus together. The ring bus is commonly used in a generating station with a large number of relatively high-capacity generating units. In most cases, current-limiting reactors are used in the bus section between generators. Outgoing feeders are connected to the same section as the generators. The short-circuit kva on the main bus, the generator leads, or the outgoing feeders is kept to a reasonable value by proper reactors in the main bus. Bus reactors normally have by-passing switches to cut out the reactors under certain operating conditions. By the use of double-winding generators the high reactance between generator windings eliminates the necessity of bus reactors.

It is questionable whether the ring bus generally can be justified either theoretically or by operating experience. In many cases it will be found that the ring is not normally closed and that all necessary transferring of power from one section of the bus to another can be carried out without closing the ring.

Star or Synchronizing Bus.—The star system consists of a synchronizing bus to which each generating bus is connected through a reactor. It thus becomes several separate generator busses connected to a common point through suitable reactors. Normally two switches would be used, one between the generator and its main bus, and one between the reactor and the synchronizing bus. If complete flexibility is desired some form of transfer bus is required. An additional generator switch allows the generator to be connected through the transfer bus to the reactor bus, thus enabling any section of the main bus to be taken out of service without affecting operating conditions.

The star arrangement is somewhat more flexible than the ring bus due to the fact that every generator or set of cables is equally available to every other one and power transfer can be carried out with a minimum voltage variation. During short-circuit conditions the voltage of the star bus as a whole can be maintained at a higher value than with other busses.

Function of Switching

With the elements of a switching arrangement carefully defined and thoroughly understood it ought to be possible to take the building of a switching scheme out of the realm of guess or personal whim; in other words, it ought to be possible to rationalize the entire procedure of combining the elements of a switching arrangement to form a unit that is sound technically and economically.

While no hard and fast line can be drawn between underground and overhead switching systems there are certain factors which affect the schemes used on both types.

Underground.—In the underground system it is common practice to have one spare cable on all important circuits to permit replacement of a cable in which failure occurs. Ordinarily the spare circuit would not be in service; however, due to thermal limits, if the peak load exists for a short time the procedure probably will be to add the spare cable rather than overload the cables in service and chance exceeding their heating limits.

Overhead.—With an overhead system, on the other hand, it is not general practice to provide a spare circuit, although there are cases where two circuits are employed and normally lightly loaded. In those cases, however, both circuits would normally be in service for reliability and one would not be open normally as in the case of the underground system. It is usually possible to overload considerably lines exposed to the air without causing undue heating.

As a result of inherent characteristics of the two systems, the switching may be somewhat different. For example, the presence of a spare cable on the underground system may eliminate the necessity of by-passing oil circuit breakers for inspection or repair. On the overhead system, however, it may be necessary to inspect the breaker while keeping the circuit in service and that, of course, calls for some by-passing arrangement.

Generation Requirements

The first step in building up a rational switching arrangement is to ask these questions: How important is the load to be served? Is it absolutely necessary that service be maintained at all times? What expansion is expected in the load and what arrangement must be made so that the initial scheme will work best with the ultimate condition?

In the case of generating stations thought must be given to the relative importance of each station in system reliability so as to determine how much can be spent on the switching scheme. Obviously, it would not be good practice to use an extremely complicated and costly switching scheme on a small and relatively unimportant station. On the other hand, future expansion plans may dictate additional expenditure in the beginning. On a large and important generating station or substation a reasonable expenditure is justified to insure reliability. However, size must be judged relatively to the size of the system and in every case a thorough study should be made to simplify the scheme as much as possible.

In the case of a generating station where reliability is exceptionally stressed, it is particularly necessary that the importance of the station be gaged relative to the whole system when laying out the switching scheme. For example, if a hydro station of say 40,000 kw is connected to a system of 1,000,000 kw capacity, the station represents such a small percentage of the total installed capacity that only the simplest form of bus could be justified economically. Because of this the station may be lost at times and some water wasted due to the outage, but the system could still be operated more economically with such an outage than if the original cost were heavily increased by the use of a very complicated switching scheme for the comparatively small generation component.

In any important station, by-passing of circuit breakers is absolutely necessary unless it is possible to take the feeder out of service. The reliability of a circuit breaker depends upon its condition, which means frequent inspection. Since the reliability of the system depends upon circuit breakers, it becomes very important that means of inspection and repair be provided without interrupting any important circuits. This generally means the use of a transfer or inspection bus with or without a special transfer breaker.

The importance of the station may warrant the use of both a main and reserve bus. If such is the case, it is important to remember that only one bus should normally be energized. If both are energized, there is in reality not a main and reserve bus but a split main bus. If conditions will warrant it, the reserve bus may be utilized either as such or as a transfer bus, but it cannot function simultaneously as both. If the main and combined reserve and transfer busses do not meet the requirements of safety and cost of operation, and if the size of the station warrants the additional expenditure, then it is logical to use three busses, each separate and distinct, namely: a main, a reserve, and a transfer bus.

Selector Breakers or Disconnects.—In the selection of switching schemes two questions very often arise—shall selector-breakers or selector-disconnects be used on the double-bus arrangement, and shall the busses be sectionalized?

A review of present schemes shows that both types of selectors are used on both low- and high-voltage layouts depending on past experience, local conditions or individual preferences. The most flexible plan makes use of oil circuit breakers, which can be justified under certain conditions on lower voltage layouts and even up to 66 kv. However, when the cost of the breaker installed is of the order of $75,000 (not unusual on 220-kv systems) cost justification becomes a serious question. One argument sometimes advanced for the use of circuit-breaker selectors at high voltages is that the cost per kilowatt can be kept the same. However, the aim should be to reduce materially the cost per kw at the higher voltages for only through such an attitude can the electrical industry keep costs down and thus insure its continued growth and development.

Bus Sectionalization.—With reference to the question of sectionalization of busses, it will be found that in cases of large concentration of power there will usually be some means for sectionalizing. Important loads may have feeders from each bus section. It is practically universal practice now to use bus-differential or fault-bus protection which in case of trouble will clear one section and drop some feeders. However, the remaining feeders will probably have sufficient capacity to care for the entire load. Here again the character of the load will determine whether or not the sections should be connected by oil circuit breakers or disconnecting switches. If interruptions are permissible, it is paracticable to use a disconnecting switch between sections of the bus; if one section is in trouble the bus can be deenergized and the disconnect opened until the bus has been repaired. A good criterion for a switching arrangement of any system is an arrangement that will prevent the concentration of extraordinarily large blocks of power on any one bus section. The magnitude of short circuit should be held down and its effect should be confined to as little of the system as possible.

Examples of Rationalization

1. A typical load center which is small but relatively important is shown in Fig. 1. Two sources of feed were available. The present size of the load warranted only one bus, but expected growth indicated the probable desirability of two busses. Hence, the future need for a second bus was indicated in the diagram as a warning to the designer to leave room for it. The load of the station being well within the capacity of one line, by-pass arrangements were not provided on the line breakers; system requirements did not call for 100% service from each of

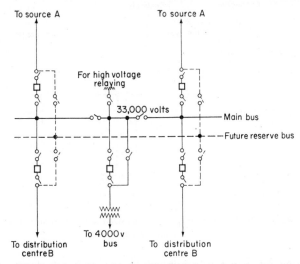

FIG. 1. Rationalized switching for typical small but relatively important load center.

the two lines going to separate sources. This was not the situation in the case of the transformer breakers; with only one transformer available by-pass features were absolutely necessary. To permit any necessary work on the bus, manual sectionalizing switches were provided because nothing more elaborate could be justified.

2. A medium-size substation serving an important load area is shown in Fig. 2. The basis for the design of the switching arrangement follows:

For the size of load involved (30,000 kw) two separate and distinct busses were felt necessary and justifiable. Hence the use of a main and reserve bus; owing to the length of lines it was desirable to provide by-pass features on all breakers, even where double-circuit lines existed. Because the station was not large enough to warrant a third bus, the transfer bus was combined with the reserve bus.

Two banks of transformers having been provided it was logical to connect them to different portions of a sectionalized bus. A sectionalizing oil circuit breaker could be justified since it permitted clearing one-half of the bus in case of transformer or bus trouble and still supply most of the load centers fed from the station since duplicate or parallel feeders were divided between the two bus sections.

By-pass facilities having been decided upon for the feeders, inspection arrangements could be made through a transfer or inspection oil circuit breaker. Because of the comparatively low voltage of the bus and, therefore, the relative inexpensiveness of the oil circuit breakers, it was felt justifiable to provide duplicate breakers for each of the two transformer banks. This arrangement not only provided two inspection breakers but also the additional flexibility and speed of duplicate breakers on the two sources of supply to the main and reserve busses.

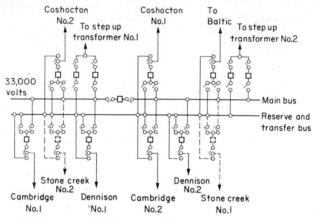

FIG. 2. Rationalized switching for medium-sized substation serving important load area.

3. The same reasoning was applied to a moderately large steam-turbine station with 240,000 kw capacity, as indicated in Fig. 3. This station, the most important one within a radius of 100 miles, serves some very important load centers and obviously must be designed for maximum reliability. With this in mind it was deemed necessary to have two busses available at all times—a main and a reserve bus that should be absolutely independent of each other. At first sight there appear to be two transfer busses but electrically they form one. The arranging of a transfer bus on each side of the main structure was to make possible the use of a single bay for the switching of two circuits, one going to the north and the other to the south. If this scheme had not been used the entire bus structure would have been lengthened unreasonably to take care of all circuits.

Selector-disconnect switches were used on each circuit for selecting the main or reserve busses. At this voltage (132 kv) the use of selector-circuit breakers could not be economically justified. The concentration of power generated by this station necessitated the sectionalization of the main and reserve busses; considering the size of the station oil circuit breakers were the obvious selection.

Effect on Fundamental System Plan

The broad outline of the switching scheme and the fundamental system plan are so closely related that one can almost be defined in terms of the other. When the switching scheme has been decided upon, the system plan has really been decided, and conversely when the system plan is determined a certain definite switching scheme is called for.

All system plans may be broadly classified into three fundamental plans.

Tightly Linked System.—Normally this considers the entire system as a unit. Lines are bussed and switched on the high-voltage side and require oil circuit breakers of high interrupting capacity. Troubles may cause severe shocks to the system, the severity being dependent to some degree on the protection used. The

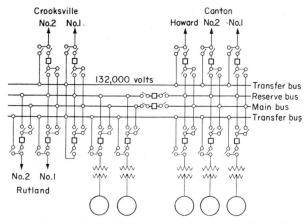

FIG. 3. Rationalized switching for important generating station.

plan makes maximum use of transmission lines and minimizes the reactance between station busses, thus allowing maximum synchronizing power between stations. This plan is especially adaptable for transmission systems involving long lines and a network covering large areas because of the amount of power than can be carried over any given set of lines.

One modification of this plan is to use the low-voltage bus instead of the high-voltage bus with switching done on the low-voltage side and the line and transformers operated as a unit. This arrangement in most cases will be more costly than the preceding one due to reserve transformer and line capacity required. Also it is necessary to pay particular attention to the balancing of loads on lines if it is desired to use balanced-relay protection. This system which limits voltage dips in cases of system disturbances, has been defended on the grounds that its additional reactance limits short-circuit currents thereby greatly reducing their shock to the system and materially lessening the interrupting capacity of oil circuit breakers. The latter point is, of course, true but the former is true only when the time for clearing faults is relatively long. By using available high-speed breakers and relays faults may be cleared in 12 to 15 cycles (in terms of a 60-cycle system). Under this condition the system having lines bussed on the high side usually shows considerable advantages over the low-voltage bus scheme in the amount of power that may be transmitted over a given set of lines.

Low Voltage or Network Synchronization.—In this plan the generating sources are normally paralleled only through the low-voltage network which may be at 4000 v or even through a 120/208-v network. The fundamental idea is to bring as many separate sources of supply to any given load area as can be justified so that the loss of any feeder will not cause an interruption to any load area. The idea may be carried so far as to split up the generating station bus and run separate feeders from each generator to the various load areas. At first thought it would appear that such a plan would give insufficient synchronizing power between generating sources to hold the system in step during disturbances. However, calculations backed up by operating experience seem to indicate that faults do not affect system stability appreciably and result in only very moderate voltage dips for any load area during disturbances.

The network-synchronization plan undoubtedly offers a very satisfactory solution for certain types of systems such as densely populated city districts where it is of utmost importance to maintain service. Such a plan obviously could not be carried out to any material extent in sparsely settled cities or districts where relatively long distances prevail.

Loosely Coupled System.—The fundamental idea is that the system is divided into districts each of which is self contained with regard to generating facilities, but the various districts are interconnected by so-called "interlinking cables or lines". The term "loose coupling" refers to the relaying rather than to line capacity. Each district is so arranged that if uncontrollable trouble occurs which the normal relays will not clear, then the back-up relays on the interlinking cables or lines will isolate the various districts. This means that to insure that an interruption in one district will not drag down all districts, each district must have sufficient generating capacity to take care of itself in case the interlinking cable or lines to the neighboring districts are opened during trouble. One arrangement in which this plan is used consists of a low-voltage system in the outskirts of the city. The city generating plants use synchronizing busses which are connected by interlinking cables between the synchronizing bus of the plants in the low-tension section and the step-down substations of the high-tension system. In the high-tension system, too, definite pull-apart points are provided.

By using this arrangement it is possible to limit fairly definitely the oil circuit breaker duty. Also it has certain definite advantages when the generating stations can be distributed so as to fulfill the essential requirement of self-contained load areas.

Consideration of these fundamental plans brings up the question of what line in the development switch gear and relaying should be stressed. If the loosely coupled system and the low-tension switching scheme are to predominate then development of circuit breakers of very high interrupting capacity will not be necessary. However, if the tightly coupled system is to be used more then development should be concentrated along the line of higher interrupting capacities, extremely high-speed opening and relays of comparable speed.

While each of these schemes is available, the proper way to obtain the best-balanced plan is not to pick a fundamental switching arrangement and then carry it through regardless of consequences but to carry out each plan only so far as its physical demands are within moderate limits. For the tightly coupled system it seems entirely logical to set up limits that would be altered when oil circuit breakers of 2,000,000 kva interrupting capacity will no longer suffice; at present this duty for oil circuit breakers is entirely feasible economically, whereas, ten years ago it would have been considered impossible.

A well balanced system will probably make use of each of these fundamental ideas using each on that portion of the system where it proves the most advantageous.

Relationship of Relay Scheme

No transmission system is better than the positiveness of its relay system in clearing system faults. Quite often when an attempt is made to coordinate the relaying with a system plan, it is necessary to change the transmission and switching schemes considerably and in come cases split up the system and use an entirely different plan.

In any case it is unfair to adopt a transmission and switching plan and then put it before the relay engineer to work out the proper protection. The only reasonable method is to consider all factors at the same time and work out the best over-all proposition.

One practical case involved connecting two systems by a long 132-kv transmission line. The economic situation was such as to justify interconnecting the two territories because it allowed two inefficient steam plants to be shut down normally, permitted economic transfer of power between the territories, and gave greater reliability due to the possibility of mutual help during times of system troubles. Conditions were such that a double-circuit line could not be justified economically and success of the plan, involving a large investment, depended upon the satisfactory operation of a single-circuit line. The chief difficulty anticipated was relaying. This interconnection included four intermediate substations between the two generating sources. The use of induction-type overload relays meant that long time settings would be involved for clearing short circuits at certain sections of the line because the relay settings must be cascaded to give selective action. Using overload relays at the generating stations and a combination of overload and reserve power relays at the substations, the time setting at the generating stations is very high resulting in slow clearing of faults and possibility of severe shock to the system during the short circuit. The amount of power that could be transmitted over the line was limited due to the necessity of making the overload relays selective.

The relay scheme selected was the best obtainable at that time and with a knowledge of its very definite limitations. However, at the time it was adopted the use of carrier-current relaying was anticipated and it was felt that its use would very easily overcome the difficulties pointed out. Troubles not anticipated were encountered in the development of this type of relaying with the result that it was not believed that it would result in proper protection.

As an alternative study was made of the reactance type of relay. Although not ideal it offered certain very definite advantages over the induction-type overload relay. By using reactance relays this section of the line is not entirely independent of other parts of the system with reference to its relay settings; for example, the reactance relays on the main line must be selective with relays on the low-voltage side of the substations along the line and also must be selective with relays beyond the section having reactance relays under certain fault conditions.

Tests have been carried out recently on this system with oil circuit breakers that open in 8 cycles, and it will, therefore, be possible in the near future by using these high-speed oil circuit breakers (and perhaps faster relays) to reduce very materially the time now required with the distance relays for clearing of faults.

These cases that have been cited illustrate forcefully the interdependence of the physical transmission arrangement, the switching scheme and the relay scheme.

5. EFFECTIVENESS OF CARRIER-CURRENT RELAYING†

WITH the general development of the 5- to 8-cycle high-voltage breaker and the modernization of many older type breakers to give substantially the same speeds, the necessity for bringing the relay system into harmony with this faster breaker speed has become manifest.

To obtain high-speed relaying a number of schemes have been evolved, each more or less limited, and many rather expensive, particularly when their limitations is taken into account. Among them are the instantaneous over-current relay set to clear faults within a definite distance of the outgoing bus; balanced-line relays, which of course, will not function when one circuit is out or when both circuits of a double-circuit line are in trouble; and reactance and impedance distance relays of various types. The distance relays are by far the most elaborate; yet they offer the possibility of satisfactory phase-to-phase protection for only approximately 80% of the particular section they are supposed to protect. Lack of complete section protection is also an objection to using instantaneous current-type relays, although they can be used for both phase and ground protection. Obviously, as long as current is the only basis of selection, complete protection cannot be obtained without some system of back-up relays.

The only relay method ever proposed for complete and rapid protection to any line section over its entire length, regardless of the type of fault and independently of the number of line sections between any two sources of supply, is the pilot-wire scheme. This, however, has never met with any great favor in the United States on account of the prohibitive cost of the pilot-wire circuits, particularly on long lines, and because the scheme is no more reliable than the pilot wire. Furthermore, experience has shown that its reliability is not quite at the 100% level desired for this service.

Solution of this difficulty was the substitution of a carrier-current for the pilot wire; a practical scheme embodying this idea was developed several years ago. But this initial scheme went completely out of its way to avoid placing reliance on any heretofore standard relay arrangement or combination of relays; the fundamental idea was to place reliance primarily on over-current (i.e. fault current) and on carrier. Hence a number of complications were introduced which militated against its wide adoption.

The soundness of the principle, and its complications, were brought out in a series of tests carried out in 1927 and 1928 on the system of the Ohio Power Company. For a thorough understanding of the simplified scheme it may be well first to describe the principal weaknesses encountered during these tests:

(a) The use of the common plate transformer introduced too many magnetic difficulties and made it substantially impossible to obtain proper plate voltage for all conditions of line-to-line and line-to-ground faults. In practice this limited the arrangement to either phase-to-phase of phase-to-ground protection.

† *Electrical World* (with C. A. Muller), September 10, 1932.

(b) The heavy volt-ampere burden placed on the secondaries of the transformers required a minimum of approximately 300-amp primary current to get satisfactory operation of the carrier-current equipment where bushing transformers were employed. Where this was not feasible instrument-type current transformers had to be used.

(c) The principle of current actuation ruled it out on a line where the normal load current from one terminal is in excess of the short-circuit supply from the second terminal without resorting to power-directional relays; installation of which, however, violated the very basis of the entire scheme.

(d) The scheme of operating the filaments with reduced voltage required an unusually long time setting for the tripping relays, since time had to be allowed for bringing the tubes up to full emission. The speed was of the order of 30 cycles which did not appear fast enough for a system as elaborate as this.

(e) The complexity of the equipment with its multiplicity of reactors, transformers, glow tubes, etc., was an innate deterrent in that it promised excessive maintenance to insure reliability of operation.

Nevertheless a complete installation of this scheme was made in 1929 on one of the large 220-kv lines with phase protection waived and carrier-current relied on for ground protection only. Although this original arrangement seem to fall short of what an ideal universal protection scheme ought to be, the idea seemed to be so inherently sound that work on it was continued culminating in the simplified system of protection now described.

Simplified Carrier-current Pilot System

In the simplified arrangement it was recognized that the prime function of carrier was to replace the pilot wire and lock out the relays at the ends of a section when trouble occurs outside the section. If this action could be depended upon definitely relay arrangements could be employed to provide very fast action under conditions of internal fault.

The elements of this scheme and the manner in which it operates are as follows:

The sole function assigned to carrier apparatus is to receive or transmit carrier-current and when received to actuate a single receiver relay.

Power required to operate the carrier-current equipment, both transmitter and receiver, is obtained from a small inverter operated from the station battery and is in no way connected with the current-transformer circuit.

Directional-power relays are employed at each station to determine, in combination, whether a fault is internal or external; no phase relationship of carrier currents is necessary for this determination.

The transmitter-tube filaments are operated continuously from ac potential at rated voltage and the life of the tubes is protected by employing a proper negative bias on the grid; this is removed when the transmitter begins to function in response to the starting relay. The receiver is prepared to receive under all conditions.

By means of directional-power relays carrier is allowed to be transmitted by only one station at a time, that station being the one at which power flow is into the bus. This recognizes the fact that in a given transmission-line section employing power-directional relays, the relay acting under power flow into the bus does its own blocking by its power-directional element. At the station where power flow is away from the bus blocking is not permissible if the power flow at the other end is away from the bus but is necessary if it is into the bus.

Speed of operation is greatly increased due to operating the tube filaments at normal voltage and employing circuit-opening relay contacts for starting the transmitters.

A single carrier frequency is used at both terminals. The equipment consists of a simple transmitter and receiver assembly coupled to the transmission line through the usual coupling capacitors and tuning unit. All plate, filament and bias voltages for both the transmitter and receiver tubes are obtained indirectly from the station battery by using a small dc to ac converter and a plate-supply rectifier. No grounds are placed on the station battery.

The receiver is prepared to receive signals at all times and does not introduce any appreciable time lag in this circuit. No harm is done if the receiver picks up signals from its own transmitter. When the transmitter is operated it is not necessary to receive signals from a distant station or to prevent energizing of the receiver relay. However, the receiver is protected against excess voltage from its own transmitter. The arrangement results in a simple, relatively insensitive, receiver capable of supplying adequate power to operate the receiver relay.

Freedom from interference due to static or arcing grounds is obtained by using adequate power in the transmitter, together with a very insensitive receiver and adequate margin on the receiver relay. However, the transmitter power required is normally low because the transmission losses between terminals are low and the use of line-trap units eliminates losses outside the section of line being protected.

Field Tests

Although this new arrangement had withstood all laboratory tests and sufficient experience had been gathered with carrier separately and with the relays over many years, experience with previous arrangements had left a definite feeling that no positive assurance could be placed in anything but a complete field test. Accordingly a complete schedule of tests was outlined covering all the various types of faults which might be encountered within and external to the section to be protected, and at both ends of the section. It was sought by mean of these tests to prove that for all types of external faults carrier is received to prevent tripping at the station where power flow is from the bus into the line, but that for all types of internal faults carrier is not received at either station. Also the tests were expected to determine the maximum speed allowable for tripping relays without faulty operations on through-faults and to determine the effect of arcing short circuits on the reception of carrier.

It was not felt that there was anything else to be learned with regard to the behavior of directional relays under various conditions of line potential, the reliability of the three-phase directional relay under all conditions except dead short circuit at the station bus, and of the ground-directional relay which had been definitely established over many years of operating experience.

Information Gained

For internal faults carrier was not transmitted. The total time from the occurrence of a short circuit to energizing the breaker trip coil ranged from 3.0 to 5.5 cycles.

Results of the tests make it apparent that the time for the starting relay to open its contacts can be made approximately one cycle. The time required for the receiver relay to open its contacts after receiving carrier is approximately 0.2 cycle, thus requiring a total of 1.2 cycles for the trip circuit to be blocked for an external fault. Hence the total time for the tripping relay to close its contacts might be made 2 cycles. However, for reasons of safety it probably is desirable for the present to make it a minimum of 3 cycles.

The development of this simple, workable and reliable system of pilot protection will give the system planner and designer freedom for the first time to plan lines without any regard to switching or other complications which might hinder proper relay protection under present relay practice. Further, this carrier system permits almost any number of sectionalizing points between any two main sources of feed since each line section provides its own protection. The speed of that protection will be the same for all sections and will be independent of the total number of sections operated. By insuring high-speed relaying over the entire range of a line, carrier relaying will bring improved system stability. In those sections where the probability of lightning damage is particularly great, this system of protection ought to reduce that source of system difficulties to an almost negligible quantity. It ought to prove of particular benefit on double-circuit lines where two-line trouble is a frequent occurrence, since all present relay schemes, except pilot, are particularly weak in this regard.

In its present form the carrier arrangement has definitely proved the feasibility of positive locking out on through-faults and at the same time positive energization of trip coils of the breakers involved in an internal fault in a maximum of 3 cycles. Until breakers are developed further to speeds that are considerably better than those obtainable at the present time and at least equal to the carrier relay time of three cycles, it appears that nothing could be gained by further speeding up the relay arrangement. However, there does not appear to be much doubt that the present arrangement can be speeded up to permit complete action within $1\frac{1}{2}$ cycles. The next move, if there is any economic advantage to be derived from faster speeds, would appear to be definitely up to the breaker.

6. CARRIER-CURRENT RELAYING SIMPLIFIED†

THE principle of carrier-current for pilot-wire protection of transmission lines to give complete and rapid protection to any line section over its entire distance, regardless of the type of fault and independently of the number of line sections, has long been recognized. Many schemes have been tried out but the most practical and reliable method employs directional power relays at each station to determine, in combination, whether the fault is internal or external and to control transmission and reception of carrier current.

In the original installation the power supply for the carrier-current equipment was obtained from a small rotating inverter connected to the station battery. All tubes were operated from the ac supply of the inverter while the plant and bias voltages were obtained by means of a step-up transformer, rectifier and filter system. Since the rotating inverter had to operate continually, it was necessary to install duplicate machines, one to act as a spare in case the other should fail—a rather cumbersome arrangement.

Consequently, a power-supply system was worked out to permit all tubes to operate directly from the station control battery, thereby eliminating the rotating inverters and rectifiers. This simplification, Fig. 1, was made possible by the

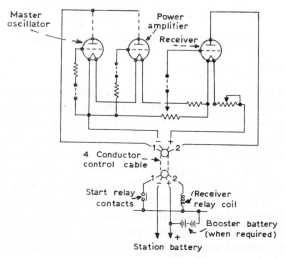

FIG. 1. Simplified power supply system permitting all tubes to operate directly from station control battery.

increased efficiency of the coupling circuit obtained by locating the carrier-current-cabinet outdoors near the coupling capacitors, whereby the losses which had been occurring previously in the long carrier-frequency leads had practically been eliminated.

On the first installation interphase coupling, using two 0.001-mfd capacitor assemblies, was employed, which resulted in a rather expensive coupling. It

† *Electrical World* (with C. A. Muller), October 28, 1933.

was found possible, however, to use phase-to-ground coupling and also reduce coupling capacity, thereby simplifying and reducing the installation coupling cost. With this arrangement, Fig. 2, it was necessary to employ only one 0.0005-mfd capacitor assembly, thus decreasing the cost of coupling to high-voltage lines by almost 75%.

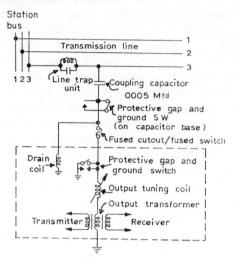

FIG. 2. Phase-to-ground coupling reduces cost.

Relaying Long High-voltage Tie

During the past year ten carrier-current relay installations were put in operation on the double-circuit section between Howard and Fort Wayne sectionalizing stations on the 132-kv system between Philo and Twin Branch generating stations, which, in turn, tie in very extensive 132-kv networks totaling some 4 million kw capacity. On this 300-mile sectionalized tie line it was essential that all types of faults, such as ground, phase-to-phase and simultaneous faults on two circuits, be cleared as rapidly as possible to keep within stability limits. To obtain high-speed clearing of faults all the 132-kv oil circuit breakers were converted to high-speed operation (6 to 8 cycles), and balanced duo-directional protection, instantaneous overcurrent protection and carrier-current relaying on the double-circuited section were installed. Only a carrier-current system offered fast relaying in case of simultaneous trouble on two circuits, and at the same time insured, in case of either single-line or two-line faults, that all breakers on the section in trouble would trip simultaneously. Cascading of breakers would result in too slow clearing.

Because of a number of unusual conditions in this type of tie line the arrangement originally employed with carrier-current relaying could not provide the desired protection. One of the conditions was the value of the normal load current, which, especially in case of single-line operation, is nearly equal to the short-circuit currents obtainable. To prevent continuous operation of the transmitter and receiver for long periods of time it would be necessary to set the carrier-current starting and tripping relays for phase protection at high pick-up values, thereby decreasing the sensitivity of protection considerably.

Another difficulty was introduced by the load swings, sometimes reaching very large values, that occur over such a tie line in both directions, carrying with them the possibility of opening the line from either the carrier-current relays or back-up protective relays. A third problem was introduced by the small values of ground currents that were obtainable on certain sections of the line, making it difficult for the standard ground starting and tripping carrier-current relays to function rapidly.

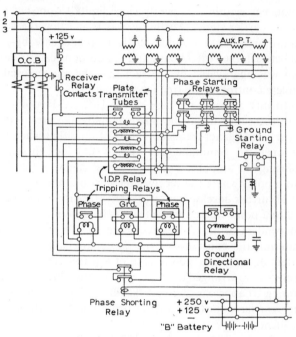

FIG. 3. Phase-to-phase protection enables ground starting and tripping relays to operate as fast as corresponding phase relays.

To take care of these conditions the carrier-current relay arrangement shown in Fig. 3 was used for phase-to-phase protection. The current-starting relays have been replaced by instantaneous undervoltage starting relays, which normally short circuit the current coils of the phase-tripping relays for both the carrier-current and back-up relays, and also control the starting of carrier transmission. By means of the voltage starting relays, it is possible to set the various phase-tripping relays as sensitively as desired regardless of the normal load currents that may exist on the tie line. On this particular installation the voltage starting relays were set to operate when the phase voltages dropped to approximately 80% of normal. To obtain fast operation on ground faults the directional element of a type IB ground directional relay was employed with the zero-sequence voltage stepped up three-to-one by means of auxiliary

22*

potential transformers. With this arrangement it was possible to make the ground starting and tripping relays operate as fast as the corresponding phase relays.

Uniform Carrier Transmission

During the final installation tests on the Howard-Fort Wayne lines it was found that the amplitude of the carrier signal transmitted and received differed under various system setups sufficiently to affect the reliability of the carrier-current relaying. The master oscillator, it was discovered, did not control the generated frequency within sufficiently close limits to prevent a variation in amplitude of the carrier signal. Further it was found that the variation of the master oscillator frequency was due partly to inadequate by-passing of the control leads between the switchboard and the outdoor carrier-current cabinet. This situation was aggravated by the fact that the master oscillator did not exercise complete control of the frequency generated.

Another difficulty arose in the case of the Howard-Fostoria-Lima circuits, which have three sectionalizing terminals employing carrier-current relaying operated on the same carrier frequency. It was found that when any two trans-mitters were operated simultaneously they would lock in step on the same frequency, causing the amplitude of the received signal at the third station to be reduced to a point where the receiver relay would not function.

As a result of further development work, the stability of the master oscillator was increased to a point where the control of the generated frequency was sub-stantially independent of the above-mentioned factors. This was accomplished largely by altering the constants of the master oscillator circuit. To prevent two transmitters drifting into step in case of a three-terminal line section, the three transmitters were arranged to operate on slightly different frequency assign-ments amounting to approximately 0.5%, which, while adequate, did not pre-vent all other carrier adjustments from being made on the basis of a single operating frequency.

The results of previous tests, subsequent development work, and operating experience described here have definitely demonstrated that the system of carrier-current protection as now developed is both simple and rugged and will furnish universal and fast protection of the simplest line or the most compli-cated network. Further, by simplification and refinement it has become possible to incorporate the protection scheme as part of the standard switchboard. This, in turn, has resulted in such lowered costs that from an economic stand-point it has become possible to apply such protection to, and justify it on almost any high-voltage line. In short, carrier-current protection has become a fully practical engineering development with a wide possible field of applica-tion.

7. TESTS FOR SELECTION OF SUSPENSION INSULATORS†

FREQUENTLY a section of transmission costing millions of dollars can be put out of commission for periods up to possibly several days by the failure of a single disk insulator, among the many tens of thousands.

Considering the complex structure of the suspension insulator, it is surprising that a study of the several elements entering into its composition and their assembly has not been pursued with the same thoroughness as that applied to many structures of a much simpler nature. Admittedly there has been vast improvement in the design and manufacture of insulators in recent years, but it does not follow that a product uniformity has been reached that will assure satisfactory operating reliability.

Nearly all insulator manufacturers periodically change details of design and methods of assembly which are such that they do not always involve a change in catalog numbers, nor are the customers always advised that changes have been made. It may be assumed generally that these changes result in an improvement of the product, but occasionally some slight modification will affect performance adversely. This may not always be apparent from a laboratory test, and it may take several years of field experience to show up the defect.

This means that the selection of insulators to give the best results from a complete engineering standpoint, involving cost, performance, and deterioration, is difficult if one attempts to do it with any degree of precision. It is necessary to know not only that the design of the insulator is correct but that the product is uniform. For example, insulators manufactured according to the same design, but during different years, for one reason or another, may vary in performance.

The authors, having been connected for the past 15 years with the construction and performance of some 3000 miles of steel-tower transmission lines, mostly 132 kv, believe that some of the methods they have used in the selection and checking of insulators may be of interest.

On the basis of the data obtained they have reached the following conclusions:

1. The present AIEE standard tests for suspension insulators, which are mainly factory acceptance tests, are not in themselves sufficient to furnish a basis for detecting all insulators of inferior design and manufacture, or for comparing competitive makes of insulators.

2. Using results obtained by tests conducted by the American Gas and Electric Company, laboratory tests can be devised that will furnish a basis for proper selection of insulators and permit the weeding out of inferior units.

3. Field testing is essential, and when carried on over a sufficiently long period, can serve as a reliable index of the condition, rate of deterioration, and life expectancy of insulators.

† AIEE Summer Convention (with V. A. Mulford and E. L. Peterson), Hot Springs, Va., June 25, 1934.

4. If proper selection tests are made in advance, the life performance and deterioration of insulators can be predicted with a fair degree of accuracy.

5. Modern suspension insulators of proper design and manufacture, properly selected and correctly applied can be expected to give satisfactory performance over a long period of time, possibly to the extent of 50 years.

Routine Factory Tests

Practically all high-voltage insulators are put through various factory routine tests to eliminate any units that would be unsatisfactory for service. Tests included in this routine inspection vary among the manufacturers, but usually include a 60-cycle flashover of the unassembled bodies, and both 60-cycle and high-frequency flashover of the assembled units, although sometimes the latter test is omitted. Porosity tests generally are made on samples selected from each firing batch of porcelain, and many companies run temperature-change tests on samples of the assembled units from time to time. Some manufacturers include regularly a mechanical test on their suspension insulators at a certain percentage of their rated strength. Routine factory tests usually follow the standards adopted by the AIEE and ASA, which were developed for use as standard factory acceptance tests.

Although factory routine tests are of great assistance to the manufacturers in keeping their products uniform, and the acceptance tests are important to the purchaser, it is questionable whether they give complete indication of the life that may be expected of the insulators. It is believed, however, that supplementary tests can be performed that will indicate the life performance and relative merits of various insulators. In confirmation of this, it has been the authors' experience that insulators that have performed best in the laboratory tests also have performed best in service.

AGE Tests

The mechanical test devised for insulators consisted merely of subjecting them, while carrying normal line voltage, to the maximum physical loads for which these lines were being designed. This maximum design load, amounting to 6000 lb, occurred on single suspension strings when the conductors were loaded with excessive ice. It was assumed that this condition might last for a period not exceeding 3 days. The 72-hr time-load test thus devised was followed by an electrical test to detect any cracks that might have developed. This electrical test consisted of flashing over each unit individually by applying a damped-wave high-frequency voltage for 30 seconds.

Some of the units passing the time-load and electrical tests were subjected to a thermal-change test wherein the units were immersed first in boiling water for 10 minutes, then immediately transferred to ice water for an equal period and followed by a short high-frequency test. Five complete cycles of this test were made. Porosity tests consisted of immersing broken porcelain samples in a water solution of eosin dye at a pressure of 260 lb per sq in. for 14 hours.

Subsequent tests included an ultimate electrical and mechanical test on all units passing the previous test. In this test each insulator, while subjected to a potential of 60 kv, was pulled to destruction. A mechanical load starting at 6000 lb, was increased in 500-lb steps every 30 sec until electrical failure, at which time the electrical potential was removed and the test continued to mechanical failure. In addition, the thermal-change test for all units passing the time-load test was revised to allow 10 cycles between boiling and freezing temperatures with high-frequency flashovers following the fifth and tenth cycles only. In the more recent tests, half the units being put through the thermal-change test were subjected at the same time to a mechanical load of 3000 lb, which is approximately the working load of the insulator. The procedure in the porosity test has been varied radically, the most recent test being at a pressure of 10,000 psi for 6 hr in an alcoholic solution of fuchsine dye.

During the test period checks also have been made at certain intervals on wet, dry, and impulse flashover values of the units and puncture values under oil. These tests, however, never have been considered a part of the standard test procedure.

The purpose of conducting the temperature-change tests on units under load was to duplicate as nearly as possible conditions that had arisen in the field causing failures of insulators, duplicates of which previously had passed the usual temperature-change tests.

Field Testing

The first systematic field test by the company on suspension insulators on which a record was kept was conducted in 1922 on some of the lines of the Ohio Power Company with insulator test sticks improvised by its engineers. The test sticks used were of two types—a condenser stick for testing with the line alive, and a spark-gap stick for testing with the line dead. The condenser stick consisted of two condensers in series with an adjustable spark-gap in multiple with one of the condensers; the spark-stick tester consisted merely of a dry battery, spark coil, and adjustable gap.

Field tests were continued on suspension insulators utilizing standard commercial testers, but not until 1929 were accurate records of such tests kept. At present, the individual companies lay out their own testing programs, using results from previous reports as a guide so that the lines having the greater number of insulator failures are tested more frequently. An effort is made, however, to test at least 10% of the insulators on each line every year.

In general the field tests conducted on suspension insulators have shown them to be performing in an excellent fashion. Of 367,963 suspension insulators tested to date only 2664 have been found defective, or about three quartes of one per cent.

Deterioration

The question of life expectancy of any piece of engineering equipment should be raised every time that its design, manufacture, or purchase is considered, if the problem is to be handled on a rational and engineering basis. Obviously the rate of depreciation to be allowed on a piece of equipment depends on life expectancy, although obsolescence frequently plays the major part in that question. In the case of steel transmission towers, it is the general opinion among engineers that where no unusual climatic conditions exist that would accelerate rusting, the life of such a structure can be taken as a rather long period, possibly 50 years or more.

Therefore, if the structure and conductor are good for a 50-year life the logical question is can the insulator be expected to last the same length of time?

Considered by themselves insulators may not have outlived their usefulness where the rate of deterioration is, say, only 10%. Yet when such a condition is reached the insulation of the line as a whole may have become totally unreliable. Thus, if the criterion of usefulness of a line is its ability to carry power continuously, except for causes extraneous to the line itself, then the useful life of the insulation on that line has substantially reached its end when the majority of interruptions to the service are caused by insulator failures.

The percentage of insulator failures determining the end of the usefulness of the insulation of a transmission line cannot be stated generally, but must be determined for each line individually according to the importance of uninterrupted service to that line. The limit possibly may have been reached when the total failures to date are between 5 and 15% of the total insulators on the line.

8. HIGH-SPEED RELAYING†

ONE needs only to go back as little as a decade to reach a time when clearing of faults in 3–5 sec was considered fast. One-second clearing at that time was considered super-fast. However, since then progress has been such that the term high-speed relaying is now considered to apply to time intervals of the order of 6 cycles maximum to one cycle minimum between the instant of fault occurrence and trip-coil energization. These speeds are in terms of a 60-cycle system, so that in absolute time these invervals would be from 0.10 sec maximum to 0.017 sec minimum. The corresponding breaker time in a modern transmission system would be from 7 cycles maximum down to 2 cycles minimum, or 0.12 sec maximum down to 0.03 sec minimum.

Benefits derived from high-speed clearing of faults are well understood today and are summarized below:

Apparatus Damage.—Extensive experience with high-speed relaying on bus structures, for example, has shown that the damage resulting even from a dead short-circuit with maximum concentration of power at the point of fault will be negligible and will result in nothing more than switch action if the time of switch energization and switch action is short enough.

Line and Insulator Damage.—Burning of high-tension copper conductors ranging from 100,000 cir mils to 500,000 cir mils, or their aluminum equivalent, is reduced to very slight pitting; insulator damage on these lines is reduced to mild markings or discoloration of the glaze; and the burning of lines in two has been almost completely eliminated where the relay and switch action has been made high speed.

Service Outages.—High-speed clearing of faults renders remote the likelihood of a long interruption because instantaneous reclosure is made possible. Furthermore the damage being extremely limited, the likelihood of the switch staying closed upon such reclosure is very much greater.

System Stability.—In an extensive transmission network, the increased stability resulting from rapid clearing of faults is considerable. Thorough studies of system stability recently carried out have shown increases in the power limits of as much as 20% when fault clearing time was decreased from 6 cycles to 3 cycles.

Effect on Load.—The possibility of reclosing rapidly makes even the most sensitive equipment operating on a transmission system insensitive to a switch opening. Even if flashover occurs the equivalent of continuous service can be obtained in most cases if reclosures can be made within 10 cycles. For such service it is thought today that the waiting time between the extinction of the arc and the time of reclosure has to be 6 cycles to prevent restriking of the arc,

† Conférence Internationale des Grands Réseaux Electriques à Haute Tension (CIGRE) (with C. A. Muller), Paris, France, Paper 357, June 27, 1935.

due to incomplete deionization. It is further thought that 10 cycles is the limit that can be allowed between initiation of a fault and the re-establishment of full potential for such sensitive equipment as heavily loaded motor-driven reciprocating compressors. With such equipment, extremely fast relay and switch action is needed to come within the maximum allowable time interval.

Product Damage.—The reduction of time of relay and switch action results in improved products in many processes and in some, such as cotton weaving, the very lowest time attainable is almost mandatory.

Customer Satisfaction.—Lighting service and satisfaction with central-station service are materially enhanced by the reduction of the time taken to clear a fault.

Summed up very concisely, high-speed relay and switch action is necessary for maintenance of proper service and is an almost necessary part of modern electric service.

Methods for Obtaining High Speed

High-speed relaying for single line faults on two or more parallel lines has been accomplished by using current-balanced protection at the source end and duo-directional balanced protection at the load end of the lines. Instantaneous-current, mechanically-balanced relays of the plunger types have been used which operate within one cycle. For duo-directional balanced protection two three-phase power-directional relays with voltage restraint in conjunction with three instantaneous over-current relays are employed, operating within one cycle. Voltage restraint is employed on the power-directional relays normally to hold their tripping contacts open. In case of an unbalanced current sufficient to operate the over-current relays, voltage restraint is removed from the power-directional relays by the circuit-opening contacts of the over-current relays, allowing them to operate as sensitive power-directional relays. The circuit-closing contacts of the over-current relays connected in series with the power-directional relay contacts complete the trip circuit. For current-balanced and duo-directional balanced protection, the trip circuits of the circuit breakers are interlocked by auxiliary switches on the breaker proper to cut out the balanced protection when one of the circuit breakers is opened.

Obviously balanced protection is limited to single line faults on two or more lines and will not provide adequate protection for simultaneous faults on two or more parallel lines, nor for faults on single line circuits.

There was an economical way of obtaining fast relaying for simultaneous trouble on two parallel circuits and for trouble on single line circuits, consisting of a simple scheme of protection costing less than one-tenth the cost of distance relays, which did not require the retirement from use of the previously developed relay installations. Further, the arrangement gave instantaneous protection for ground faults, as well as phase-to-phase faults for approximately 90% of the cases of trouble that occurred on our transmission systems.

This scheme consists of three instantaneous over-current relays, two of which are connected as phase relays and the third as a ground relay working in conjunction with the already existing and installed directional-power relays and induction-type time-delay over-current relays on each circuit. These instantaneous phase and ground over-current relays are set to pick up at a value of phase or ground current which is slightly greater than the maximum current that would flow on any line section from the bus into the line on a through fault. With the instantaneous over-current relays so set, relay action is obtained instantaneously (1 cycle) for internal faults up to 90% of the line section.

In the great majority of cases after the breaker at one end opens, the relay settings on the breaker at the other end, due to the increased current flow over that breaker, cover over 100% of the line section and immediately trips. With this installation of instantaneous over-current relays, the settings of the time-delay over-current relays can be materially decreased since they need only to cascade with each other for a fault at the end of a line section instead of at the beginning of a line section. It is generally found in a loop circuit containing five sectionalizing stations radiating from a generating source, that the reduction in relay settings of the time-delay over-current relays on the breakers at the generating station will be from 60 to 90 cycles for a fault at the end of the line section if instantaneous over-current relays are installed at each of the stations in the loop.

Well over 90% of the faults occurring on our lines have been cleared by the instantaneous relays. In practically every case of simultaneous trouble on two circuits of a double-circuit line, all four breakers have tripped from the instantaneous relays. Installation of this instantaneous over-current protection on our transmission systems has almost entirely eliminated customers' complaints of motors dropping off the line from voltage surges. In addition, maintenance costs of transmission systems have been greatly reduced since emergency repairs of transmission lines have been reduced to almost zero.

Ultra High-speed Protection

Although the relay schemes employing balanced-line relays and instantaneous over-current relays definitely set to clear faults within a given distance of the outgoing bus do give high-speed relaying for the greater majority of the faults that occur on a transmission system, they do not give complete and rapid protection to any line section over its entire distance, regardless of the type of fault. On important line sections, such as short parallel lines radiating out from generating stations and on single- or double-circuit tie lines between two generating sources with a series of intervening sectionalizing stations, it is particularly essential that all types of faults be cleared as rapidly as possible to keep within the stability limits of the system. To give complete high-speed protection for these important line sections, we have used carrier-current pilot-wire relaying extensively in addition to balanced-line relays and instantaneous over-current relays.

The standard scheme of carrier-current pilot-wire relaying employed by the authors consists of one three-phase duo-directional power relay without voltage restraint, three instantaneous over-current relays, three definite-time induc-

22a*

tion over-current relays, and one receiver relay at each station. The power-directional relay is so connected that for power flow from the bus into the line its tripping contacts close, and for power flow from the line into the bus its transmitting contacts close.

The three instantaneous over-current relays, usually called the starting relays, two of which are connected as phase relays and the third as a ground relay, have their contacts connected in parallel but in series with the transmitting contacts of the power-directional relay to apply plate voltage to the transmitter. Hence, carrier current will be transmitted only after a predetermined value of current flows in the line section and only at that station where the direction of power flow is from the line into the bus. The three definite-time over-current induction relays, usually called the tripping relays, two of which are connected as phase relays and the third as a ground relay, close their contacts upon a predetermined value of current in 3 to 4 cycles. Their contacts are connected in parallel but in series with the tripping contacts of the power-directional relay and the receiver-relay contacts to the trip coil of the circuit breaker. The receiver-relay contacts are normally closed but on reception of carrier current open and block the trip circuit. Hence, no tripping operation will occur unless a predetermined value of current flows into the line section, and no carrier current is received unless power flows from the bus into the line section.

The operation of the system for an external fault to the protected line section is as follows: The instantaneous over-current starting relays at both stations close their contacts; at the station where the power flow is from the line into the bus, the transmitting contacts of the power directional-relay close to transmit carrier current which is received at the other station (where the flow is from the bus into the line) to actuate the receiver relay, thereby opening the trip circuit. But at the same station where the power flow is from the line into the bus, the tripping contacts of the power-directional relay open, thereby preventing tripping of its circuit breaker. Thus, at both stations the tripping of the circuit breakers is prevented and the line section holds intact.

The operation of the system for an internal fault to the protected line section is as follows: At both stations the power flow will be from the bus into the line section. The transmitting contacts of the power-directional relays will open, preventing the transmission of carrier at either station, and the tripping contacts of the power-directional relays at both stations will close. After the contacts of the definite-time over-current tripping relays at both stations close, the circuit breakers at both stations will trip.

In applications where ground-directional relays are necessary, the relay arrangement is modified as follows: the contacts of the ground definite-time tripping relay are connected in series with the tripping contacts of the ground-directional relay and the contacts of the receiver relay. The circuit-closing contacts of the instantaneous over-current ground starting relay are connected in series with the transmitting contacts of the ground-directional relay to apply plate voltage to the transmitter. In addition the ground starting relay has a set of circuit-opening contacts which are connected in series with the transmitting contacts of the three-phase power-directional relay and the phase starting relays. Hence, in case of a ground fault the control of carrier-current transmission is taken away from the three-phase power-directional relay and given to the ground-directional relay.

For phase-to-phase faults the operation of the system is exactly the same as above described. In case of a ground fault, the operation also is similar except that the control of carrier transmission is taken over by the ground-directional relay and no carrier current can be transmitted by the three-phase power-directional relays, regardless of their action.

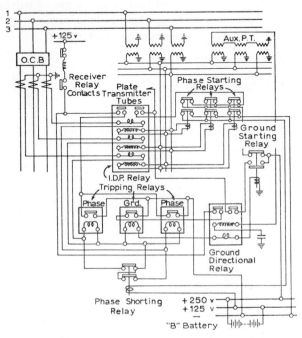

Fɪɢ. 1. Under-voltage relays start carrier and tripping relays.

Another modification of the relay arrangement, Fig. 1, that we have employed is to replace the instantaneous phase starting relays by instantaneous under-voltage starting relays. In this case the three instantaneous undervoltage starting relays are connected across the three-phase voltages, one set of normally open contacts are connected in parallel and in series with the transmitting contacts of the three-phase power-directional relay to apply plate voltage to the transmitter and the other set of normally open contacts are connected in parallel to energize an auxiliary dc relay having two sets of circuit-opening contacts. These contacts normally short circuit the current coils of the carrier-current phase-tripping relays and back-up phase relays.

This modified relay arrangement is applicable where the normal load current, especially in case of single line operation, may be nearly equal to the short-circuit currents obtainable. Otherwise, in order to prevent continuous operation of the transmitter and receiver for long periods of time, it would be necessary to set the carrier-current starting and tripping relays for phase protection at high pickup values, thereby decreasing the sensitivity of protection considerably.

One-cycle Carrier-current Relaying

Those responsible for the operation of the high-tension systems with large concentrations of power have for some time realized that the relay speeds heretofore attainable on carrier are no longer adequate. Where instantaneous reclosure is desired, or where system stability is an economic consideration, a maximum of one cycle for relay action was all that could be permitted on a 60-cycle system. During the past two years the authors have carried out extensive research and development work on a remedial arrangement.

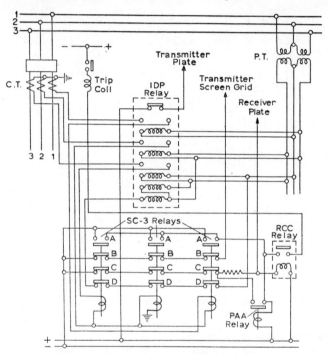

FIG. 2. High-speed carrier relay arrangements.

Shown in Fig. 2 is a diagram of the high-speed arrangement without ground-directional relays the outstanding feature of which is the employment of three simple, multi-contact, instantaneous, over-current relays (SC-3) in combination with one three-phase, power-directional relay (IDP) with voltage restraint, and one receiver relay (RCC). All contacts of the instantaneous over-current relay are circuit opening except the tripping contacts which are circuit closing. This combination gives a circuit having the equivalent of a continuous carrier circuit with the battery taking the place of carrier under normal conditions, while eliminating all the disadvantages of continuous carrier. Because of the use of a single relay the system lends itself to a maximum reduction in the time required for tripping, which is determined by the combined time required to operate the circuit-opening contacts of the power-directional relay, and to close the contact of the receiver relay based on the assumption that the multi-contact instantaneous relay time is less than the opening time of the power-directional relay. This has proven not to be difficult since such a relay can be made to function in 0.1 cycle. The time for circuit opening of the power-directional relay used in the present arrangement is between 0.4 and 0.5 cycle, and the time of closing of the receiver relay is approximately 0.3 cycle, making the overall minimum time approximately 0.8 cycle.

Under normal conditions, the IDP directional relay contacts are held closed by voltage restraint applying plate voltage to the transmitter; but the transmitter does not operate due to the normally closed contacts "B" of the SC-3 current relays applying a negative bias to the screen grid of the transmitter. Further, the RCC receiver relay contacts are held open by the closed contacts "C" of the SC-3 relays energizing receiver relay coil from the station battery.

In case of an external fault, the instantaneous SC-3 relays operate at both stations to remove voltage restraint from the directional relays, bias from the screen grid of the transmitters, and local battery supply from receiver-relay coils. The instant that the bias is removed from the screen grids, the transmitters at both stations operate, thereby holding receiver relay contacts open. At the station where the power flow is from the bus into the line, the directional-relay contacts open, stopping the local transmitter from operating, but at the other station the transmitter is operating, causing carrier to be received, preventing the receiver-relay contacts from closing. After 3 cycles, the PAA lockout-relay contacts at both stations close to prevent the receiver-relay contacts from closing on sudden reversals of power flow after the fault is cleared. Therefore, at both stations the receiver-relay contacts remain open, thus preventing tripping of the circuit breakers.

In case of an internal fault the operation is similar except that the directional-relay contacts will open at both stations, stopping transmission of carrier and causing the receiver-relay contacts to close at both stations and trip their respective circuit breakers.

9. DIFFERENTIAL BUS PROTECTION†

DIFFERENTIAL protection of buses as an economical means of obtaining reliable electrical service has not received the widespread recognition it deserves, possibly because it is somewhat strange and new.

Its essentials are extremely simple. Basically the total incoming and outgoing currents in any section of bus are equal and in opposite directions under normal conditions. If a fault develops on the bus the perfect balance is upset and a relay current calls for the immediate clearing of the section in trouble. If this relay action is sufficiently fast it generally is possible to clear the equipment before any serious damage results. Thus differential protection is not a technical fad, but an essential tool and is comparateve inexpensive for the service it performs.

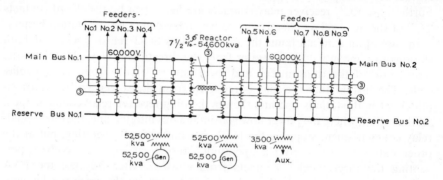

FIG. 1. Differential relays applied to five sections of 66-kv double bus.

Bus differential relays have, as shown in Fig. 1, been applied to the five sections of a 66-kv double bus of a large modern steam station in which all breakers and buses are normally energized. Only because of differential protection is it possible to do this safely and accomplish as well the following:

(a) Employ all four buses continuously; in case of a bus fault occurring on any bus section no interruption will occur to any of the circuits.

(b) Bus short circuits or troubles are cleared by relay action of approximately one cycle. Absence of long-time arcs means the unlikelihood of any appreciable bus damage.

(c) Having the reserve bus energized continuously (which can be done with this differential scheme) means greater certainty of availability of one bus whenever the other bus is in trouble.

There are a number of difficulties which boil down to fear of an incorrect operation that would result in disconnecting important apparatus when no trouble exists. While this fear has not materially retarded the installation of

† *Electrical World*, May 9, 1936.

differential protection on generators, large motors, synchronous condensers and power transformers it has had a paralyzing effect with respect to busbars.

The reasons underlying this fear are tangible enough:

1. Inequality in ratios of current transformers: with heavy overload currents this can produce sufficient unbalance current to operate the differential relay.

2. Open circuits in the current transformers may cause differential relay action either with normal load current or through short circuit.

3. Relay testing accidents: an accidental opening of any one of the current transformers' secondary circuits which might result in improper relay action.

4. False operations due to some abnormal switching set-up which has the effect of disturbing the normal differential set-up.

Remedial Measures

Because all these possibilities are likely, preventive measures have been worked out and found effective for more than 10 years.

Specifically, to overcome possible difficulties due to any one of the four contingencies enumerated above, there can be applied the following:

1. To prevent inaccuracies of current transformers from causing false operations on through short circuits, it has been found advisable to choose their ratios so that the maximum through short-circuit current obtainable will produce a secondary current of less than 50 amp in any current transformer; most transformer designs generally hold their ratio up to that secondary value. This practice has been reinforced by installing a terminal cabinet in the case of outdoor yard practice, centrally located with respect to the various circuits, so that the lengths of secondary leads are substantially the same; thereby minimizing unbalance due to unequal burdens. At this cabinet the various transformer leads are totalized and one set of common wires is taken to the switchboard for the differential relays.

2. To prevent false operation in case of an accidental opening of a current transformer, it has been advisable to set the differential relays so that they will not operate on a current less than twice the normal load current taken by any circuit forming part of the differential group.

3. To prevent false operations due to more or less permanent open circuits in the current-transformer secondary leads a 1-amp ammeter and test switch are provided, as shown in Fig. 2. The latter normally disconnects the meter from the differential relay, but can be flipped to the test position to measure the unbalance current at any time. At manually operated substations this testing is carried out by each shift, the operators being required to read the unbalance current on every differential ammeter the first thing when they come on duty. If the unbalance current is more than normal they are expected to cut the differential relays out of service and immediately call the relay engineer. In automatic stations the reading is taken periodically by trained test men so that the unbalanced condition is detected before a through short circuit occurs and possibly cause faulty action.

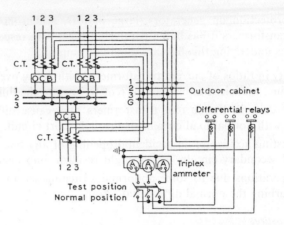

FIG. 2. Ammeter with test switch to measure unbalance of differential current.

4. To prevent false operation of differential relays due to abnormal switching setups, it has been found advisable to so interlock the differential relays that under any abnormal setup they are cut out of service automatically. This obviously causes definite loss of the differential protection until normal switching conditions are restored but it gives so much added confidence in the perfect functioning of the differential scheme that its loss for the short period that the abnormal switching situation exists is well worth the price paid.

Proven Results

These precautions have been put to practical test on the system with which the author is associated thereby enabling the following conclusions to be drawn:

Differential protection of buses greatly improves performance and reliability of electric supply systems. Not only is damage minimized thereby but customers' equipment is subjected to minimum disturbance. Once bus trouble has been cleared by differential protection, the equipment should be in shape to go back into service because of reduced damage even though some repairs of a permanent nature might have to be made later at some convenient time. We have averaged approximately two cases of bus trouble per year. Yet in every case, due to the almost instantaneous (one cycle) bus-differential protection, the trouble has been cleared so rapidly that no extensive or substantial damage resulted and the bus could be re-energized immediately.

The benefits from this bus differential protection are not only the intangible benefits of better service and good will, but the very tangible benefits of de-creased damage to equipment and, therefore, decreased maintenance costs. From this it seems clear to the author that no important high-tension bus or station can afford to be without it. When all theoretical considerations and extensive practical experience definitely show an adequate return from every standpoint on the cost of providing differential protection, there seems to be no reason for deferring such essential protection on any bus supplying a load area of any consequence.

10. ULTRAHIGH-SPEED RECLOSING OF HIGH-VOLTAGE
TRANSMISSION LINES†

THE problem posed was the attainment of such a rapid circuit-breaker reclosure that the net effect on load would be the same as if no interruption had taken place—that no load would be lost. Thorough exploration of the possibilities of ultra-rapid reclosure indicated that the following were essential parts of the solution to the problem:

1. Relaying must be provided which will locate the fault and communicate the trip impulse to all breakers involved in the shortest possible time, preferably not exceeding one cycle.

2. Reclosing mechanisms must be provided which will open and reclose the breakers in a total time somewhere between 0.25 and 0.50 sec (15 to 30 cycles), preferably near the higher speed.

3. The circuit breaker must extinguish the arc in the least possible time, preferably not to exceed 5 or 6 cycles.

4. The arc causing the interruption must not restrike when voltage is restored.

Some of these requirements were already met, or the facts about the phenomena involved were definitely known. For example, standard breakers could be obtained to operate in not to exceed 8 cycles, and special breakers in not over 3 cycles. Experience had been obtained previously with a relay system functioning in a period not exceeding one cycle. However, the problem of getting a mechanism that would reclose a standard breaker in 7 to 8 cycles, or 15 cycles after initiation of a fault, had not been solved. Although the action of the arc upon re-establishment of potential had been explored previously by others, insufficient and inconclusive data were available for a satisfactory solution of the problem. Nevertheless, it was decided that the time had arrived when it was necessary to attempt to solve this particular problem, taking advantage of what was available, and developing or exploring those parts that were available from an equipment standpoint or unknown from a theoretical standpoint.

Carrier-current Relaying

Choice of relay equipment was determined by the desirability of high-speed operation combined with nearly simultaneous energization of the trip circuits of the breakers at each terminal. Carrier-current relaying is the only practical system which provides high-speed operation for any fault at any location on the circuit being protected. When it came to a decision as to the type of carrier relaying to be employed, it was a matter of plain arithmetic that a one-cycle system was almost essential. It was obvious if 6 cycles is allowed for arcing and 8 cycles is permitted for reclosure after the arc has been extinguished then, if the total time of 15 cycles is to be adhered to, the time available for relaying is only one cycle. If a larger amount of time was satisfactory for reclosure, some leeway was available in the relaying time but the one-cycle system was ideally suited for this problem.

† *Electrical Engineering* (with D. C. Prince), January 1937.

In the one-cycle system the trip circuit is held open normally by the receiver relays and it is closed only by the dropping out of these relays when no carrier signal is received. Since the relays controlling the trip circuits at both ends are simultaneously de-energized by the stoppage of the carrier signal, the actual closing of the trip circuits will occur at nearly the same instant.

Because of the importance that the relaying plays in the entire scheme, it may be pertinent to review the salient parts of the relay plan, their function and performance.

Under normal conditions the directional-relay contacts are held closed by voltage restraint, and applying plate voltage to the carrier transmitter, but the transmitter does not operate because of the normally closed contacts of the fault-detector relays which apply a negative bias to the screen grid of the transmitter. Furthermore, the receiver-relay contacts are held open by the closed contacts of the fault-detector relays. In case of an internal fault the directional relays will operate to stop the transmission of carrier at both ends of the line. When this occurs the receiver relays are de-energized and they both drop out, completing the trip circuits. The receiver relay operates in an average time of 0.35 cycle. The directional relay opens its contacts in an average time of 0.40 cycle so that the fastest time possible is approximately 0.75 cycle.

Tripping and Reclosing Mechanisms

Most standard mechanisms require about 0.5 sec to close the circuit breaker. This has been accomplished by connecting two mechanisms together through a walking beam, one of which is at all times at rest with its latch reset. When tripped the circuit breaker opens until the energized mechanism builds up enough force to overcome the opening springs and momentum.

It was obviously necessary to secure a much higher speed than was obtainable in this way. Accordingly, one of standard mechanisms was replaced by a spring mechanism of the stored-up energy type with two sets of springs, as shown in diagram, one set to reinforce the breaker opening springs, the other set to close. When charged this mechanism performs an opening and closing operation, its normal position being closed.

The spring mechanism has a high-speed latch of the flux-shifting type which operates with the opening springs to give a contact parting time of from 2.5 to 3.5 cycles. After the contacts have moved 8 in. a link attached to the opening mechanism trips the auxiliary opening springs and releases the closing springs. These then reclose the breaker.

During the reclosing stroke the tripping circuit is transferred to the motor mechanism so that a permanent fault will be tripped off at once. As adjusted for test, one of these mechanisms gave an overall time from initial trip impulse to contact make of 15 cycles. Since it proved rather difficult to hold this fine adjustment throughout the service aimed at was nearer 18 cycles.

The mechanism is set for reclosure upon the completion of 8 in. of stroke without waiting for arc extinction, so that the circuit breaker had to extinguish the arc on the completion of that allowed stroke, with enough time to spare for the tripping relays to reset.

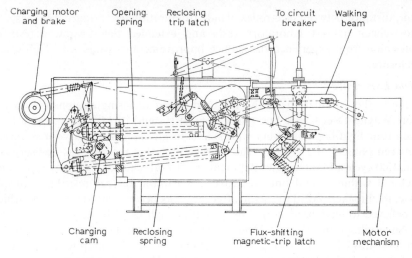

Ultrahigh-speed spring reclosing mechanism for circuit breakers.

Basic Ideas of UHS Reclosing

As previously pointed out, one of the essential requirements for ultrarapid reclosing was that the actual arc causing the service interruption would not restrike when voltage was restored. The usual fault against which protection is sought is caused by a lightning stroke. In such cases the circuit cannot be re-energized until the ionization produced by the stroke has been dissipated enough that the restored voltage may be withstood. Griscom and Torok have investigated this problem, testing with currents from 800 to 1500 amp, voltages from 66 kv to 130 kv, and electrode spacings from 12 to 93 in. They concluded that:

The restriking of arcs is a random phenomenon.

The probability of restrike increases with operating voltage for spacings ordinarily used.

The probability of restriking is only slightly affected by large variations in current, and

For a given recovery voltage above a certain spacing, the probability of restriking is but slightly affected by the spacing of electrodes; below this value, spacing plays an important part.

They found that the restriking times vary from about $1\frac{1}{2}$ cycles at 66 kv to a little over 3 cycles at 130 kv. The authors of this paper made some preliminary tests in 1935 under conditions similar to those used by Griscom and Torok but with currents of the order of 5000 amp and found a probable restriking time of nearly 5 cycles.

What would be the practical effect of multiple strokes of lightning? Would field experience show a larger or smaller proportion of successful reclosures than obtained in fair-weather tests? It was decided to undertake tests along these lines: Additional laboratory tests to clarify further the effects of voltage, current, distance between electrodes, wind, arc duration, etc., staged field tests to cross-check against laboratory tests and extended field-operation tests to determine from experience the service improvement possible with high-speed reclosure.

Field Experience

Since the installation of the ultrahigh-speed reclosing equipments on the Fort Wayne–Deer Creek line (AGE central system) in May 1936, there have been four lightning flashovers on this line. In all cases, the flashover occurred between phase-3 and ground. The average time that the line was de-energized was 13.0 cycles. In three cases the breakers at both ends of the line tripped and reclosed within a total time of 22.0 cycles without the arc restriking. In one case, the breakers failed to stay in after reclosing because the arc restruck, which is believed to have been caused by a multiple lightning stroke.

An oscillogram taken at Deer Creek during a flashover from phase-3 to ground showed the following: The fault current of 1165 amp was interrupted in 6.0 cycles: the breaker contacts reclosed in 21.5 cycles after the trip circuit was energized and the line was de-energized for a period of 12.5 cycles.

Future Possible Applications

Application of ultrahigh-speed reclosing may make it possible in some instances to reduce the line investment now required to give uninterrupted service to large industrial loads. With the further development of this idea duplicate lines may give way to single lines with the possibility of giving as good service as is now given with multiple feeds.

The field of application of UHS reclosing will be determined largely by the additional cost of the mechanisms compared with standard mechanisms, assuming quantity production will be possible. In any event it appears that a new and ingenious tool has been made available to the transmission-system designer and operator that gives promise of improved service and lower costs.

Conclusions

Perhaps it is premature to draw final conclusions as to the significance of this development, although it would appear that the following have been fairly well established:

1. On high-voltage lines outages caused by lightning and possibly other factors can be materially reduced, perhaps to the extent of 75%, by ultrarapid reclosing.

2. While a great deal of benefit can be obtained in reducing outages resulting from phase-to-ground troubles, phase-to-phase outages need to be investigated either to bring them within the range of treatment successfully applied to phase-to-ground faults or to reduce their frequency.

3. Finally, as in so many other cases of research and development, it appears that the by-product of the search for higher-speed circuit-breakers and higher-speed relay protection has given most fruitful results, perhaps equal to those obtained from the original line of research, and with promises of a great deal more.

11. HIGH-SPEED MULTIBREAK 138-kv OIL CIRCUIT BREAKER†

HIGH-SPEED oil circuit breaker development was really started in the period 1929–30; the first tests with high-speed 138-kv breakers were made at Philo in July 1930. While it was definitely expected at that time that operating benefits would be obtained from the higher speed the problem was not pressing.

Anteceding that was the development in 1926 of high-speed relaying, in what proved to be an impractical form. At that time the aim was for a universal relay system that would give an impulse to the breaker trip circuit within 20 cycles.

The two developments went along in parallel although practical achievement in breaker speeds was attained much earlier than in relaying. However, by 1935 both had been so improved that it was routine to get a 6- to 8-cycle 132-kv breaker of fairly standard construction and get a relay system that would give universal protection in 1 cycle or less. Development in breakers had been carried through not only in the 138-kv classes but also in the lower-voltage classes. For special services, such as the Boulder Dam job, breakers of much faster speed were needed and 3-cycle breakers of the impulse type were developed and built.

Solution of typical stability problem was made possible by an 8-cycle breaker in combination with a 1-cycle relay system. However, that is no guarantee that similar or other more involved problems in the future will not call for even faster time. If it is granted that 1-cycle relay time is about as fast as can be practical, it is obvious that a much higher breaker speed than 8 cycles is needed.

Such a need arose in connection with the development of ultrahigh-speed reclosing of transmission lines. Although such reclosing was successfully accomplished with a breaker having very little better time than 8 cycles, it was evident that the amount of waiting time necessary to allow deionization after the line was de-energized was not only a function of the arc current and atmospheric conditions, but also was dependent upon the duration of arcing. Obviously if the arcing time could be reduced, the total time necessary for reclosure would be reduced not only by that time but also by the reduced waiting time.

Realization that the solution of many transmission problems was possible only by high-speed interruption led to a decision some 2 years ago to apply the knowledge gained in the development of the impulse breaker to the development of a breaker having substantially higher speeds than appeared possible with the conventional oil-blast breaker; and with no insurmountable economic handicaps.

In short, it was hoped to get a breaker having an interrupting time close to 4 cycles in a sufficiently simple design that would not involve an appreciable, if any, increase in cost. Having such designs available on an economical basis when they were actually needed would aid considerably in maintaining reliability and keeping system growth within economic limits.

† AIEE Summer convention (with H.P. St. Clair), Washington, D.C., June 20, 1938.

Conclusions

Field tests of the new high-speed interrupter demonstrated that no matter how skilled the designers nor how exhaustive the factory testing may be, it is still necessary to look to high-capacity field tests to bring out possible weaknesses which otherwise may escape detection. It is no reflection whatever on the designers that unforeseen weaknesses were disclosed in the field tests; on the contrary, it is proof of the soundness of the design that only comparatively minor changes based upon the tests were necessary. The final design went through what is believed to be the most severe series of tests yet made on a high-voltage circuit breaker, and it did so with perfect performance.

It is also interesting to look at this development in the light of what has recently been said and written in considerable volume on the subject of oilless and oil-poor circuit breakers. Many engineers have felt that such breakers might eventually become the final answer to the circuit-breaker problem. However, here we have in an oil circuit breaker of the conventional type the culmination of many years of orderly development which have resulted in a 1,500,000-kva unit that has demonstrated its ability to interrupt repeatedly and with no apparent effort short circuits up to a maximum of 2,000,000 kva, with speeds consistently under 4 cycles, Viewed from this angle, we cannot help feeling that this development is truly a tribute to the skill and ingenuity of our American designers.

12. EXPERIENCE WITH ULTRAHIGH-SPEED RECLOSING †

ALTHOUGH the effects and phenomena behind the principles of ultrahigh-speed reclosing of high-voltage transmission lines are generally known, there are still a number of unknown factors that have not been explored fully. Their effects need to be known to make a thorough and scientific application of these principles. For example, further data are needed on the effects of breaker time and elapsed time between clearing of the arc and re-energization of the circuits on the deionization of the arc and re-establishment of the insulation strength of the surrounding area. Data are needed on the probable extent of the total elapsed time and intervals between multiple strokes. Knowledge is also needed concerning the effects of the type of load and the likelihood of an arc restriking, or the failure of lines to hold when re-energized because the systems drift away from synchronism, and finally on the dead time needed after phase-to-phase faults to prevent restriking.

While waiting for the results of such further investigation there is no reason why high-voltage lines of any importance should be designed and installed without ultrahigh-speed reclosing. Since the presentation of the first data on this subject, three additional line sections, that is six terminals, have been equipped with ultrahigh-speed reclosing and a number of others are under way.

Elements of UHS Reclosing

An ultrahigh-speed reclosing setup consists of three elements—the high-speed breaker, the ultrahigh-speed mechanism and the high-speed relay system.

The circuit breaker should be capable of interrupting a fault and de-energizing the line in the least possible time. So-called high-speed breakers available heretofore in the Unites States have required approximately 8 cycles to operate after the trip coil was energized. This does not refer to special designs. Within the last year more or less standard breakers having a total time not to exceed 5 cycles have been developed; although faster speeds are expected, they are not available today in the American market. The work described herein has been carried out on breakers operating in approximately 8 cycles.

The reclosing mechanism of the breaker has to be so designed that it is capable of performing its function in no greater time than the minimum necessary to assure complete deionization of the arc. This is essential to avoid its restriking and to assure minimum probability of loss of synchronism or loss of load. With 8-cycle breakers it has been felt that the reclosing time has to be of the order of 7 cycles but this has not been obtainable heretofore on standard breaker mechanisms without running into stresses beyond those desirable or practical. So far this time has been from 10 to 14 cycles.

The relay system must operate positively to clear the circuit on both ends in minimum time. With the limits of breaker and reclosing speeds immediately contemplated the relay time should not exceed 1 cycle. No difficulty has been experienced so far in getting that with the utmost reliability.

† AIEE Winter Convention (with C. A. Muller), New York, N.Y., January 23, 1939.

In the earlier designs of ultrahigh-speed oil-circuit breaker mechanisms, two were provided for closing the breaker. One was a standard closing mechanism for normal operation and the other for high-speed reclosing. These mechanisms were attached to opposite ends of a walking beam the center of which was attached to the rod operating the breaker contacts. Energy for the high-speed reclosure was provided by a heavy spring which was reset automatically after each reclosure by a small motor. Each mechanism was provided with a trip coil but all relays which would normally trip the reclosing mechanism were automatically transferred to the standard mechanism after a high-speed reclosure and remained there until the reclosing spring had been reset.

The high-speed reclosing mechanism was also provided with a spring to speed up the opening of the breaker contacts. When the reclosing trip coil was energized this spring pulled the breaker contacts through a travel of about 8 in., at which point a latch released the closing spring, thus completing the reclosing operation. It was found that these mechanisms could be adjusted to give reliable operation with 18 cycles total elapsed time from trip coil energization to reclosure of breaker contacts.

Because two mechanisms were required for each breaker, the total space needed for the complete equipment was considerably increased over that required by a standard electrically-operated breaker. Further, the space occupied by the high-speed reclosing part of the device was much greater than that used by the standard closing mechanism, so that the total space occupied by the complete equipment was more than doubled.

It was felt that this space requirement should be reduced if possible and at the same time the mechanism should be simplified in its operation. Keeping these objectives in mind, a new mechanism was designed and built which used the same motors for both normal closing and high-speed reclosing, thus eliminating the need for powerful reclosing springs, reducing the space required for the equipment, and considerably simplifying its operation.

New Ultrahigh-speed Reclosing Mechanism

A schematic diagram of the new ultra-high-speed reclosing mechanism is shown. As in the earlier equipments these new mechanisms are provided with a standard trip coil and a high-speed reclosing trip coil. The closing motors drive a cam which operates the breaker output crank through a roller on its surface. At the end of the closing operation a prop falls into place and holds the breaker contacts closed. At the same time the cam is prevented from returning to the open position by the presetting prop. When the reclosing trip coil is energized the prop is removed and the breaker contacts begin to open. After the contacts have opened a predetermined amount, an auxiliary switch energizes the closing motors starting the cam revolving in a direction to close the breaker. The breaker opens until the cam roller comes in contact with the cam surface; then the contact motion is reversed, and the breaker recloses. Operation of the normal trip coil releases a trip-free toggle opening the breaker without a high-speed reclosure. As in the case of the earlier mechanisms, these new mechanisms permit breaker adjustment to reclose contacts in 18 cycles after the trip coil is energized but with considerably less strain.

Operating Experience.—During a portion of 1938 a total of eight 132-kv breakers so equipped were operating and serving four line sections. Since their installation there has been a total of 15 operations on these sections. In all

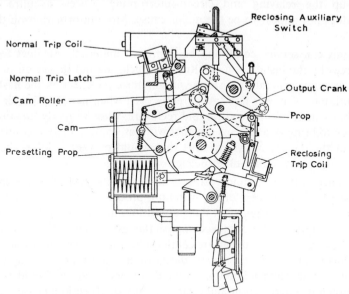

Fig. 1. Diagram of new ultrahigh-speed breaker reclosing mechanism.

cases flashovers occurred between a phase wire and ground, 12 of the flashovers being between Phase-3 which is the top conductor and ground, one between Phase-3 which is the middle conductor and ground and two between Phase-1 which is the bottom conductor and ground. In 13 of the 15 cases the breakers on both ends of the line tripped and reclosed without the arc restriking and without any of the normal deleterious effects of a feeder outage being felt in any case.

In 2 of the 15 cases the breakers failed to stay in after the first initial ultra-rapid reclosure. In at least one of these cases analysis showed that multiple lightning strokes were the cause; it is quite likely that the other failed to stay in for the same reason. However, it is interesting that in the case of one double-circuit line, where approximately 25% of the lightning flashovers involved both circuits prior to the installation of a one-cycle relaying and ultrarapid reclosing, not a single double-circuit flashover occurred in more than 2 years that have elapsed since the installation of the ultrarapid equipment, although no steps of any kind were taken on the line.

Conclusions

The results obtained so far definitely warrant the following conclusions:

1. On high-voltage overhead lines outages caused by lightning can be reduced materially by ultrarapid reclosing. The originally expected figure of 75% seems

conservative in the light of the additional experience obtained since that time.

2. Apparently two-circuit flashover is reduced materially on double-circuit transmission lines when properly equipped with ground wire and with ultra-rapid-reclosing breakers at both ends. This is in a large measure the result of speeding up the relaying and circuit-interrupting process as ultra rapid reclosure does not appear to be the major cause. More information on that is needed.

3. Not only can service continuity be improved materially on lines feeding isolated areas by the installation of ultrarapid reclosing but its use on tie lines between two major generating systems will minimize the effect of line flashover on the continuity of power flow. Apparently even when handling power close to the stability limit of a line, ultrarapid reclosure, when properly functioning, can be made to bring two systems together without waiting for synchronization and without the danger of system drift that would prevent holding the two systems together, or produce other deleterious effects.

4. Experience with ultrarapid reclosure indicates the need for a complete re-examination of the concept of the switching process on any important high-voltage circuit, including relaying, de-energization, and re-energization of the circuit. If ultrarapid reclosure will accomplish the results here indicated in a vast majority of cases, it would appear logical that ultrarapid reclosure to the extent of one reclosure ought to become standard practice and that a waiting period longer than the absolute minimum required for deionization should be permitted between de-energization and reclosure only in exceptional cases.

5. Work on high-voltage systems indicates that very similar results can be obtained on intermediate and low-voltage lines by utilizing the same principle. Exploration of that field is a problem for the future. The authors hope to be able to contribute something to that.

13. GLAZE ON HIGH-VOLTAGE LINES†

OF ALL the hazards in the path of continuous transmission-line performance, glaze or "sleet", with its concomitant problems, undoubtedly is the greatest that as yet remains unsolved. Lightning, which once constituted the major difficulty, has been guarded against to the point where in some cases it is only 10% of the problem it formerly was, and with work now going on there will be a further reduction. But in the case of glaze there has been no such progress.

Compared with lightning, which is a recurring annual phenomenon, glaze may not present any difficulties in a particular territory for 5 or 6 years, and sometimes longer. But when a major sleet storm does take place the effects are so severe that its influence on service may be felt for days, weeks and sometimes even months.

The theory of producing sufficient heat in an ice-covered conductor to break-up glaze is well understood and it is possible to calculate quite accurately the time required to prevent or break-up glaze for any combination of ice thickness, ambient temperature, and wind velocity within the range of permissible currents. The main weakness has been the economic difficulty of isolating a section of line in trouble and connecting it to a segregated generator either directly or through specially tap-provided transformers. Where widely varying lengths of trans-mission line have been involved, special transformers of considerable size with a wide range of taps and connections had to be employed. While this has not prevented the practice of glaze melting it has restricted it to comparatively few locations, usually where glaze is almost a continuous year-to-year occurrence.

Where an area is served by a single transmission line, even though it is double circuited, and where a load of from 40,000 to 50,0000 kw is involved, the loss of a line even for several hours, and not more often than once every 5 years, still constitutes a break in service that cannot be looked at with equanimity. Hence to meet the threat of glaze attempts were made many years ago to prevent ice formation by shifting load. However, because of the limited load that generally can be transferred, and because of the delay in starting the shift, owing to the impossibility of gaging which of a number of lines need first attention, this practice did not prove very effective.

As a later development, steps were taken on the American Gas and Electric system to work out connections on one section of the 132-kv transmission system, so that it could be connected to an available 27-kv bus for sleet melting. But this proved satisfactory only in the case of two lines, each having a length of the order of 30 miles.

Full Line Voltage

Analysis of the problem disclosed that on an integrated system where both power and reactive supplies were available, and where almost no important load was being handled by anything but a double-circuit transmission line, the most effective method was the utilization of the main transmission voltage for sleet-melting current.

† *Electrical World* (with G. G. Langdon and V. M. Marquis), May 20 and August 12, 1939.

To get a proper current, it was necessary to interconnect enough transmission sections between power supply and grounding points to give a short-circuit current that would bring about melting within about one-half to one hour. The current and kvas needed for sleet melting on 397,500-circ mil ACSR conductor for lines of different lengths are shown in Fig. 1.

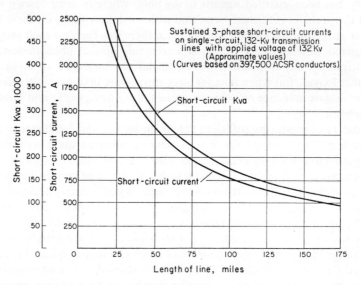

Fig. 1. Current and reactive needed for sleet melting for lines of different lengths.

The magnitudes of power and kva involved generally cannot be obtained economically and easily except on a large interconnected and integrated system. The losses needed for sleet melting require tying up a generator of no mean size, and that demands reserve capacity that cannot usually be found except on a fairly large system. However, once these conditions are satisfied the method described offers a means of combating glaze with a facility and quickness that have never been equaled. Further, it does so with a minimum of additional expenditure for special facilities such as transformers or similar devices. This work emphasizes again the great value of an interconnected and integrated system under emergency conditions.

If melting of glaze is to be done in time to insure continuous performance of transmission lines during sleet storms, it is essential that a dependable means of detecting the presence and severity of sleet conditions on any section of line be provided to serve as an accurate guide for melting. Once glaze is formed on the conductor a light wind may bring about "dancing" of conductors, which often results in conductor burn-downs. Therefore, if trouble or loss of line is to be avoided, it is really necessary to melt glaze early before it gets too heavy. Furthermore, if ice is still forming, delay in melting means a greater thickness on the conductor and hence a longer melting time.

Detection of Glaze

At nine stations and junctions on the AGE system, the carrier-relaying equipment is being used successfully to give an indication of the presence and degree of accumulation of ice on the conductors. A relatively simple addition to the carrier-receiving apparatus reveals the progressive attenuation in sections where the glaze is forming. The technique has been useful in improving the service continuity of the 132-kv lines during the glaze season.

It is not particularly difficult to determine when to melt if conductor glaze occurs uniformly throughout the length of a line and if it can be seen at the terminals of the line. However, experience has shown that sleet is frequently confined to a relatively restricted area and, while there may be no evidence of glaze formation at the terminals of the line, there may be a severe sleet condition at remote points. Quite often glaze formation occurs at night when roads may be impassable so that sleet detection by inspection in remote sections is unsatisfactory, if not impossible, until night passes. Even by day bad roads offer a handicap to successful inspection. Though sleet may be found by inspection, it may have put telephone circuits out of commission or telephones may be so remotely located that there is bound to be delay in getting word to the dispatchers. The use of two-way space radio on mobile units, while offering some help in overcoming this difficulty, is only partially satisfactory.

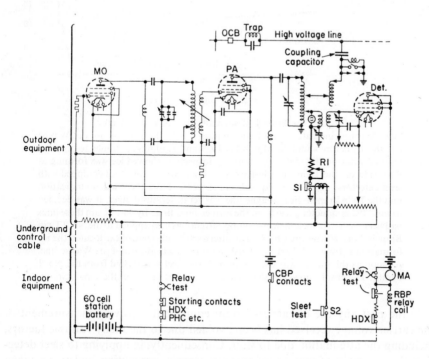

FIG. 2. Sleet detector circuit of carrier receiver.
(Equipment added to standard receiver shown in heavy lines).

Carrier Equipment Used

The idea of using carrier-current transmission as a means of detecting sleet was evolved because an analysis of experience with carrier showed its transmission over high-tension lines is affected appreciably by the presence of glaze. Several sleet detectors, Fig. 2, were built and used with highly satisfactory results during the last sleet season, see Fig. 3. The initial installation was made in 1938 on the section of the 132-kv system where sleet is most prevalent.

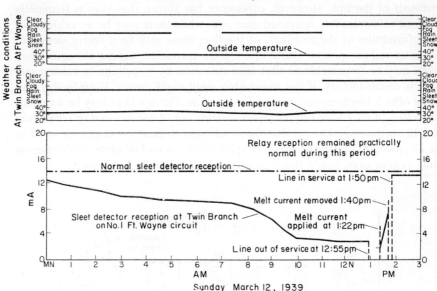

FIG. 3. Case history of successful use of sleet detector
Sleet detectors indicated slight ice formation on the Twin Branch–Fort Wayne line shortly before midnight March 11, 1939 and readings continued to decrease gradually during the night. By morning, inspection showed ice was forming at Riverside on the Twin Branch–Riverside lines (which were not equipped with sleet detectors) and continued to become worse. Since there was no indication that the Twin Branch–Fort Wayne line was yet in serious danger, it was decided to melt the conductor glaze from the former lines first. By 11 : 31 am these lines had been cleared of ice and sleet-melting current was next applied to the No. 2 Twin Branch–Fort Wayne circuits. Shortly afterward a radio patrol car located the ice formation between Elkhart and Goshen on the Twin Branch–Fort Wayne line. As soon as this circuit was back to normal the sleet was melted from the No. 1 circuit. The graph shows the sleet detector readings taken on the No. 1 circuit.

In the past any use of carrier-current equipment involving measurement of the carrier voltage received has been avoided due to the many variable factors, including the attenuation due to sleet. Consequently, in applying to sleet detection the principle of measuring received signals, precaution had to be taken to minimize all variables except carrier attenuation due to sleet within any line section. While the use of separate transmitters and receivers as well as the incorporation of sleet detection on carrier-current telephone and carrier-current metering equipment was considered, it was not attempted on the system with which the authors are associated primarily because of the existence of carrier-current relaying equipments which adapted themselves more readily to sleet detection, due to the following advantages:

The line-trap units which are a necessary part of carrier-current relaying tend to minimize the effect of switching conditions outside the particular section involved, consequently a measurement of received signal level should indicate conditions within the section;

Carrier relaying operates at a relatively high energy level, which makes a highly sensitive detector unnecessary;

Routine testing with sleet detectors is simplified and the routine carrier-relay tests provide a cross-check on most of the equipment used in sleet detection;

The additional equipment required is inexpensive and easily installed.

Operating Experience

The sleet storms which occurred during the past winter provided at least one practical test on each of the detectors that had been installed and on some of them several tests. Advantage was taken in each of these cases to correlate the detector readings with actual sleet conditions on the lines. It was realized that the detectors on different lines would perform differently under similar glaze conditions due to such variables as length of line, carrier frequency, detector adjustment, type of set and other factors, and that in all probability each detector must be calibrated by actual sleet experience. While a complete experience record under varying sleet conditions is not absolutely necessary for detecting the presence of sleet, any such information should be valuable in giving the operating man a better means of interpreting readings in the future. It was considered desirable to determine for each detector any variations caused by conditions other than sleet, snow, etc. Therefore a routine was set up for taking and recording sleet detector readings at the same time the routine carrier-relay tests were made.

Heavy frost was found to give indications of the same order of magnitude as sleet, but its presence was recognized by the fact that weather conditions were outside of the range productive of sleet. In no case did the detector fail to indicate the presence of sleet on a line; therefore it is felt that the indication of such non-hazardous conditions as frost is on the safe side and not a serious disadvantage.

Summary

While it is felt that the use of carrier-current waves for sleet detection is still in its infancy, it is believed that this method gives a direct means for detecting the presence of sleet, snow or frost on transmission lines, even though present only in small amounts or on a small portion of the line. Taking weather conditions into consideration, the detector readings, when properly interpreted,

23 VEP

give the operator a more reliable guide for determining when conductor glaze should be melted than he has heretofore had. Further, it is believed that this method will eliminate the necessity of visual inspection for determining the presence of sleet on transmission lines, although it should not be inferred that visual observations are not useful.

While in no instance did the detectors fail to indicate the presence of sleet on the transmission lines, experience has shown that in general sleet-detector readings taken when conductor glaze is present will be different for each line. Therefore it is necessary to accumulate experience through actual line inspections and to correlate them with detector readings taken under varying glaze conditions before detector readings on a given line can be utilized to the greatest advantage. For this reason additional operating experience will be required to determine definitely the full possibilities as well as the limitations of the detector and to serve as a guide for the most satisfactory design characteristics.

14. FIVE YEARS' EXPERIENCE
WITH ULTRAHIGH-SPEED RECLOSING†

ULTRARAPID reclosure has not only been widely extended but our operating data are now extensive enough to furnish a basis for some well-founded conclusions.

All installations of the reclosing equipment with the exception of four, representing two line sections, are on the central system of the American Gas and Electric Company. This is an integrated system operating in seven states and comprising at this time 1,455 miles of 132-kv transmission, substantially all of double-circuit steel-tower construction, making the total circuit mileage 2483. The last-recorded maximum one-hour integrated peak was 997,000 kw. On this system the generation and transmission are integrated and co-ordinated to the highest degree. Important foreign interconnections on several portions of the system through which diversity and other power exchanges are made necessitate frequent and in many cases sudden changes in power flow over many of the important transmission lines. The concentration of generation in stations at points where combined optimum fuel and water conditions prevail makes the heavy transfer of power over most of the transmission lines economically desirable and necessary. Under these conditions, the importance of continuity of supply is apparent. Hence, the first and so far the widest application for ultra-high-speed reclosing has been found on this system.

Our record covers a total of 33 terminals involving 16 line sections; which include approximately 840 circuit-miles of line, all but 126 of which operate at 132 kv. In the period May 1936 to November 1940, covering substantially five lightning years, 72 cases of flashover occurred resulting in line openings. Of these, 63 were single-line trouble and nine were two-line trouble. In these cases 65 reclosures were successful and seven resulted in restrikes. This is a record of 9.7% unsuccessful reclosures, but in one case permanent physical damage of a tower had previously resulted, so that successful reclosure was a physical impossibility. Disregarding this one case, the unsuccessful reclosures were slightly over 8%. In not a single case did the equipment fail mechanically; reclosure was successful in every case where the circuit condition permitted re-energization of the line.

Among the cases of flashover, 55 were phase-to-ground faults on one line resulting in 53 successful reclosures and only two restrikes, which were probably due to multiple lightning strokes. Three cases of phase-to-phase and ground faults on one line occurred; in each case the breakers reclosed successfully. Five cases of three-phase faults on one line occurred with three successful reclosures and two restrikes. Three cases of phase-to-ground faults on two lines occurred; all were reclosed successfully. Four cases of phase-to-phase and ground faults on two lines occurred with three successful reclosures and one restrike. One case of three-phase fault on two lines occurred which restruck.

---

† AIEE Winter convention (with C. A. Muller), Philadelphia, Pa., January 27, 1941.

23*

Additional Installations

The accumulated operating experience has been so beneficial and the results obtained so much better than the authors had anticipated that when the problem of reliability of supply had to be met on other sections of the system, ultrarapid reclosure was extended and continues to be extended. Thus at the present time there are twelve additional terminals in process of installation covering six additional line sections involving 224.2 miles of 132-kv circuit.

Conclusions

Based upon five years' experience the authors believe that the following conclusions are clearly indicated:

1. In general, on a high-voltage system properly insulated and provided with ground-wire protection, at least 90% successful reclosure can be expected by the use of ultrarapid reclosing.

2. The probability of restrike on double-circuit lines of the type existing on the central system is approximately four times as great as in the case of single-circuit flashovers.

3. Ultrarapid reclosure has established itself as an indispensable protective tool in high-tension integrated power systems. A start has been made in extending it to lower-voltage lines but further development is needed and will depend upon the successful removal of the economic limitation to such extension.

4. There is reason to believe that the present 8.5% unsuccessful operation can be reduced, particularly the 25% failure to reclose successfully on two-line faults. The means for carrying this out so far indicated are by reducing the total percentage of two-line flashovers by further increasing the speed of fault clearance and thus permitting a corresponding increased waiting time before reclosure is attempted.

15. HIGH-SPEED 138-kv AIR-BLAST BREAKER TESTS†

RECENTLY a 138-kv air-blast breaker of novel design was given a series of field tests for normal interrupting duty and for ultrahigh-speed reclosing service. Circuit-interrupting ability at least equal to that expected of any modern oil breaker of conventional design was obtained. In addition, an unusual super-speed reclosing performance points the way to a possible liberalization of existing derating factors for this kind of service.

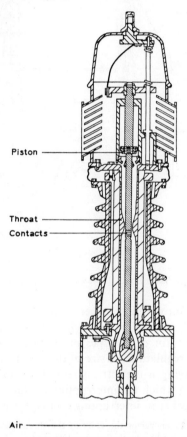

Piston

Throat

Contacts

Air

FIG. 1. Cross-section of 138-kv air-blast interrupter.

While the air-blast breaker is still somewhat of an innovation in the high-voltage field, it is believed that these tests are an important step forward in developing its possibilities for this class of service.

This new unit, known as the conserved-pressure type, as built for 138 kv, has two interrupting units of the axial or longitudinal-blast type in series per pole, Figs. 1 and 2. Two such columns plus a disconnecting member comprises

† *AIEE Pacific Coast Convention* (with H. E. Strang), August 27, 1941.

a pole unit. Each pole has a storage tank in which sufficient air for two complete close-open operations is stored at 350 psi. The three tanks are held together by a common header, and connected to a central air compressor through a double-acting check valve. Each pole has its own electrically operated blast valve to control the flow of air to the contacts. One pneumatic cylinder, with electrically actuated control valves, operates the three isolating switches through an enclosed system of push-pull rods between phase units.

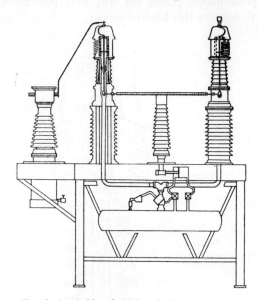

FIG. 2. Assembly of 138-kv air-blast interrupter.

The interrupting action of this breaker is unique in that the arc is drawn into a space deliberately maintained at high pressure, instead of into the free air as has been common for other types of air breakers. This back pressure, which is maintained by regulating the size of the vent from the arcing chamber, provides a medium having a dielectric strength several times that of air at atmospheric pressure, in which the interrupting contacts are separated.

The action of the breaker in interrupting a circuit is as follows:

1. The protective relay energizes the coils of the three blast valves, causing them to admit air to the passage leading to the interrupting units.

2. Contacts are separated by action of the pistons in each unit.

3. The arc is drawn into the insulating orifice through which air is passing where it is extinguished; the moving contact continues into the area of high pressure and high dielectric strength which prevents the arc from restriking.

4. A definite time after the blast valves have been energized (they are interlocked pneumatically to require action of all three valves) air is admitted to the disconnect-actuating cylinder.

5. After the disconnect has started to open, blast air is cut off, allowing the interrupting contacts to return to the normal closed position after the isolating switch has opened.

The closing action is performed entirely by the isolating switch, its action being fast and controlled by a positive driving force, so that it is capable of closing repeatedly against currents as high as 6000 to 10,000 rms amp at 132 kv without harmful effects.

The development of an outdoor, oil-less, high-voltage circuit breaker, which to some may appear to be of only academic interest at the present time, may emerge as a timely, significant and much needed undertaking. It is not at all inconceivable that restrictions of one form or another may be encountered in the use of oil for future breakers. This leads pertinently to a discussion of some of the advantages and disadvantages of oil as used in conventional oil circuit breakers.

Pros and Cons of Oil for Breakers

It cannot be denied that oil has proven to be an excellent medium for circuit breakers, or that it has helped to bring them to the present high state of development. Some of the advantages of oil are; high insulating value, uniformity, good arc-quenching or cooling medium, particularly effective when properly controlled, and has a background of decades of experience and development.

The disadvantages of oil are likewise quite real, and by comparison with air may in the course of time appear even greater.

Oil is a definite fire hazard, though for out-door breakers this hazard is not regarded as very serious. It presents a maintenance problem of sizable proportions from the standpoint of cost and time to condition the oil, including filtering and drying out, and the equipment required for that purpose, and to handle the oil, both for conditioning and for maintenance of the breaker, as well as the pumping, piping, and storage facilities required.

Viewing the comparison from the standpoint of possible future war conditions, oil circuit breakers may present other serious disadvantages. Also the normal fire hazard from oil circuit breakers may be increased substantially by greater duties accompanying the rapid growth of systems, and possibly by the necessity for larger physical concentrations or greater crowding of oil circuit breakers.

The value of high-speed reclosing on high voltage is becoming more widely recognized and the demand is increasing for breakers capable of performing this duty. To find wide application, therefore, any new type of circuit breaker must be capable of clearing a fault promptly, reclosing preferably in less than 20 cycles, and then clearing again in the event the fault on the line still persists.

Significance of Air-blast Breaker

Considering the long time which has been spent in developing oil circuit breakers to their present state of satisfactory performance, it is remarkable that in such a comparatively brief time an entirely new principle of arc interruption has been developed and incorporated in a successful breaker which gave substantially 5-cycle performance on its first real system test.

Even more striking, we have never before been able to complete an initial series of tests on a new oil circuit breaker design without disclosing some difficulty in the breaker, not of basic importance but still sufficient to prevent completion of the tests until such difficulties were remedied. These tests on the air-blast breaker were the first ever undertaken on the American Gas and Electric system in which the breaker in its original condition successfully completed an entire schedule series without any adjustments, difficulties or inspections.

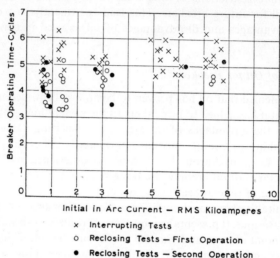

Fig. 3. Test performance of 1500 Mva air-blast breaker.

Conclusions

1. The high-voltage air-blast breaker tested has demonstrated an interrupting performance at least equivalent to that expected from any modern oil breaker of conventional design. It may reasonably be expected that its mechanical design and construction will undergo changes during the next few years, but a successful interrupting principle has been established beyond doubt.

2. The present air-blast breaker demonstrated the inherent adaptability to reclose at speeds at least equal to, and at short-circuit duty far beyond, any similar tests on oil breakers. Based on prevailing standards, the interrupting ratings of 138-kv oil breakers are subject to a reduction of 15 to 25% when applied on 20-cycle reclosing service. Although it may be too early to form definite conclusions, even a conservative interpretation of these tests results points the way to a probable downward revision of such derating factors as applied to air-blast breakers.

3. The current-transformer problem is decidedly more complex with the air-blast than with the conventional oil circuit breaker since the relatively simple and economical procedure of applying bushing current transformers to oil circuit breakers cannot be used with the air-blast breaker. Therefore it would be highly desirable, in the interests of the future development of this type of

breaker, if a more economical solution of this problem than the use of separate current transformers could be found.

4. The success of this air-blast breaker on 138 kv points encouragingly to the prospect of its developments for higher voltages, such as 230 kv, or even higher. The economic picture here might be even more favorable to the air-blast breaker considering the physical dimensions and large quantities of oil required in conventional breakers for such voltages.

5. As regards the possible difficulties in the use of air and equipment for handling it, there is no doubt that practical experience is needed. The only way to get this is through trial installations.

Fortunately, the existence of reliable oil circuit breakers makes it possible to carry out such a program at least in the moderately high-voltage class like 138 kv, systematically and without delay, but unhurriedly.

16. NINE YEARS' EXPERIENCE
WITH ULTRAHIGH-SPEED RECLOSING†

SINCE the last publication of operating experience with ultrahigh-speed reclosing of high-voltage transmission lines, four more years of operating experience have been gathered on lines of the American Gas and Electric Company. This experience is so much more extensive than that previously available that it is now possible to check more thoroughly the ideas and conclusions previously entertained but not entirely proved. It appears that sufficient data are now available to serve as a basis for definite and final conclusions on a number of important aspects of ultrarapid reclosure and the benefits inherent therein.

The equipment under discussion is installed on the central system of the company which consists of 45 circuit miles of 154-kv, 2818 circuit miles of 132-kv, 64 circuit miles of 110-kv, 304 circuit miles of 88-kv, and 695 circuit miles of 66-kv high-voltage circuits and serves a maximum one-hour integrated peak of more than 1,450,000 kw. Most of the 132-kv circuits are of double-circuit steel-tower construction.

Although transmission distances are comparatively small, it has been necessary to maintain maximum net-work continuity because of sudden changes in power flow and especially because of the industrial character of the load. This in turn results in rendering the most reliable service, the most co-ordinated operation, and utilization at all times of the most efficient combination of generating sources regardless of their location.

Ultrahigh-speed reclosing equipment has been installed at 91 terminals to protect 43 line sections comprising approximately 1634 circuit miles of line, 1360 miles of which operate at 132 kv. In the period May 1936 to November 1944, covering substantially nine lightning years, 635 cases of flashover resulted in line opening. Of these, 531 cases were single-line trouble and 104 cases were two-line trouble. In those 635 cases 570 reclosures were successful and 65 (or 10.2%) were unsuccessful. In eight of the unsuccessful reclosures a permanent fault existed on the line so that successful reclosure was a physical impossibility, since nothing but a permanent outage could clear the trouble. These permanent faults were caused either by lines being on the ground, physical damage to steel tower, failure of a lightning arrester, or a string of insulators pulled loose because of ice loading. If one chooses to disregard these eight cases, the total unsuccessful reclosures were 57, representing 9.0%.

In 104 cases of two-line trouble, 21 unsuccessful reclosures were experienced giving an unsuccessful 20.2% for simultaneous trouble on two lines. The method of scoring employed for recording unsuccessful reclosures for two-line trouble was this: In cases where only one line of the two reclosed successfully, the operation was credited with one-half unsuccessful reclosure, and in cases where neither line reclosed successfully, the operation was credited with a full unsuccessful reclosure. Since in 80% of the cases of reclosures on two-line trouble

† AIEE Winter Technical Meeting (with C. A. Muller), New York, N.Y., January 22, 1945.

which were not fully successful one of the two lines reclosed successfully, it is evident that the scoring method adopted reflects, if anything, a pessimistic version of the effect of reclosure. Two Roanoke–Reusens lines experienced 48 cases of two-line trouble of which $13\frac{1}{2}$ resulted in unsuccessful reclosures during the period from May 1940 to November 1944. The unsuccessful reclosures consisted of 19 cases where only one of the two lines failed to reclose successfully and four cases where both lines failed to reclose successfully. An investigation of the large percentage of unsuccessful reclosures experienced on these lines for two-line trouble indicated that apparently cascading of breaker-tripping operations was occurring.

Conclusions

The data obtained in the course of nine years' experience with ultrarapid reclosure of high-voltage circuits are complete enough to warrant the following definite conclusions:

1. On high-voltage lines, properly insulated and provided with ground-wire protection, 90% successful reclosure has been and can be obtained by the use of ultrarapid reclosing as now developed.

2. While the experience obtained with double-circuit lines indicates a record of apparently unsuccessful reclosure approximately double that of the average, the significant fact is that in these cases 80% of the apparently unsuccessful reclosure resulted in successful reclosure of one of the two circuits.

3. Ultrarapid reclosure has proved itself a tool of major importance in planning and building any overhead high-voltage transmission system or circuit. Without question it is the most economical and dependable means of improving high-voltage-transmission reliability.

4. The confirmed and highly successful result obtained on high- and medium-high-voltage lines again points clearly to benefits obtainable from application to lower-voltage lines. The only barrier to the extension of such application is economic: Simpler and lower-cost reclosing mechanisms and, if possible, relaying are needed.

5. There is excellent reason to believe that single-circuit, and particularly double-circuit, reclosure performance can be improved by decreasing the time of fault duration and thus speeding up the reclosure cycle. The recent successful development of 3-cycle breakers, making possible 12-cycle reclosure, will open the way for exploring this possibility.

17. HEAVY-DUTY HIGH-SPEED BREAKER FIELD TESTS†

FIELD tests of two 138-kv, 800-amp, 3500-Mva, 3-cycle circuit breakers—one oil the other air-blast—have been made with 20 interruptions on each breaker.

Development of these breakers was undertaken because of need for breakers of such high rupturing capacity on a number of interconnected power systems in the United States, including the central system of the American Gas and Electric Company. The search for faster opening speeds is part of a program undertaken close to ten years ago to improve high-voltage transmission-system performance. The fact that rupturing duties of close to 3500-Mva are required under present conditions would be surprising were it not that expansion has taken place at a much more than normal rate in response to demands for power during the war. Rupturing duties of a totally different order than any considered likely to be needed heretofore have been encountered and there is reason to believe that further growth in short-circuit duty is bound to take place as long as transmission net-works continue to expand.

Objectives and Results of Tests

While specifically the purpose of the tests was to determine the interrupting capacity and reliability of the two breakers, from a broader standpoint they were: To explore the limits of the self-generating pressure-type oil interrupter; to test the ability of an air-blast design to meet the highest rupturing-duty performance reached by the oil breaker and to explore the possibilities of a faster reclosing cycle on a 138-kv system than that offered by the oil design even in its most improved form.

Both breakers came through the entire series of tests without any visible evidence of distress. In the oil breaker, the occurrence of an interrupting operation was evidenced only by wisps of smoke emerging from the separating chamber vents. Dielectric tests on the condition of the oil before and after showed an over-all deterioration of only 30 kv to 26 kv. As compared with the oil breaker, the air-blast breaker gave much greater outward evidence of interrupting performance by the high-velocity blast of air, as well as visible ejection of incandescent gases and smoke from the interrupter vents. Performance of the air-blast breaker was consistent throughout, and no visible difference appeared between the opening and close-opening tests in the first series, nor between the first and second shots of the high-speed reclosing tests. However, high-speed motion pictures taken during the heavy-duty reclosing tests, showed a considerably larger emission of gases and smoke on the second interruption than on the first. This would be expected as a result of the accumulation of some vaporized contact material remaining in the vicinity of the breaker contacts between the first and second shots.

† AIEE Winter Technical Meeting (with H. P. St. Clair), New York, N.Y., January 22, 1945.

As in the case of the non-reclosing tests, the air-blast breaker came through all of the reclosing tests with an interrupting time of 3 cycles or under, whereas one pole of the oil breaker required slightly more time on several tests, the maximum being $3\frac{1}{3}$ cycles. The other two poles of the oil breaker showed consistent performance at 3 cycles or below on all tests except on the preliminary shot at less then 10% of rating.

Reclosing tests in the second series showed a minimum of $15\frac{1}{2}$ cycles reclosing time on the oil breaker, and down to 13 cycles on the air-blast breaker. While the principal objective in making the tests on these breakers was to obtain data on interrupting-capacity performance under conventional opening–closing combinations and also under more or less standard high-speed reclosing duties, much faster reclosing speeds were hoped for and sought. From that standpoint the reclosing speed obtained was gratifying, particularly on the air-blast breaker at 13 to $13\frac{1}{3}$ cycles. In the case of the oil breaker, although the decrease in time from 19 cycles to $15\frac{1}{2}$ cycles for the last test in the series was obtained by means of certain adjustments that could be made in the field, it appears that this is about as far as it is feasible to go with the present design; any further marked reduction in time would require some redesign. However, with the air-blast breaker, there appears to be no question but that 12-cycle reclosing definitely is attainable. This may perhaps represent one of its most attractive features, which include such other advantages as the stored-energy interrupting principle and the elimination of oil.

From the standpoint of mechanical and electrical performance of the test breakers, this program of tests was accomplished with the least difficulties and with the highest degree of satisfactory performance of any series of tests yet carried out on the system with which the authors are associated.

From a system standpoint, the consistently high-speed performance on all tests was undoubtedly responsible for the almost complete absence of system voltage complaints, notwithstanding the fact that many industries normally sensitive to voltage disturbance were in full operation during the tests. Furthermore, the program produced no operating disturbances, such as faulty relay operations, which may be due principally to the very high speeds obtained on all the tests and in part to constant vigil to improve and perfect relay operation. It has always been felt that one of the valuable by-products of staged system tests of this kind is the opportunity given to discover and correct imperfections in relay systems if and when they fail to accomplish the protective functions for which they are designed.

Conclusions

For the first time a full 3-phase short-circuit of 3500 Mva was successfully interrupted with a speed of 3 cycles. In comparing the performance of the two types of breakers, the following conclusions seem warranted:

The tests on the oil circuit breaker showed performance closely approaching 3500 Mva at a 3-cycle interrupting time, even though the 3-cycle time was slightly exceeded on one pole in some cases. The air-blast breaker, however, gave unquestioned performance at 3 cycles or less on all of the tests.

Judging by the performance of both breakers it seems apparent that a goal of 5000 Mva is attainable with both oil and air-blast designs, and unquestionably with the latter.

It was demonstrated, though not fully accomplished, that 12-cycle reclosing is entirely feasible in the case of the air-blast breaker. The deficiency is quite small, only $\frac{1}{2}$ to 1 cycle, and it is believed that this can be eliminated by relatively minor design changes.

Field tests at short-circuit values as high as 3500 Mva can be conducted smoothly and safely on a large and extensive interconnected power system without serious operating disturbances.

Again was demonstrated the value of full-capacity field tests to prove the adequacy of circuit breaker designs.

MADE IN GREAT BRITAIN